High-Speed and Lower Power Technologies

Electronics and Photonics

T0314351

Devices, Circuits, and Systems
Series Editor
Krzysztof Iniewski

For more information about this series, please visit: https://www.crcpress.com/Devices-Circuits-and-Systems/book-series/CRCDEVCIRSYS

High-Speed and Lower Power Technologies

Electronics and Photonics

Edited by
Jung Han Choi
Krzysztof Iniewski

CRC Press
Taylor & Francis Group
Boca Raton London New York

CRC Press is an imprint of the
Taylor & Francis Group, an **informa** business

CRC Press
Taylor & Francis Group
6000 Broken Sound Parkway NW, Suite 300
Boca Raton, FL 33487-2742

First issued in paperback 2020

© 2019 by Taylor & Francis Group, LLC
CRC Press is an imprint of Taylor & Francis Group, an Informa business

No claim to original U.S. Government works

ISBN 13: 978-0-367-65609-6 (pbk)
ISBN 13: 978-0-8153-7441-1 (hbk)

Library of Congress Cataloging-in-Publication Data

Names: Choi, Jung Han, editor. | Iniewski, Krzysztof, 1960- editor.
Title: High-speed and lower power technologies : electronics and photonics /
Jung Han Choi and Krzysztof Iniewski, editors.
Description: Boca Raton : Taylor & Francis, a CRC title, part of the
Taylor & Francis imprint, a member of the Taylor & Francis Group,
the academic division of T&F Informa, plc, 2018. | Series: Devices, circuits,
and systems | Includes bibliographical references and index.
Identifiers: LCCN 2018026596| ISBN 9780815374411 (hardback : alk. paper) |
ISBN 9781351242295 (ebook : alk. paper)
Subjects: LCSH: Low voltage integrated circuits. | Integrated optics--Power supply. |
Electronic apparatus and appliances--Energy conservation. |
Very high speed integrated circuits.
Classification: LCC TK7874.66 .H54 2018 | DDC 621.3815--dc23
LC record available at https://lccn.loc.gov/2018026596

Visit the Taylor & Francis Web site at
http://www.taylorandfrancis.com

and the CRC Press Web site at
http://www.crcpress.com

Special thanks to my parents

for

their endless love and support

Contents

Preface

Advances in semiconductor technologies over the last few years have led to such rapid and considerable improvements with respect to speed, power, integration density, and cost, that the semiconductor device, especially the Si CMOS gate length, is now approaching the theoretical quantum limit of its feature size. Many academic institutions and industries are aggressively investing their efforts in discovering next-generation technologies in every area: for example, nano devices using carbon-based materials, new structure-materials, heterogeneous material systems to form optics in Si, and high-temperature superconducting materials.

In addition, technological evolutions like quantum computing and quantum communications, which it was believed would take at least a decade to make practically available, are now already on the horizon. People are starting to bring their attention to those areas to work out feasible future applications combined with IoT, 5G, and beyond 5G. One distinct area that has recently attracted a lot of interest from many research institutions is artificial intelligence, particularly since AlphaGo, developed by Google Deepmind, won a Go match against 18-time world *human* champion, Lee Sedol. This historic event inspired many engineers to dig into the enormous possibilities of artificial intelligence in ever-developing electronic machinery and all sorts of engineering areas. Machines embedded with artificial intelligence even began mimicking human intelligence in part and people seek to imagine applications in daily life. It is strongly anticipated that many electronics will embed AI in the near future.

Also, the explosion of data traffic has triggered the development of higher-speed and lower-power intra-/inter-server communication modules for both data centers and communications. According to the survey by Cisco in 2016, global IP traffic will show 24% compound annual growth rate (CAGR) between 2016 and 2021, reaching 278 EB per month (EB = one billion gigabytes).* It is reported that global electricity consumption by data centers is estimated at about 1.4% of global electricity consumption with a CAGR of 4.4% from 2007 to 2012†. It also contributes significantly to CO_2 emissions. The significant message we have to take from these statistical surveys is that we have to produce lower-power systems compared to old ones, supporting higher speed and lower complexity. One of the key elements in the data center as regards lower power-consumption is the interconnection

---

* Cisco, "White Paper: Cisco Visual Networking Index: Forecast and Methodology, 2016–2021." Available at: www.cisco.com/c/en/us/solutions/collateral/service-provider/visual-networking-index-vni/complete-white-paper-c11-481360.html.
† M. Avgerinou, et al., "Trends in Data Centre Energy Consumption under the European Code of Conduct for Data Centre Energy Efficiency," *Energies* 2017, 10, 1470; doi:10.3390/en10101470.

technology between servers and/or data centers. A lot of research and development activity is under way to support higher speed and lower energy-consumption with regard to devices, circuits, and systems. There are lots of R&D initiatives to reduce the cost of the optical transceiver modules for both short- and long-range communications using Si photonics.

This book aims to present recent research and development activities focusing on lower-power and higher-speed electronics and optics, against the background of current technology trends mentioned above. We have tried to bring together relevant and heterogeneous topics being investigated now in order to predict future research directions. Each chapter pursues its topic from basics to feasible applications, discussing relevant theories and examples.

Chapter 1, the introductory chapter of this book, suggests an answer to the open question of how to deal with the end of the Moore's law and looks at how the semiconductor academia and industries need to prepare for the beginning of a new paradigm. The author tries to introduce general ideas about overcoming recent challenges and discusses the limit of the Moore's law. Chapter 2 presents how machine learning can be combined with modern CMOS technology. The authors discuss this topic in detail, giving concrete examples like physically unclonable functions (PUFs) and time-to-digital converters (TDCs). This chapter is a good example how the impact of AI will affect concurrent technologies. Chapters 3 and 4 look at the recent development of SiGe devices: as well as the scaling of CMOS devices, SiGe is also improving fast in terms of scaling (Chapter 3) and complementary PNP-SiGe (Chapter 4). Development engineers can see recent achievements in industry and academia and are capable of predicting future applications using SiGe devices. Chapter 5 considers graphene as a potential future interconnect material for deeply scaled technology. The authors provide results comparing the proposed multilayer graphene material system with conventional Cu wires, in terms of energy-delay-product and bandwidth density improvements.

Chapter 6 presents possible future directions of the laser device, considering practical communication protocol like forward error correction (FEC). Highly efficient lasers in general lack output power, thus power-hungry FEC will necessarily be employed, degrading the power efficiency of the system. This chapter discusses this aspect in detail, considering the future development direction of the laser. Chapter 7 presents two examples of low-latency, energy-efficient, automatically controlled integrated photonic interconnects for computing platforms using silicon photonic platforms. Two architectures are experimentally tested in static and dynamic conditions and their performances are analyzed. Chapter 8 is about the design methodology, the so-called co-design, of the InP Mach-Zehnder modulator and its driver-IC in order to achieve low power. Details of this method are presented, helping engineers working in this area of technology to exploit the proposed concept in their development of optical transmitters.

Chapter 9 presents an efficient and low-power network-on-chip (NoC) router architecture using two novel methods of buffering and arbitration to improve the performance and hardware metrics of NoC systems. Chapter 10 addresses how power-aware mobile devices can use the rapid SRAM hardware enhancement technique to extend their in-battery system's operating time. Chapter 11 proposes direct conversion X-ray quantum-counting detectors to solve the noise problem associated with photon weighting, instead of the currently available technique of integrating the X-ray quanta (photons) emitted from the X-ray tube. Chapter 12 presents almost all sorts of high-efficiency power amplifiers, presenting pros and cons, comparing each with the others. If readers want to get a comprehensive overview quickly without losing details about the power amplifiers, then this chapter is recommended and should be useful material for practical design, as well.

The editors would like to thank all the chapter contributors for their active communications and valuable contributions toward helping complete this book. Without their help, this book could not have been finished. I would like to give my deep thanks to my wife, Mrs. Hye Won Nam, for her warm support in completing this book, even though she was pregnant during its editing and gave birth to our daughter, Yuna Choi.

Dr. Jung Han Choi
Kladow, Berlin, Germany

MATLAB® is a registered trademark of The MathWorks, Inc. For product information, please contact:

The MathWorks, Inc.
3 Apple Hill Drive
Natick, MA 01760-2098 USA
Tel: 508 647 7000
Fax: 508-647-7001
E-mail: info@mathworks.com
Web: www.mathworks.com

Series Editor

Krzysztof (Kris) Iniewski is managing R&D at Redlen Technologies Inc., a start-up company in Vancouver, Canada. Redlen's revolutionary production process for advanced semiconductor materials enables a new generation of more accurate, all-digital, radiation-based imaging solutions. Kris is also a President of CMOS Emerging Technologies (www.cmoset.com), an organization of high-tech events covering Communications, Microsystems, Optoelectronics, and Sensors. In his career, Dr. Iniewski held numerous faculty and management positions at University of Toronto, University of Alberta, SFU, and PMC-Sierra Inc. He has published over 100 research papers in international journals and conferences. He holds 18 international patents granted in USA, Canada, France, Germany, and Japan. He is a frequent invited speaker and has consulted for multiple organizations internationally. He has written and edited several books for IEEE Press, Wiley, CRC Press, McGraw-Hill, Artech House, and Springer. His personal goal is to contribute to healthy living and sustainability through innovative engineering solutions. In his leisurely time, Kris can be found hiking, sailing, skiing, or biking in beautiful British Columbia. He can be reached at kris.iniewski@yahoo.ca.

Editor

© Fraunhofer HHI

Jung Han Choi received BS and MS degrees in electrical engineering from Sogang University, Seoul, Korea, in 1999 and 2001, respectively, and the Dr.-Ing. degree from the Technische Universität München, Munich, Germany in 2004. From 2001 to 2004, he was a research scientist in the Institute for High-Frequency Engineering at the Technische Universität München. During this time, he worked on high-speed device modeling, thin-film fabrication, network analyzer measurement and circuit development for high-speed optical communications. From 2005 to 2011, he was with the Samsung Advanced Institute of Technology and the Samsung Digital Media & Communication Research Center, where he worked on the RF bio-health sensor, nano devices and RF/millimeter-wave circuit design, including 60 GHz Si CMOS ICs. In 2011 he joined at the Fraunhofer Institute (Heinrich-Hertz Institute), Berlin, Germany and holds tenured position. Now he is working on ultra-low power high data bitrate transmitter and receiver circuits for optical communications, microwave devices, electromagnetic simulations, and network analyzer measurement up to 170 GHz.

He was awarded the EEEfCOM (Electrical and Electronic Engineering for Communication) Innovation prize in 2003 for his contribution to the development of the high-speed receiver circuit. His current research interests range from microwave active/passive devices and IC, electromagnetic simulation and analysis, and metamaterials. He holds 19 international registered and 22 pending patents in the area of semiconductor device, circuits, and systems for high-frequency engineering.

Contributors

George Ph. Alexiou
CEID – University of Patras
Patra, Greece

Nicola Andriolli
Scuola Superiore Sant'Anna
Pisa, Italy

Wayne Burleson
Department of Electrical and
    Computer Engineering
University of Massachusetts,
    Amherst
Amherst, Massachusetts

Isabella Cerutti
Scuola Superiore Sant'Anna
Pisa, Italy

Jung Han Choi
Fraunhofer Heinrich-Hertz Institute
Berlin, Germany

Il-Sug Chung
School of Electrical and Computer
    Engineering
Ulsan National Institute of Science
    and Technology
Ulsan, South Korea
and
Department of Photonics
    Engineering
Technical University of Denmark
Kongens Lyngby, Denmark

Alfonso Conesa-Roca
Universitat Politècnica de Catalunya.
Barcelona, Spain

Adam Grosser
Redlen Technologies Inc.
Saanichton, British Columbia, Canada

Themistoklis Haniotakis
ECE – Southern Illinois University
Carbondale, Illinois

Kris Iniewski
Redlen Technologies Inc.
Saanichton, British Columbia, Canada

Gul Khan
Electrical and Computer
    Engineering
Ryerson University
Toronto, Ontario, Canada

Raghavan Kumar
Intel
Portland, Oregon

Suhas Kumar
Hewlett Packard Labs
Palo Alto, CA

Shuo Li
Department of Electrical and
    Computer Engineering
University of Massachusetts,
    Amherst
Amherst, Massachusetts

Odile Liboiron-Ladouceur
Department of Electrical and
    Computer Engineering
McGill University
Montreal, Quebec, Canada

Yiheng Lin
Department of Materials
    Engineering
University of British Columbia
Vancouver, British Columbia, Canada

Herminio Martínez-García
Universitat Politècnica de Catalunya
Barcelona, Spain

Atul Kumar Nishad
School of VLSI Design and
    Embedded Systems
National Institute of Technology
Haryana, India

Edward Preisler
TowerJazz
Newport Beach, California

Rohit Sharma
Department of Electrical
    Engineering
Indian Institute of Technology
    Ropar
Punjab, India

Theodoros Simopoulos
CEID – University of Patras
Patras, Greece

Chris Siu
British Columbia Institute of
    Technology (BCIT)
Burnaby, British Columbia, Canada

Nicolas Sklavos
CEID – University of Patras
Patras, Greece

Guillermo Velasco-Quesada
Universitat Politècnica de Catalunya
Barcelona, Spain

Guangrui (Maggie) Xia
Department of Materials
    Engineering
University of British Columbia
Vancouver, British Columbia, Canada

Xiaolin Xu
Department of Electrical and
    Computer Engineering
University of Florida
Gainesville, Florida

1

The End of Moore's Law and Reinventing Computing

Suhas Kumar

CONTENTS

Why do we care to build better computers? It is because the problems faced by our societies have become prohibitively complex and varied, and solving such problems has become necessary to sustain our survival. We need better computers to do that.

Consider for example the efforts into decoding the causes of cancer. A magic wand that could address this issue would be a computer that can solve computationally hard problems like gene sequencing in polynomial time. It turns out that problems such as gene sequencing, solving Sudoku, resource optimization, vehicle route scheduling, etc., are a specific class of problems, which become exponentially more complex as the size of the problem increases. For example, to perform gene sequencing, with three units of the gene, there are three possible combinations to choose from, while with 15 units we face 43 billion possibilities. We need exponential improvements in computing capabilities to address such outstanding issues and none of the best supercomputers of today can come even close to solving these problems. There are several examples of other comparable issues, such as increasing the supply of drinking water by simulating networks of water-filtering nanomaterials, alleviating poverty by effective resource-sharing, predicting weather more accurately by solving chaotic equation systems, all of which embody computationally difficult problems of massive scale, and critically depend on better computing resources.

The manufacturing of integrated circuits using electronic components such as the transistor, which started in the 1960s, is probably the most important revolution to date in the field of computer science. Gordon Moore predicted in 1965 that electronic component dimensions would get smaller following a trend [1]. It is accepted today as Moore's law, a rule of thumb that the number of transistors packed into an integrated circuit doubles approximately every two years. People innovate to stay ahead of the competition by making devices smaller to attain several advantages, including packing and information-processing density [2]. This race has enabled continual dramatic improvements in computer performance, leading to the hand-held computers of today that are far more capable than those that put man on the Moon. Despite the significant improvements in computer technology over several decades since Moore's early prediction, we now face physical or technological limits that will prevent us from scaling down forever.

In this introductory chapter, we examine the fundamental limits to Moore's law and discuss directions that could effectively help us circumvent these limits.

1.1 Fundamental Limits to Moore's Law

1.1.1 Thermodynamic Limits

To perform useful computation, we need to irreversibly change distinguishable states of cell(s) containing information. The thermodynamic entropy to change the state of n information cells within m states is

$$\Delta S = k_B \ln\left(m^n\right), \tag{1.1}$$

where k_B is the Boltzmann constant. From the second law of thermodynamics, $\Delta S = \Delta Q / T$, where ΔQ is the energy spent and T is the temperature. So the energy required to write information into one binary bit is $E_{bit} = k_B T \ln 2$. This is known as the Shannon-von Neumann-Landauer (SNL) expression. This tells us that we need at least 0.017 eV of energy to process a bit at 300 K.

From Heisenberg's Uncertainty Principle, $\Delta E \Delta t \geq \hbar$, for E_{bit} of 0.017 eV, where ΔE and Δt are the uncertainties in energy and time, respectively. Thus the minimum time to change/process information (t_{min}) is 0.04 ps. From Heisenberg's Uncertainty Principle, represented as the uncertainty in the position $\Delta x \geq \hbar / \sqrt{(2mE)}$, the minimum feature size (x_{min}) corresponding to an electron as the carrier is 1.5 nm. The power per area, $P = n \times E_{bit} / t_{min}$, where $n = 1/x_{min}^2$ is the packing density (~4.7×10^{13} devices/cm^2), which is about 3.7 MW/cm^2. For comparison, the surface of the Sun produces about 6000 W/cm^2. These are not the limits, however. In the next section, we will correct these formulas by considering tunneling.

1.1.2 Inclusion of Quantum Tunneling

Consider a quantum well system as shown in Figure 1.1. The probability of thermionic injection of the electron over the barrier height is $G_T = \exp(-E_b/k_B T)$. The probability of tunneling through the barrier is $G_Q = \exp(-2a\sqrt{(2mE)}/\hbar)$ [3]. For the two states to be distinguishable, the limiting case is $G_{error} = G_T + G_Q - G_T G_Q = 0.5$. Solving, we get the minimum energy:

$$E_{bit}^{min} = k_B T \ln 2 + (\hbar \ln 2)^2 / (8ma^2). \tag{1.2}$$

The power dissipation corresponding to an area A having n devices operating at a frequency f is

$$P_{max} = f(n/A)\left[k_B T \ln 2 + (\hbar \ln 2)^2 / (8ma^2)\right]. \tag{1.3}$$

1.1.3 Thermal Limits

How much we can allow the power dissipation to rise depends on how much rise in temperature the chip can stand (typically upto 400 K) and on how fast we can remove the heat from the chip. Newton's law of cooling governs heat removal as

$$Q = H(T_{Dev} - T_{sink}). \tag{1.4}$$

H is the heat transfer coefficient, which is determined by the material constants like specific heat, viscosity, thermal conductivity, heat capacity, etc., separate from the geometry of the cooling structure [4,5]. T_{Dev} and T_{sink} are the temperatures of the device and its heat sink, respectively. When $T_{Dev} < T_{sink}$, it appears from the first section that E_{bit} gets more tolerable. But Carnot's theorem says that the work needed to remove heat Q is

$$W = Q(T_{sink} - T_{Dev})/T_{Dev}. \tag{1.5}$$

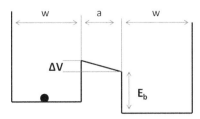

FIGURE 1.1
Model for a quantum well computation system.

Therefore,

$$E_{bit}^{total} = E_{bit} + E_{bit}\left(T_{sink} - T_{Dev}\right)/T_{Dev} \tag{1.6a}$$

$$= k_B T_{sink} \ln 2 + \left(T_{sink}/T_{Dev}\right)\left(\hbar \ln 2\right)^2\big/\left(8ma^2\right) \tag{1.6b}$$

E_{bit}^{total} and power are plotted against a and temperature in Figure 1.2. From Figure 1.2, we see that:

1. cooling the system does not help lower E_{bit} at all,
2. power is ridiculously high for features less than a nanometer, and
3. E_{bit} required is also very high below a feature size of 2 nm, while it is about $k_B T \ln 2$ for larger features.

Notice that, as we approach smaller features, E_{bit} and power are far better behaved at higher temperatures than at lower temperatures.

1.1.4 Compton Wavelength

The Compton wavelength $\lambda_c = h/mc$ (~0.00243 nm for $m = 9.1 \times 10^{-31}$ kg) is the characteristic dimension of an electron, which has been proposed as a fundamental absolute limit of the size of an electronic device [6]. At these length scales, there is a runaway divergence in power and E_{bit}, as is apparent from Figure 1.2. The reader is encouraged to plug this length scale into the

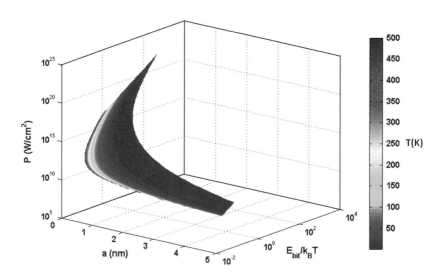

FIGURE 1.2
Plot of energy, E_{bit}, and power, P, as a function of feature size, a, at different temperatures, T.

equations presented here (and into Figure 1.2) to estimate power and E_{bit} and decide if it is sensible to even approach this limit.

1.1.5 Other Practical Aspects

Power consumption and speed are limited fundamentally by the devices, but practically by the electrical parasites, interconnects and chip architecture. This is the reason why clock speeds have stalled at <5 GHz in most of today's processors. All alternative ideas, like optical interconnects and more, would see their limit in the domain conversion (such as optical to electrical conversion) as long as the critical parts of a computing system utilizes electronic components, which is limited by the thermodynamics discussed here. The 2 in $k_B T \ln 2$ can be increased to, say, z, but that only pushes the limits by a factor of $\ln(z)/\ln 2$. [7,8]

From the estimates above, we predict that Moore's law will begin tapering out by 2021, by when we should have serious investments in the research and development of alternate computing and information processing technologies. These technologies will be mainstream in 20 years from now.

1.2 Beyond Moore's Law

The question remains of what approaches, from new physics to new architectures, can get us to dramatically improve computing, despite the impending end of Moore's law [9]. The approaches that supported the electronics industry during the era of Moore's law were:

1. Heroic device characterizations
2. Aggressive transistor scaling
3. Conformation of all advances to Boolean logic and the von Neumann architecture.

The era of Moore's law was limited by the von Neumann bottleneck, which is that the speed of computing is limited by the separation of information processing and storage, a fundamental property of the von Neumann architecture. Secondly, digital or Boolean logic performs high-precision floating point operations, but compromises on the speed of convergence. And then we have the Boltzmann tyranny, which sets a minimum energy required to process a unit of information. These issues did not seem critical or prohibitive for most practical purposes until about a decade ago. But the entry of big-data, brain-inspired computing, novel machine learning algorithms, etc., have necessitated the overhauling of all the approaches we have been taking toward building computers. The end of Moore's law has transformed the above three approaches into:

1. Discovery of new physico-chemical properties, of possibly new electronic materials as well;
2. Inventing devices using new materials, such that a single device can functionally replace hundreds of transistors;
3. New device functions that have enabled and necessitated new hardware architectures and software algorithms.

Hence, the end of Moore's law is far more exciting than the beginning of Moore's law. We are now faced with open directions for research and innovations to create the future paradigms of all computing technologies. In order to take this route and overcome the limitations we are increasingly experiencing with the era of Moore's law, there are probably hundreds of parallel approaches being explored in the industry and academia alike. These include using the properties of (possibly new) new materials to construct problem-specific analogue computer accelerators, learning from the brain how to develop better hardware architectures within neural networks, and using non-linear phenomena such as chaotic and stochastic processes to emulate natural processes.

The scope of this book is to give an overview of the last stages of Moore's law, while giving insights into some of the fundamental developmental efforts that may form the paradigms of the world of computers beyond Moore's law. In these efforts to design better components that can extend Moore's law in terms of speed and power efficiency, there is a very clear emphasis on materials discovery. Using exotic materials like graphene for electronics and exploring III-V materials for optical interconnects exemplify this direction. There is also a clear realization that most of the computing speed today is limited by communications delays due to the von Neumann bottleneck. Thus nearly a third of the chapters are dedicated to finding faster and power-efficient ways of communicating across chips. A significant part of the book is also dedicated to discussing better ways of storing information – from new techniques to write and access information, to using non-transistor approaches such as two-terminal resistive memories to gain energy efficiency and speed. Finally, this book also sheds light on recent advances in low-power design of electronics at several scales – from components, through FPGAs, to personal computing systems. The efforts to break free of the limitations of Moore's law are very apparent in all the advances described here, while giving a sense of the technical depth required to run the last lap of Moore's law.

References

1. G. E. Moore, Cramming more components onto integrated circuits, *Electronics* 38, No. 8, 114 (1965).
2. C. A. Mack, Fifty years of Moore's law, *IEEE Trans. Semicond. Manuf.* 24, 202 (2011).
3. A. P. French and E. F. Taylor, Eds., *An Introduction to Quantum Physics* (W.W. Norton & Company, 1978).
4. I. Mudawar, Assessment of high-heat-flux thermal management schemes, *IEEE Trans . Comp. Packag. Technol.* 24, 122 (2001).
5. C. LaBounty, A. Shakouri, and J. E. Bowers, Design and characterization of thin film microcoolers, *J. Appl. Phys.* 89, 4059 (2001).
6. J. R. Powell, The quantum limit to Moore's law, *Proc. IEEE* 96, 1247 (2008).
7. L. B. Kish, End of Moore's law: Thermal (noise) death of integration in micro and nano electronics, *Phys. Lett. A* 305, 144 (2002).
8. V. V. Zhirnov, et al., Limits to binary logic switch scaling – A Gedanken model, *Proc. IEEE* 91, 1934 (2003).
9. R. S. Williams, What's next?, *Computing in Science & Engineering* 19, 7 (2017).

2

When the Physical Disorder of CMOS Meets Machine Learning

Xiaolin Xu, Shuo Li, Raghavan Kumar, and Wayne Burleson

CONTENTS

While the development of semiconductor technology is advancing into the nanometer realm, one significant characteristic of today's complementary metal-oxide-semiconductor (CMOS) fabrication is the random nature of process variability, i.e., the physical disorder of CMOS transistors. It is observed that with the continued development of semiconductor technology, physical disorder has become an important factor in CMOS design during the last decade, and is likely to continue to be in forthcoming years. The low cost of modern semiconductor design and fabrication techniques benefits the ubiquitous applications, but also poses strict constraint on the area and energy of these systems. In this context, many traditional circuitry design rules should be reconsidered to mitigate the possible negative effects caused by the physical disorder of CMOS devices. As the physical deviation from nominal specifications becomes a big concern for many electronic systems [1], the performance of electronic blocks that place high requirements on the symmetric design and fabrication process is greatly impacted. For example, the resolution of time-to-digital (TDC) will be decreased by the physical process variations, which leads to deviations in the fabricated delay elements from the designed (nominal) delay length. On the one hand, such process variations introduce uncertainty into the standard CMOS design [2], while on the other hand, this opens up the possibility of leveraging such properties for constructive purposes. One major production of this philosophy is the development of physically unclonable functions (PUFs) in the literature, which extract secret keys from uncontrollable manufacturing variabilities on integrated circuits (ICs).

As electronic designs like PUF and TDC are impacted by the process variations of fabrication, many issues including variability, modeling attacks and noise sensitivity also need to be reconsidered and addressed. Due to the microscopic characteristics of such physical disorder, it is either infeasible or very expensive to measure and mitigate it with tractional electronic techniques. For example, any physical probing into a PUF instance would change the original physical disorder and therefore the measured results are not necessarily correct. In this context, employing a *non-invasive* way to characterize the internal physical disorder becomes a promising solution. This chapter presents some recent work on advancing this physical disorder modeling with the help of Machine Learning techniques. More specifically, it shows that through modeling the physical disorder, Machine Learning techniques can benefit the performance (i.e., reliability improvement) of PUFs by filtering out the unreliable challenge and responses (CRPs). As for a fabricated TDC circuitry with given internal physical disorder, it is demonstrated that a backpropagation-based Machine Learning framework can be utilized to mitigate the process variations and optimize the resolution.

This chapter is structured as follows. Section 2.1 briefly describes the sources of process variations, i.e., physical disorder in modern CMOS circuits. An introduction to common PUF terminologies and performance metrics used throughout this chapter is provided in Section 2.2. Section 2.3 provides a brief overview of some of the most popular silicon-based PUF circuits along with their performance metrics. Next, the Machine Learning modeling method is reviewed as a threat in attacking PUFs in Section 2.4. Our discussion continues on applying Machine Learning techniques to help with understanding and mitigating the physical disorder of CMOS circuitry in Sections 2.5 and 2.6. More specifically, Machine Learning techniques can also be employed in a constructive way to improve the reliability of PUFs and the resolution of TDCs. Finally, we provide concluding remarks in Section 2.7.

2.1 Sources of CMOS Process Variations

The sources of process variations in ICs are summarized in this section. The interested reader is also referred to the book chapter by Inyoung Kim et al. [3] for more details on the sources of CMOS variability. From the perspective of circuits, the sources of variations can be either *desirable* or *undesirable*. The desirable source of variations refers to process manufacturing variations (PMV) as identified in [3]. The environmental variations and aging are undesirable for some functional circuits like PUFs.

2.1.1 Fabrication Variations

Due to the complex nature of the manufacturing process, the circuit parameters often deviate from the intended value. The various sources of variability include proximity effects, chemical-mechanical polishing (CMP), lithography system imperfections and so on. The process manufacturing variations consist of two components as identified in the literature, namely *systematic* and *random* variations [4].

2.1.1.1 Systematic Variations

The systematic component of process variations includes variations in the lithography system, the nature of layout and CMP [4]. By performing a detailed analysis of the layout, the systematic sources of variations can be predicted in advance and accounted for in design step. If the layout is not available for analysis, the variations can be assigned statistically [4].

2.1.1.2 Random Variations

Random variations refer to non-deterministic sources of variations. Some of the random variations include random dopant fluctuations (RDF), line edge roughness (LER) and oxide thickness variations. The random variations are often modeled using random variables for design and analysis purposes.

2.1.2 Environmental Variations and Aging

Environmental variations are detrimental to PUF circuits. Some of the common environmental sources of variations include power supply noise, temperature fluctuations and external noise. These variations must be minimized to improve the reliability of PUF circuits. Aging is a slow process and it reduces the frequency of operation of circuits by slowing them down. Circuits are also subjected to increased power consumption and functional errors due to aging [5].

2.2 Terminologies and Performance Metrics of PUF

Some terminologies and performance metrics like CPR, reliability, uniqueness and unpredictability used to evaluate PUFs are briefly summarized in the following sections.

2.2.1 Challenge-Response Pairs (CRP)

The inputs to a PUF circuit are known as *challenges* and the outputs are referred to as *responses*. A challenge associated with its corresponding response is known as a *challenge-response pair (CRP)*. In an application scenario, responses of a PUF circuit are collected and stored in a database. This process is generally known as *enrollment*. In the *verification* or *authentication* process, the PUF circuit is queried with a challenge from the database. The response is then compared against the one stored in the database. If the responses match, the device is authenticated.

2.2.2 Performance Metrics

Three important metrics are usually used to analyze a PUF circuit, namely *uniqueness*, *reliability* and *unpredictability*.

2.2.2.1 Uniqueness

One important application of PUF devices is to generate unique signatures for device authentication. Thus, it is desirable that any two PUF instances can be easily distinguished from one another. Toward this end, a typical measure used to analyze uniqueness is known as *inter-distance* and is given by [3]:

$$d_{inter}(C) = \frac{2}{k(k-1)} \sum_{i=1}^{i=k-1} \sum_{j=i+1}^{j=k} \frac{HD(R_i, R_j)}{m} * 100\%. \qquad (2.1)$$

In Equation 2.1, HD(R_i, R_j) stands for the Hamming distance between two responses R_i and R_j of m bits long for challenge C, and k is the number of PUF instances under evaluation. For a group of PUF instances, the desired inter-distance (i.e., uniqueness) of them is 50%. By carefully looking at Equation 2.1, one can correspond the inter-distance $d_{inter}(C)$ to the mean of the Hamming distance distribution obtained over k chips for a challenge C. While designing a PUF circuit, inter-distance is often measured through circuit simulations. A common practice is to perform Monte Carlo simulations over a large population of PUF instances. In simulations, care must be taken to efficiently model various sources of manufacturing variations in CMOS circuits, as they directly translate into uniqueness. During the simulations, manufacturing variations are modeled using a Gaussian distribution. In such cases, mean and standard deviation of the Gaussian distribution under consideration must correspond to either inter-die or inter-wafer variations' statistics.

2.2.2.2 Reliability

As PUFs are built on microscopic process variations, a challenge applied to a PUF operating on an integrated circuit will not necessarily produce the same response under different operating conditions. The reliability of a PUF refers to its ability to produce the same responses under varying operating conditions. Reliability can be measured by averaging the number of flipped responses for the same challenge under different operating conditions. A common measure of reliability is *intra-distance*, given by [3, 6]:

$$d_{intra}(C) = \frac{1}{s} \sum_{j=1}^{s} \frac{HD(R_i, R'_{i,j})}{m} * 100\%. \qquad (2.2)$$

In Equation 2.2, R_i is the response of a PUF to challenge C under nominal conditions, s is the number of samples of response R_i obtained at different operating conditions, $R'_{i,j}$ corresponds to jth sample of response R_i for challenge C and m is the number of bits in the response. Intra-distance is expected to be 0% for ideal PUFs, which corresponds to 100% reliability. The terms intra-distance (d_{intra}) and reliability have been used interchangeably in the rest of this chapter. Given d_{intra}, reliability can always be computed (100 d_{intra}(%)). As an important feature of PUF performance, there exist several contributions in the literature to improve the reliability of PUFs from circuit and system perspectives [7–13].

2.2.2.3 Unpredictability

For security purposes, the responses from a PUF circuit must be unpredictable. Unpredictability is a measurement that quantifies the randomness of responses from the same PUF device. This metric can be evaluated using NIST tests [6, 14]. Silicon PUFs produce unique responses based on intrinsic process variations that are very difficult to clone or duplicate by the manufacturer. However, by measuring responses from a PUF device for a subset of challenges, it is possible to create a model that can mimic the PUF under consideration. Several modeling attacks on PUF circuits have been proposed in the literature [15]. The type of modeling attack depends on the PUF circuit. A successful modeling attack on a PUF implementation may not be effective for other PUF implementations. Modeling attacks can be made harder by employing some control logic surrounding the PUF block that prevents direct readout of its responses. One such technique is to use a secure one-way hash over PUF responses. However, if PUF responses are noisy and have significant intra-distance, this technique will require some sort of error correction on PUF responses prior to hashing [8].

2.3 CMOS PUFs

This section provides an overview of some PUF instantiations proposed in the literature. Based on the challenge-response behavior, PUFs can be classified into two categories, namely strong PUFs and weak PUFs. Strong PUFs have a complex challenge-response behavior and it is very hard to physically clone them. They are often characterized by CRP pairs exponential to the number of challenge bits. So, it is not practical to read out all CRP pairs within a limited time using measurements. On the other hand, weak PUFs accept only a reduced set of challenges and in the extreme case, they accept just a single challenge. Such a construction requires that the response is never shared with the external world and only used internally for security operations. We present an overview of some popular constructions of PUFs, namely arbiter PUFs in Sections 2.3.1, SRAM PUFs in Section 2.3.2 and one of its variants called DRV (data retention voltage) PUF in Section 2.3.3.

2.3.1 Arbiter PUF

The concept of arbiter PUFs was introduced in [16, 17]. An arbiter PUF architecture is shown in Figure 2.1. An n-bit arbiter PUF is composed of n stages, with each stage employing two 2:1 MUXs, as depicted in Figure 2.1. A challenge vector $\mathbf{C} = \{c_1, c_2, \ldots c_n\}$ is applied as the control signals for all stages to configure two paths through the PUF toward the arbiter; at each stage the paths are configured to be either straight or crossing. Thus, a rising edge applied at the input of the first stage gathers the delay mismatch from the paths through each stage while propagating toward the arbiter. The arbiter is a latch that digitizes the response into "1" or "0" by judging which rising edge is the first to arrive. It is important to note that the number of CRP pairs is exponential to the number of challenge bits. So, arbiter PUFs fall under the category of strong PUFs.

2.3.2 SRAM PUF

The concept of SRAM PUFs was first introduced in [18–20]. Static random-access memory (SRAM) is a type of semiconductor memory, which is composed of CMOS transistors. SRAM is capable of storing a fixed written value "0" or "1", when the circuit is powered up. An SRAM cell capable of storing a single bit is shown in Figure 2.2. Each SRAM bit cell has two cross-coupled inverters and two access transistors. The inverters drive the nodes q and qb as shown in Figure 2.2. When the circuit is not powered up, the nodes Q and $\bar{Q}$ are at logic low (00). When the power is applied, the nodes enter a period of metastability and settle down to either one of the states ($Q=0$, $\bar{Q}=1$ or $Q=1$, $\bar{Q}=0$). The final settling state is determined by the extent of device mismatch in driving inverters and thermal noise. For PUF operation, it is desirable to have device mismatch dominant over thermal noise, as thermal noise is random in nature. The power-up states of SRAM cells were used for generating unique fingerprints in [19, 20]. Experiments were conducted over 5,120 64-bit SRAM cells from commercial chips and d_{inter} was found to be around 43%. Similarly, d_{intra} was found to be around 3.7%. Similar results were obtained in [18]. An extensive large-scale analysis over 96 ASICs, with

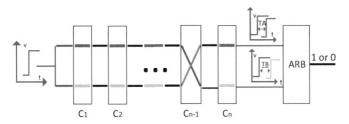

FIGURE 2.1
Schematic of an arbiter PUF. Challenge **C** controls the propagation paths of rising edges that gather delay mismatch as they propagate toward the final arbiter.

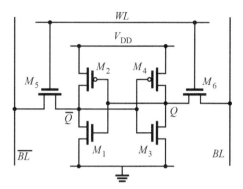

FIGURE 2.2

An SRAM cell using standard CMOS transistors. *BL* is the bit line, *WL* refers to the word line, Q and $\overline{Q}$ are the state nodes for storing a single bit value. The transistors that make up inverters are shown in black and access transistors are shown in grey.

each ASIC having four instances of 8-kB SRAM memory, was performed in [21]. The experimental results from [21] show that SRAM PUFs typically have high entropy and responses from different PUF instances were found to be highly uncorrelated.

2.3.3 DRV PUF

In SRAM study, data retention voltage (DRV) stands for the minimum supply voltage at which the written state is retained in an SRAM. In [22], DRV is proposed as the basis for a new SRAM-based PUF that is more informative than SRAM PUFs [23]. For example, DRV can uniquely identify circuit instances with 28% greater success than SRAM power-up states that are used in PUFs [22]. The physical characteristics responsible for DRV are imparted randomly to each cell during manufacturing, providing DRV with a natural resistance to cloning. It is shown that DRVs are not only random across chips, but also have relatively little spatial correlation within a single chip and can be treated in analysis as independent [24]. The proposed technique has the potential for wide application, as SRAM cells are among the most common building blocks of nearly all digital systems.

However, one major drawback of DRV is that it is highly sensitive to temperature, which makes it unreliable and unsuitable for use in a PUF. In [10], the idea of DRV fingerprinting is further extended to create a PUF based on DRV. Moreover, to overcome the temperature-sensitivity of DRV, a DRV-based hashing scheme that is robust against temperature changes is proposed. The robustness of this hashing comes from the use of reliable DRV-ordering instead of less reliable DRV values. The use of DRV-ordering can be viewed as a differential mechanism at the logical level instead of the circuit level, as in most PUFs. To help validate the DRV PUF, a Machine Learning technique is also proposed for simulation-free prediction of DRVs as a function of process variations and temperature. The Machine Learning model enables the rapid creation of the large DRV data sets required for evaluating the DRV PUF approach.

2.4 Machine Learning Modeling Attacks on PUFs

In this section, we review some existing attacks on PUFs; more specifically, we focus on the Machine Learning-based modeling attacks. Recently, the emergence of unreliable CRP data as a tool for modeling attacks has highlighted an important aspect to be addressed. Also, the vulnerabilities posed by strong PUFs toward modeling attacks will pave the way for an immense competition between codemakers and codebreakers, with a hope that the process will converge on a PUF design that is highly resilient to known attacks.

2.4.1 Modeling Attacks on Arbiter PUFs

Because of the additive nature of path delay, several successful attempts have been made to attack arbiter PUFs using *modeling attacks* [15, 25–27]. The basic idea is to observe a set of CRP pairs through physical measurements and use them to derive runtime delays using various Machine Learning algorithms. In [26], it was shown that arbiter PUFs composed of 64 and 128 stages can be attacked successfully using several Machine Learning techniques, achieving a prediction accuracy of 99.9% by observing around 18,000 and 39,200 CRPs respectively. These attacks are possible only if the attacker can measure the response, i.e., the output of a PUF circuit is physically available through an I/O pin in an integrated circuit. If the response of a PUF circuit is used internally for some security operations and is not available to the external world, modeling attacks cannot be carried out unless there is a mechanism to internally probe the output of the PUF circuit. However, probing itself can cause delay variations, thereby affecting the accuracy of response measurement.

Several non-linear versions have been proposed in the literature to improve the modeling attack resistance of arbiter PUFs. One of them is feed-forward arbiter PUFs, in which some of the challenge bits are generated internally using an arbiter as a result of racing conditions at intermediate stages. However, such a construction has reliability issues because of the presence of more than one arbiter in the construction. This is evident from the test chip data provided in [16]. It was reported that feed-forward PUFs have $d_{\text{intra}} \approx 10\%$ and d_{inter} 38%. Modeling attacks have been attempted on feed-forward arbiter PUFs and the attacks used to model simple arbiter PUFs were found to be ineffective when applied to feed-forward arbiter

PUFs. However, by using evolution strategies (ES), feed-forward arbiter PUFs have been shown to be vulnerable to modeling attacks [26]. Non-linearities in arbiter PUFs can also be introduced using several simple arbiter PUFs and using an XOR operation across the responses of simple arbiter PUFs to obtain the final response. This type of construction is referred to as XOR arbiter PUF. Though XOR arbiter PUFs are tolerant to simple modeling attacks, they are vulnerable to advanced Machine Learning techniques. For example, a 64-stage XOR arbiter PUF with 6 XORs has been attacked using 200,000 CRPs to achieve a prediction accuracy of 99% [28–31]. All these modeling attacks impose a strong pressing need for the design of a modeling-attack-resistant arbiter PUF.

2.4.2 Attacks Against SRAM PUFs

Since SRAM PUFs based on power-up states only have a single challenge, the response must be kept secret from the external world. Modeling attacks are not relevant for SRAM PUFs. However, other attacks such as side-channel and virus attacks can be employed, but they are not covered in this chapter.

2.5 Constructively Applying Machine Learning on PUFs

Reliability is an important feature of PUFs that reflects their ability to produce the same response for a particular challenge despite the existence of noise. So it is possible that a PUF operating in different conditions generates a different response to the same challenge vector. The PUF output is therefore a function of not only the challenge and process variations, but also of transient environmental conditions. Therefore, to make a PUF highly reliable, unstable CRPs that are easily flipped by environmental noise and aging should be corrected or even not used. Generally, the reason that PUF responses are unreliable is because supply voltage and temperature variations can overcome the impacts of process variations and flip the responses. Besides the transient noise, device aging is also an important but rarely studied source of unreliability in PUFs. Unlike environmental noise that temporarily flips PUFs' challenge- response pairs (CRPs) (PUFs work more reliably when the supply voltage and temperature return back to normal), device aging causes a permanent change in the behavior of a PUF. Device aging is usually caused by negative bias temperature instability, hot carrier injection, time-dependent dielectric breakdown, and electromigration [32, 33]. In this section, we present some methodologies that use Machine Learning-related techniques to improve the reliability of PUFs.

2.5.1 Using Machine Learning to Improve the Reliability of Arbiter PUFs

2.5.1.1 Mechanism of Arbiter PUFs

A Machine Learning-based modeling method is proposed in [34] that helps with generating reliable CRPs without extra resources other than a normal PUF circuitry. The unreliability source of a PUF is classified into two aspects: transient noise (e.g., temperature and supply voltage variations) and device aging. A Machine Learning model can be trained for PUF characterization and the model can be used for identifying and filtering out the unreliable challenge vectors, allowing higher reliability to be achieved. In an n-bit arbiter PUF (Figure 2.1), the propagation delay from the input to the first stage to the top and bottom outputs of the ith stage can be defined as D_{top}^i and D_{bottom}^i respectively. The delay mismatch between two delay paths is summed up as the timing difference between D_{top}^n and D_{bottom}^n (Figure 2.3). By mapping original challenge $c_i \in \{0, 1\}$ into $c_i \in \{-1, 1\}$, the path delay can be formulated as:

$$D_{top}^i = \frac{1+c_i}{2}\left(t_{top}^i + D_{top}^{i-1}\right)$$

$$+ \frac{1-c_i}{2}\left(t_{u\_across}^i + D_{bottom}^{i-1}\right)$$

$$D_{bottom}^i = \frac{1+c_i}{2}\left(t_{bottom}^i + D_{bottom}^{i-1}\right) \qquad (2.3)$$

$$+ \frac{1-c_i}{2}\left(t_{d\_across}^i + D_{top}^{i-1}\right)$$

where $D_{top}^0 = D_{bottom}^0 = 0$ and $t_{top}^i, t_{bottom}^i, t_{u\_across}^i, t_{d\_across}^i$ represent the four possible delays through the ith stage. Denoting the delay difference between top and bottom arbiter inputs as $T_A - T_B$, following Equation 2.3, the delay difference between two paths is $T_A - T_B = D_{top}^n - D_{bottom}^n$. The response ($r$) of an arbiter PUF of n-bit length is therefore determined by the sign of $T_A - T_B$ (Equation 2.4).

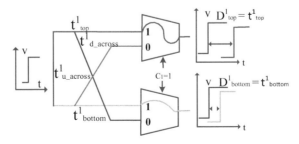

FIGURE 2.3
Propagation paths through the delay cells of an arbiter PUF.

$$r = \begin{cases} 0, & \text{if} \quad \text{sgn}(T_A - T_B) > 0 \\ 1, & \text{if} \quad \text{sgn}(T_A - T_B) < 0 \end{cases} \tag{2.4}$$

From Equation 2.4, it is clear that a PUF response is flipped when the sign of $T_A - T_B$ changes, either from positive to negative or vice versa.

2.5.1.2 Modeling the $T_A - T_B$ of Arbiter PUFs

Because a PUF operating in different conditions can generate a different response to the same challenge vector, the PUF output is a function of not only the challenge and process variations, but also of transient environmental conditions. Therefore, to make a PUF highly reliable, unstable CRPs that are easily flipped by environmental noise and aging should not be used or corrected. Toward this end, a Machine Learning-based modeling method is proposed in [34], that helps with generating reliable CRPs without extra circuitry other than a normal PUF. Based on this technique, the unreliability source of a PUF is classified into two aspects: transient noise (e.g., temperature and supply voltage variations) and device aging. A Machine Learning model can be trained for PUF characterization and utilize the model for identifying and filtering out the unreliable challenge vectors for each PUF, allowing higher reliability to be achieved.

A n-bit arbiter PUF is composed of n stages, with each stage employing two 2:1 MUXs, as depicted in Figures 2.1 and 2.3. A challenge vector $\mathbf{C} = \{c_1, c_2, \ldots c_n\}$ is applied as the control signals for all stages to configure two paths through the PUF toward the arbiter; at each stage the paths are configured to be either straight or crossing. To explore the impact of noise in more detail, simulations on a set of 64–bit arbiter PUFs show that the flipped responses are from challenges that correspond to $T_A - T_B$ in a small range:

$$\text{DD}_{\text{umin}} \leq T_A - T_B \leq \text{DD}_{\text{umax}} \tag{2.5}$$

where DD_{umin} and DD_{umax} represent the minimum/maximum delay difference between T_A and T_B of unreliable challenges. Based on this observation, it can be concluded that if only the challenge vectors satisfying either $T_A - T_B > \text{DDumax}$ or $T_A - T_B < \text{DDumin}$ are applied to the PUF, then the responses of the PUF will be reliable.

Since knowing the $T_A - T_B$ of each challenge vector makes it possible to get reliable CRPs by avoiding challenges with smaller delay difference, these challenges with smaller $T_A - T_B$ are likely to be unreliable and therefore can be discarded. However, probing inside a PUF to measure $T_A - T_B$ is not practical, since it will introduce extra bias into the circuit operation and impact the original results. A Machine Learning modeling method is proposed to accomplish this job by building a PUF model. With a PUF model, the $T_A - T_B$ value for each challenge vector can be derived. Two parameters are defined below:

$$\alpha_i = \left(t_{\text{top}}^i - t_{\text{bottom}}^i + t_{\text{d\_across}}^i - t_{\text{u\_across}}^i \right)/2$$

$$\beta_i = \left(t_{\text{top}}^i - t_{\text{bottom}}^i - t_{\text{d\_across}}^i + t_{\text{u\_across}}^i \right)/2 \tag{2.6}$$

Based on Equation 2.6, for a given challenge vector $\mathbf{C} = \{c_1, c_2, \ldots c_n\}$, the corresponding response generation can be modeled by accumulating the delay mismatches through delay stages as:

$$T_A - T_B = \alpha_1 k_0 + \ldots + \left(\alpha_n + \beta_{n-1} \right) k_{n-1} + \beta_n k_n \tag{2.7}$$

where $k_n = 1$ and $k_i = \prod\limits_{j=i+1}^{n} c_j$ reflecting the number of times that the rising edges will change tracks between the ith stage and the arbiter. Thus, knowing the challenge and α_i and β_i, $i \in (1 \ldots n)$ makes it possible to compute DD for any challenge. By denoting the delay parameters of an arbiter PUF with vector $\mathbf{p}_{\text{model}} = \{\alpha_1, \alpha_2 + \beta_1, \ldots \alpha_n + \beta_{n-1}, \beta_n\}$, and defining challenge features as $\mathbf{k} = \{k_0, k_1 \ldots k_n\}$, the model-predicted $T_A - T_B$ of each challenge vector can be denoted as:

$$T_A - T_B = \left\langle \mathbf{p}_{\text{model}}, \mathbf{k} \right\rangle \tag{2.8}$$

Based on the equations above, Machine Learning modeling technique can be used to model $\mathbf{p}_{\text{model}}$ and predict $T_A - T_B$. To accomplish this, a set of known CRPs can be used to train a support vector machine (SVM) classifier. SVM models are powerful learning tools that can perform binary classification of data; classification is achieved by a linear or non-linear separating surface in the input space of the data set. Previously, SVMs have been widely used in attacking arbiter PUFs [16, 28, 35]. Note that finding the accurate delay difference is not an explicit objective of the SVM model, as the SVM model only seeks a value of $\mathbf{p}_{\text{model}}$ that can accurately predict responses. Fortunately, because the raw PUF responses are determined by the sign of $T_A - T_B$ (Equation 2.4), a model that is good at predicting unknown responses is also accurate in quantifying the values of $T_A - T_B$. Therefore, the flow using a PUF model to enhance PUF reliability becomes: (1) train a binary classifier for a PUF, and (2) use the trained PUF model to quantify the delay difference induced by each challenge. In other words, the model-quantified delay difference can be used to infer whether or not a challenge will generate a reliable PUF response. The proposed method can be evaluated by comparing the model-predicted $T_A - T_B$ with the golden value of $T_A - T_B$ that is extracted from a PUF. If the PUF model works well for response prediction, then it is expected that the ρ in Equation 2.9 is close to 1 when the size of CRP dataset used to train the model is large enough.

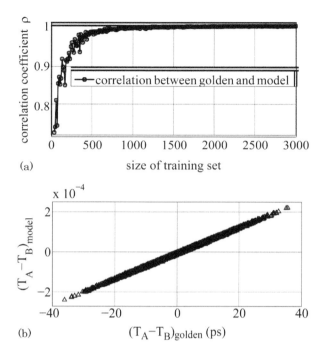

(a)

(b)

FIGURE 2.4

In Equation 2.4a, the correlation coefficient ρ between golden delay difference and model-predicted delay difference. While the PUF training size is increasing, higher ρ is achieved. In Equation 2.4b, based on the model trained with 3,000 CRPs, there is good agreement between $(T_A - T_B)_{golden}$ and $(T_A - T_B)_{model}$ for 2,000 random challenges: (a) Correlation coefficient increases with training size; (b) Good agreement between golden and model-predicted delay difference.

$$\text{corr}\left(\left(T_A - T_B\right)_{golden}, \left(T_A - T_B\right)_{model}\right)$$
$$= \text{corr}\left(\left(T_A - T_B\right)_{golden}, \left\langle \mathbf{p}_{model}, \mathbf{k}\right\rangle\right) = \rho \qquad (2.9)$$

The correlation coefficient (ρ) between $(T_A - T_B)_{golden}$ (from simulation) and $(T_A - T_B)_{model}$, which is calculated as ($\mathbf{p}_{model}, \mathbf{k}$) is as shown in Figure 2.4a. As can be seen, although $(T_A - T_B)_{golden}$ and $(T_A - T_B)_{model}$ are different in scale, the correlation between them is very high. Note that it is not required to know the exact value of delay differences to select the reliable challenges, but only the relative magnitudes of delay differences are needed. From Figure 2.4a,b, it can be inferred that the model-predicted $T_A - T_B$ values have a good agreement with that from the PUF circuit.

2.5.1.3 Improving PUF Reliability with the PUF Model

As concluded above, the flipped PUF responses are only those satisfying $DD_{umin} \leq (T_A - T_B)_{golden} \leq DD_{umax}$. However, due to the large size of a PUF CRP space, 2^n for a n-bit arbiter PUF, it is necessary to reconsider

---

Algorithm 1: Use ML model to compute the range of model-predicted delay differences that are likely to be unreliable for a given PUF. Challenges predicted to have delay differences inside this range will not be applied to the PUF, and this will improve the overall PUF reliability. Reliability can be improved by using a larger value of dr to discard more challenges.

Input: A discard ratio dr and a set of challenges and corresponding responses obtained from a single PUF at nominal supply voltage and temperature.

Output: A range $[DD_{min}, DD_{max}]$ of delay differences to consider as unreliable for this PUF.

1: Let $\mathbf{k}$ be the challenges mapped to challenge features (Equations 2.7 and 2.8)

2: $\mathbf{p}_{model} \leftarrow \mathbf{SVM}(\mathbf{k}, responses)$ {train the PUF model}

3: $\mu_p = \text{avg}\langle \mathbf{p}_{model}, \mathbf{k} \rangle$ {mean predicted delay difference}

4: $\sigma_p^2 = \text{var}\langle \mathbf{p}_{model}, \mathbf{k} \rangle$ {variance of predicted delay difference}

5: the distribution of delay differences across all challenges is modeled to be $N(\mu_p, \sigma_p^2)$

6: $DD_{min} = F^{-1}(0.5 - dr/2) = \mu_p + \sigma_p \Phi^{-1}(0.5 - dr/2)$

7: $DD_{max} = F^{-1}(0.5 + dr/2) = \mu_p + \sigma_p \Phi^{-1}(0.5 + dr/2)$ {delay difference cutoffs based on PUF model $\mathbf{p}_{model}$ and selected challenge features $\mathbf{k}$ (Equation 2.11)}

8: **return** $[DD_{min}, DD_{max}]$

---

how to build the PUF model to predict the delay cutoffs that will rule out the discarded ratio (dr) of the overall challenge space. For this purpose, there are still two questions that need to be answered:

1. How many CRPs are enough to train an accurate model to characterize the DD_{umax} and DD_{umin} cutoffs for each PUF?
2. For a given PUF model, what is the dr of CRPs that must be filtered to achieve an expected reliability level?

The first question can be answer by employing techniques in Algorithm 1. The physical features of PUFs follow Gaussian distribution, therefore the $T_A - T_B$ of each applied challenge vector will also follow a Gaussian distribution. A model can be firstly built to model this distribution, as shown in Equation 2.10. By selecting and applying the challenges randomly, the training set of a PUF ensures that it covers more unreliable range.

$$F\left(\text{DD},\mu_p,\sigma_p\right) = \frac{1}{\sigma_p\sqrt{2\pi}}\exp^{-\frac{(\text{DD}-\mu p)^2}{2\sigma_p^2}} \tag{2.10}$$

The unreliable responses are only related to challenges that are generating smaller $T_A - T_B$, therefore, even without knowing the exact range of the delay difference for these unreliable CRPS, it can be quantified by applying Quantile function. For example, by denoting the Probit function of standard normal distribution with $\Phi^{-1}(dr)(dr \in (0, 1))$, the range of the delay difference that should be discarded can be expressed in Equation 2.11 (as steps 6 and 7 in Algorithm 1), where dr stands for the ratio of challenges that should be discarded from the CRP database of each PUF. Apparently, there exists a trade-off between the value of dr and the number of usable challenges. For example, a larger dr implies that more challenges should be discarded, while a smaller dr means less challenges will be filtered out.

$$F^{-1}\left(dr\right) = \mu_p + \sigma_p\Phi^{-1}\left(dr\right) \tag{2.11}$$

By applying a challenge to the trained PUF model, if its delay difference satisfies $((T_A - T_B)_{\text{model}} \notin [\text{DD}_{\text{min}}, \text{DD}_{\text{max}}])$, then it will be marked as reliable and applied on PUFs. By formulating a set of such reliable challenges and comparing the corresponding responses with the golden data set, it is found that as the training size of a PUF model increases, the characterized cutoffs DD_{min} and DD_{max} also become more accurate, as shown in Figure 2.5. As a result, fewer challenges need to be discarded to achieve the same reliability.

2.5.2 Using Machine Learning to Model the Data Retention Voltage of SRAMs

Data retention voltage (DRV) is a commonly studied feature of SRAMs in low-power research. A SPICE simulator is usually employed to characterize the lowest DRV of different CMOS technology nodes. However, the time consumed by a SPICE simulator is relatively high to formulate the DRV of an SRAM cell. This is because, multiple supply voltage values must be applied to find the maximum voltage that induces a data retention failure. In [10], it is reported that simulating a single test voltage on a single SRAM cell for 2 ms has a runtime of 0.17 s with an Intel Xeon E5-2690 processor running at 2.90 GHz with 64 GB of RAM.

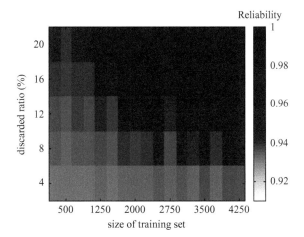

FIGURE 2.5
Validation under aging and environmental noise, across all of the simulated PUF instances. Trade-off between training size and discarded ratio can be seen in the figure. A larger *dr* is conservative and can compensate for the lower quality delay predictions of a model trained from a smaller training set.

To save the DRV simulation time, an alternative method is to predict DRV using a device model. According to [36], the DRV of an SRAM cell is determined by the environmental temperature and the process variations of the transistors. Therefore, the DRV value of an SRAM cell can be formulated as a function of physical features like temperature T and transistor width, length, and threshold voltage (W, L, and V_{th} respectively). To study this relationship, Qin et al. [36] propose an analytical model as shown in Equation 2.12, where DRV_r is the DRV at room temperature, and DRV_f is defined in Equation 2.13 with ΔT representing the temperature difference from room temperature. Terms a_i, b_i, and c in Equation 2.13 are fitting coefficients and their values are determined empirically for each CMOS technology process [36].

$$DRV = DRV_r + DRV_f \qquad (2.12)$$

$$DRV_f = \sum_{i=1}^{6} a_i * \frac{\Delta\left(W_i/L_i\right)}{W_i/L_i} + \sum_{i=1}^{6} b_i * \Delta\left(V_{thi}\right) + c * \Delta T \qquad (2.13)$$

Although this model can be used to estimate the DRV of an SRAM cell, it has two weaknesses that create the need for a more advanced model:

1. To formulate a specified DRV value with Equation 2.13, firstly the user needs to know the DRV_r for each SRAM cell. However, this physical feature cannot be expressed as a function of transistor parameters and can only be characterized through hardware measurement or computationally expensive circuit-simulation.

2. It is impractical to apply the same coefficients a_i, b_i and c for different SRAM cells. In reality, the DRV of different cells increases according to different coefficients depending on their unique process variations. This distinction is especially important in building the model.

2.5.2.1 Predicting DRV Using Artificial Neural Networks

To address the two aforementioned weaknesses, Xu et al. propose to use Machine Learning for DRV prediction [10]. The proposed technique can predict the DRV value for an SRAM cell by feeding the physical parameters into a Machine Learning model. The basis of this technique is that for a certain CMOS technology node, the values of process variations only vary over a bounded range, therefore the DRV values also fall into a bounded range $[DRV_{min}, DRV_{max}]$. Based on the modeling technique, the range of DRVs is firstly divided into K labels with each standing for a smaller range with size DRV (Equation 2.14). By training a Machine Learning model that maps the physical features into these K DRV classes, the DRV value for unknown SRAM cells can be predicted with the model, as shown in Figure 2.6.

$$\left[DRV_{min}, DRV_{max}\right] = \{[DRV_{min}, DRV_{min} + \Delta DRV) \cup$$
$$[DRV_{min} + \Delta DRV, DRV_{min} + 2^*\Delta DRV) \cup \dots \quad (2.14)$$
$$\dots \cup [DRV_{max} - \Delta DRV, DRV_{max}]\}$$

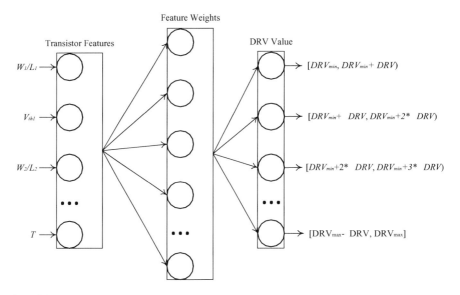

FIGURE 2.6
The layer composition of artificial neural network for DRV classification and prediction.

An artificial neural network (ANN) is a widely used Machine Learning method to solve the so-called multi-classification problems. More specifically, to model the DRV values of SRAMs, the outputs of an ANN network are used for different DRV classes. During the modeling training, the SRAM parameters are used to optimize the internal layers and neuron nodes of the ANN network and minimize the general prediction errors. A set of samples (including the DRV values and corresponding physical characteristics of SRAM cells) are collected by SPICE simulation and used to train an ANN model.

2.5.3 Performance of DRV Model

The ANN-model-predicted DRV values are compared with the SPICE-simulated ones in Figure 2.7, where the R value denotes the correlation between the model outputs and golden values. For the ANN-based DRV model, it is noted that there is a high correlation between prediction and output for all data sets. Before the ANN model proposed in [10], the linear model described in Equation 2.13 was widely used to model the DRV value of SRAM designs, which can be optimized with the linear regression (LR) method that fits data set with linear coefficients. The most common type of LR is a "least-squares fit", which can find an optimal line to represent the discrete data points. In an LR model, the same physical parameters of ANN models are defined as input training features $p=\{p_i, |p_i \in \{W_1/L_1, W_2/L_2, \dots, T\}\}$. By denoting the linear coefficients with $\theta = \{\theta_0, \theta_1, \dots, \theta_n\}$:

$$h_\theta(p) = \theta_0 + \theta_1{}^*p_1 + \cdots + \theta_n{}^*p_n \qquad (2.15)$$

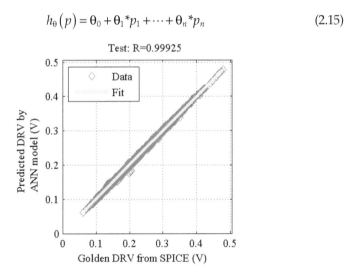

FIGURE 2.7
Training results based on neural network model, across three data sets. R denotes the correlation between golden DRV data from SPICE simulation, and predicted DRV value from our model.

where θ stands for the set of coefficients (e.g. a_i and b_i as shown in Equation 2.13). Each training sample is composed of transistor feature set p and the corresponding golden DRV value DRV_{golden} from SPICE simulation. Based on the "least-squares fit" rule, the cost function of m training examples can be expressed as:

$$J(\theta) = \frac{1}{2m} \sum_{k=1}^{m} \left(h_\theta \left(p^{(k)} \right) - DRV_{\text{golden}}^{(k)} \right)^2 \tag{2.16}$$

where $p^{(k)}$ corresponds to the training features of the kth training sample, like the transistor sizes and temperature. To obtain the optimal θ, "gradient descent" can be applied simultaneously on each coefficient θ_j, $j \in (1, 2 \ldots n)$:

Repeat{

$$\theta_j := \theta_j - \alpha \frac{\partial J(\theta)}{\partial \theta_j}$$

$$= \theta_j - \alpha \frac{1}{m} \sum_{i=1}^{m} \left(h_\theta \left(p^{(k)} \right) - DRV_{\text{golden}}^{(k)} \right) p_i^{(k)} \tag{2.17}$$

}

where α is the learning rate of linear regression model, $p_i^{(k)}$ is the ith feature of the kth training sample.

The experimental results demonstrate that the neural network model achieves smaller prediction errors than the LR model. According to [10], the mean μ and standard deviation σ of prediction error for the neural network model are 0.01 mV and 0.35 mV respectively, while those of the linear regression model are 0.041 mV and 0.9 mV. Hence, it can be concluded that the neural network model outperforms the linear model in modeling DRV of SRAMs. This is because varied weights and bias are employed in the ANN model for different feature patterns, whereas the linear model formulates all input features with the same optimized coefficients.

2.6 Using Machine Learning to Mitigate the Impact of Physical Disorder on TDC

This section presents a case study that uses Machine Learning to help with mitigating the negative impact of physical disorder on time-to-digital converter (TDC) designs. Instead of discussing the traditional TDC schemes that are based on delay line, the TDC design scheme presented in this section is the configurable compact algorithmic TDC (CCATDC) proposed in [37] and [38]. For brevity, this section does not cover the design scheme of this TDC architecture but focuses more on the usage of Machine Learning for mitigating process variations of this circuitry.

2.6.1 Background of TDC

High-resolution time measurement is a common need in modern scientific and engineering applications, such as time-of-flight measurement in remote sensing [39, 40], nuclear science [41], biomedical imaging [42], frequency synthesizer and time jitter measurement for RF transceivers in wireless communication [43]. To advance the post-processing of time signals, TDC that bridges time measurements and digital electronic devices is proposed. As a measurement system, TDC quantifies the time interval between two events by digitizing it, which greatly favors the post-processing in a digital way. Various TDC schemes have been proposed, such as the cyclic successive approximation based TDC, pulse interpolation based TDC, and delta-sigma based TDCs [44, 45].

Most conventional TDC designs are implemented with delay lines. These TDCs are usually composed of two channels for two input signals,

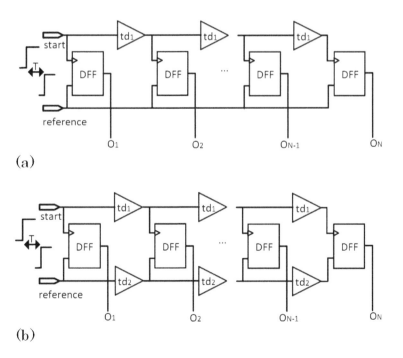

(a)

(b)

FIGURE 2.8
Schematic of traditional delay-line-based TDCs: (a) Delay-line-based TDC; (b) Vernier delay-line-based TDC.

as shown in Figure 2.8a. The time difference between the two input signals, *start* and *reference*, is digitized into binary codes. To characterize the time difference (TD) between two signals, the *start* signal is sequentially postponed by the delay elements. Theoretically, each delay element "slows down" the *start* signal by a constant time scale (td1 in Figure 2.8b), the output of each delay element is then compared with the *reference* signal to generate a binary output. The highest resolution of a delay-line-based TDC is the delay of each single element, which is limited by the CMOS fabrication technology node (more advanced CMOS fabrication technology achieves higher resolution). To improve the resolution and conversion accuracy of such TDCs, several techniques have been proposed, such as stretching pulse [46], using a tapped delay line [47] and employing a differential delay line [48].

As it becomes more difficult to control the process variations of CMOS fabrication with advanced technology nodes, the physical deviation from designed value also becomes a big concern that limits the resolution of TDC design. For example, the fabricated delay elements in a TDC circuitry usually deviate from the designed (nominal) delay length, for example, T_{d1} in Figure 2.8a. Such process variations introduce design uncertainty into the time-to-digital conversion and decreases the conversion accuracy [2].

2.6.2 Mitigating the Process Variations by Reconfiguring the Delay Elements

The formulation of physical disorder of CMOS devices come from the fabrication stage, during which the length of the TDC delay element deviates from the designed value. Therefore, one solution to mitigate such impact is designing the delay element with an adjustable length and then regulating the length of the delay chain circuitry to improve its performance. This philosophy is based on the truth that once an electronic circuit is fabricated, it is infeasible to remove the process variations. Fortunately, the actual in-path delay can be configured by using an adjustable delay line, thereby improving the robustness of CCATDC. There are many possible realizations of the adjustable delay element, which may be voltage-, current- or digital-controlled.

The schematic of an adjustable delay channel is shown in Figure 2.9, which consists of two parts: the constant delay line D_{const} and the adjustable delay line D_{adj}. Note that there still exist process variations in the constant delay line that deviate it from the designed delay length. The purpose of adding it is to roughly calibrate the reference time T_{ref} for the purpose of coarse-tuning. The high-resolution adjustment is realized with the adjustable delay line, which fulfills the job of fine-tuning. To achieve higher resolution, two parallel delay chains are utilized in two channels to form a Vernier architecture. The control signals of the adjustable delay line can be supplied by an external calibration circuit.

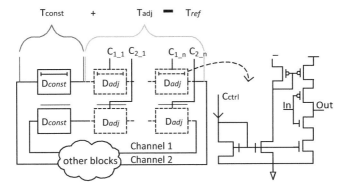

FIGURE 2.9
Schematic of the proposed configurable delay element, which is composed by two blocks, D_{const} and D_{adj}. There are many possible implementations of the adjustable delay element, in this figure a current-controlled delay element is shown as an example.

2.6.3 Delay Chain Reconfiguration with Machine Learning

An important purpose of configuring the propagation delay of each element is to minimize the deviation of process variations of CMOS transistors. However, since all components in the CCATDC design are of nanometer magnitude, it is again impractical to measure such process variations with external instrument [34]. In [34] and [10], Machine Learning technique was utilized to characterize the microscopic process variations and related physical features. This section continues this thread, looking at how Machine Learning techniques are employed to characterize and configure the delay length of CCATDC. More specifically, the backward propagation of errors (backpropagation) algorithm is used. Backpropagation is a widely used technique in training artificial neural networks, and is often used with other optimization methods such as gradient descent. Usual backpropagation training consists of two phases: *propagation* and *weight updating*. Once an input vector is applied to the neural network model, it will be propagated from the input layer to the output layer. The output value of each input vector will be obtained and compared with the desired (golden) one; a loss function will be used to calculate the error for each neuron in the trained model, and update the weight parameters correspondingly.

Within the framework, the controlling signals (for example, the controlling current inputs) of the adjustable delay line (as shown in Figure 2.9) are fed into the backpropagation model as input vectors: $\mathbf{C} = (C_0, C_1, \ldots, C_{n-1})$ in Algorithm 2. To comprehensively configure the delay chain, m different time inputs $\mathbf{T} = (t_0, t_1, \ldots, t_{m-1})$ and corresponding digital conversion outputs $\mathbf{O} = (O_0, O_1, \ldots, O_{m-1})$ are used to train the model. The accuracy of CCATDC is optimized by setting Δ_{err} as the threshold of conversion error. Before the model is trained, weight parameters are initialized as 0. During the model training, the controlling signals are fed into the HSPICE simulation and corresponding conversion outputs are obtained (line5), the simulated results are then compared with the golden values (line7). The conversion error is then backpropagated to the network (line8) and the controlling input vector is optimized to achieve higher conversion accuracy (line9). The configuration procedure will terminate while the conversion error is below the pre-set threshold Δ_{err} (line10).

2.6.3.1 Performance of Configurable Compact Algorithmic TDC

The proposed CCATDC architecture is implemented with the PTM 45nm standard cell libraries [49] and its performance is tested by applying random process variations on all transistors of the circuit, transient noise is also added in the simulation. The experimental result demonstrates that, CCATDC is robust against process variations and transient noise, the conversion accuracy is much higher if more conversion bits are utilized.

Algorithm 2: Given a CCATDC chip, configure the delay elements to improve the time digitization accuracy.
Input: a configurable compact algorithmic TDC
Input: n control signals $\mathbf{C} = (C_0, C_1, \ldots, C_{n-1})$
Input: weight vector with k parameters, $\mathbf{\Theta} = (\theta^0, \theta^1, \ldots \theta^k)$
Input: m time input $\mathbf{T} = (t_0, t_1, \ldots, t_{m-1})$ and corresponding ideal conversion output $\mathbf{O} = (O_0, O_1, \ldots, O_{m-1})$ for the designed T_{ref} and amplification gain
Input: Δ_{err}, the threshold of conversion error
　1: Initialize weight parameters $\theta^i \leftarrow 0$
　2: Initialize error parameters $\Delta^i \leftarrow \Delta_{err}$
　3: **while** max(Δ^i) >= Δ_{err} **do**
　4: **for** $i := 0$ **to** $m - 1$ **do**
　5: $\hat{O}_i = HSPICE\,(t_i, \mathbf{C})$
　6: **end for**
　7: compute conversion error $\Delta = \left(\hat{\mathbf{O}} - \mathbf{O}\right)$
　8: BackwardPropagateError(Δ)
　9: UpdateWeights($\mathbf{\Theta}$ $\mathbf{C}$)
　10: **end while**
　11: **return C**

2.7 Conclusion

In this chapter, an overview of some of the latest CMOS circuits in the literature, like PUFs and TDCs, is provided. Notably, these two circuit primitives have different relationships with physical disorder: PUF is built on the physical disorder of CMOS fabrication, while the uncontrollable process variations of a CMOS transistor degrade the resolution of TDCs. Apart from constructively leveraging these physical disorders, the modeling of them also provides a new way to study process variations and environmental variations on the performance of CMOS circuitry. Based on the case studies presented in this chapter, we can see that Machine Learning is a good tool for studying the microscopic physical disorder of CMOS circuitry and therefore provides a unique perspective to guide design. The performance metrics used for analyzing these circuits are identified and several Machine Learning techniques are employed in a constructive way to benefit the performance of these designs.

Bibliography

1. S. R. Sarangi, B. Greskamp, R. Teodorescu, J. Nakano, A. Tiwari, and J. Torrellas. Varius: A model of process variation and resulting timing errors for microarchitects. *IEEE Transactions on Semiconductor Manufacturing*, 21(1):3–13, 2008.
2. J.-P. Jansson, V. Koskinen, A. Mantyniemi, and J. Kostamovaara. A multichannel high-precision CMOS time-to-digital converter for laser-scanner-based perception systems. *IEEE Transactions on Instrumentation and Measurement*, 61(9):2581–2590, 2012.
3. I. Kim, A. Maiti, L. Nazhandali, P. Schaumont, V. Vivekraja, and H. Zhang. From statistics to circuits: Foundations for future physical unclonable functions. In A.-R. Sadeghi and D. Naccache, editors, *Towards Hardware-Intrinsic Security*, Information Security and Cryptography, pages 55–78. Springer: Berlin, Heidelberg, 2010.
4. D. Blaauw, K. Chopra, A. Srivastava, L. Scheffer. Statistical timing analysis: From basic principles to state of the art. *IEEE Transactions on Computer-Aided Design of Integrated Circuits and Systems*, 27(4):589–607, 2008.
5. W. Wang, V. Reddy, B. Yang, V. Balakrishnan, S. Krishnan, and Y. Cao. Statistical prediction of circuit aging under process variations. In *IEEE Custom Integrated Circuits Conference*, pages 13–16, 2008.
6. D. Forte and A. Srivastava. On improving the uniqueness of silicon- based physically unclonable functions via optical proximity correction. In *IEEE/ACM Design Automation Conference*, pages 96–105, June 2012.
7. V. Vivekraja and L. Nazhandali. Feedback based supply voltage control for temperature variation tolerant pufs. In *IEEE International Conference on VLSI Design*, pages 214–219, Washington, DC, USA, 2011. IEEE Computer Society.

8. M.-D. Yu and S. Devadas. Secure and robust error correction for physical unclonable functions. *IEEE Design Test of Computers*, 27(1):48–65, 2010.

9. C. Bösch, J. Guajardo, A.-R. Sadeghi, J. Shokrollahi, and P. Tuyls. Efficient helper data key extractor on FPGAs. In *Cryptographic Hardware and Embedded Systems*, volume 5154 of *Lecture Notes in Computer Science*, pages 181–197. Springer: Berlin, Heidelberg, 2008.

10. X. Xu, A. Rahmati, D. E. Holcomb, K. Fu, and W. Burleson. Reliable physical unclonable functions using data retention voltage of SRAM cells. *IEEE Transactions on Computer-Aided Design of Integrated Circuits and Systems*, 34(6):903–914, 2015.

11. X. Xu and D. Holcomb. A clockless sequential PUF with autonomous majority voting. In *Great Lakes Symposium on VLSI, 2016 International*, pages 27–32. IEEE, 2016.

12. X. Xu and D. E. Holcomb. Reliable PUF design using failure patterns from time-controlled power gating. In *Defect and Fault Tolerance in VLSI and Nanotechnology Systems (DFT), 2016 IEEE International Symposium on*, pages 135–140. IEEE, 2016.

13. X. Xu, V. Suresh, R. Kumar, and W. Burleson. Post-silicon validation and calibration of hardware security primitives. In *VLSI (ISVLSI), 2014 IEEE Computer Society Annual Symposium on*, pages 29–34. IEEE, 2014.

14. R. Maes and I. Verbauwhede. Physically unclonable functions: A study on the state of the art and future research directions. In A.-R. Sadeghi and D. Naccache, editors, *Towards Hardware-Intrinsic Security*, Information Security and Cryptography, pages 3–37. Springer: Berlin, Heidelberg, 2010.

15. G. Hospodar, R. Maes, and I. Verbauwhede. Machine learning attacks on 65nm arbiter PUFs: Accurate modeling poses strict bounds on usability. In *IEEE International Workshop on Information Forensics and Security*, pages 37–42, 2012.

16. D. Lim. Extracting secret keys from integrated circuits. Master's thesis, Massachusetts Institute of Technology, Dept. of Electrical Engineering and Computer Science, 2004.

17. G. Suh and S. Devadas. Physical unclonable functions for device authentication and secret key generation. In *IEEE/ACM Design Automation Conference*, pages 9–14, June 2007.

18. J. Guajardo, S. S. Kumar, G.-J. Schrijen, and P. Tuyls. FPGA intrinsic PUFs and their use for IP protection. In *International workshop on Cryptographic Hardware and Embedded Systems*, pages 63–80,.Springer: Berlin, Heidelberg, 2007.

19. D. E. Holcomb, W. Burleson, and K. Fu. Initial SRAM state as a fingerprint and source of true random numbers for RFID tags. In *Proceedings of the Conference on RFID Security*, 7: 2, 2007.

20. D. E. Holcomb, W. P. Burleson, and K. Fu. Power-up SRAM state as an identifying fingerprint and source of true random numbers. *IEEE Transactions on Computers*, 58(9):1198–1210, 2009.

21. S. Katzenbeisser, U. Kocabaş, V. Rožić, A.-R. Sadeghi, I. Verbauwhede, and C. Wachsmann. PUFs: myth, fact or busted? A security evaluation of physically unclonable functions (PUFs) cast in silicon. In *International Conference on Cryptographic Hardware and Embedded Systems*, pages 283–301. Springer: Berlin, Heidelberg, 2012.

22. D. E. Holcomb, A. Rahmati, M. Salajegheh, W. P. Burleson, and K. Fu. DRV-fingerprinting: Using data retention voltage of SRAM cells for chip identification. In *Radio Frequency Identification. Security and Privacy Issues*, pages 165–179. Springer: Berlin, Heidelberg, 2013.

23. J. Guajardo, S. Kumar, G. Schrijen, and P. Tuyls. FPGA intrinsic PUFs and their use for IP protection. *Cryptographic Hardware and Embedded Systems*, pages 63–80, 2007.

24. A. Kumar, H. Qin, P. Ishwar, J. Rabaey, and K. Ramchandran. Fundamental data retention limits in SRAM standby – experimental results. In *Quality Electronic Design, 2008. 9th International Symposium on*, pages 92–97. IEEE, 2008.

25. M. Majzoobi, F. Koushanfar, and M. Potkonjak. Testing techniques for hardware security. In *IEEE International Test Conference*, pages 1–10, Santa Clara, CA, USA, Oct. 2008.

26. U. Rührmair, F. Sehnke, J. Sölter, G. Dror, S. Devadas, and J. Schmidhuber. Modeling attacks on physical unclonable functions. In *ACM Conference on Computer and Communications Security*, pages 237–249, New York, NY, USA, 2010. ACM.

27. X. Xu, U. Rührmair, D. E. Holcomb, and W. Burleson. Security evaluation and enhancement of bistable ring PUFs. In *International Workshop on Radio Frequency Identification: Security and Privacy Issues*, pages 3–16. Springer: Berlin, Heidelberg, 2015.

28. U. Rührmair, J. Sölter, F. Sehnke, X. Xu, A. Mahmoud, V. Stoyanova, G. Dror, J. Schmidhuber, W. Burleson, and S. Devadas. PUF modeling attacks on simulated and silicon data. *Information Forensics and Security, IEEE Transactions on*, 8(11): 1876–1891, 2013.

29. X. Xu and W. Burleson. Hybrid side-channel/machine-learning attacks on PUFs: a new threat? In *Proceedings of the Conference on Design, Automation & Test in Europe*, page 349. European Design and Automation Association, 2014.

30. U. Rührmair, X. Xu, J. Sölter, A. Mahmoud, M. Majzoobi, F. Koushanfar, and W. Burleson. Efficient power and timing side channels for physical unclonable functions. In *International Workshop on Cryptographic Hardware and Embedded Systems*, pages 476–492. Springer: Berlin, Heidelberg, 2014.

31. U. Rührmair, X. Xu, J. Sölter, A. Mahmoud, F. Koushanfar, and W. Burleson. Power and timing side channels for PUFs and their efficient exploitation. *IACR Cryptology ePrint Archive*, 2013: 851, 2013.

32. S. Khan, S. Hamdioui, H. Kukner, P. Raghavan, and F. Catthoor. Incorporating parameter variations in BTI impact on nano-scale logical gates analysis. In *Defect and Fault Tolerance in VLSI and Nanotechnology Systems (DFT), 2012 IEEE International Symposium on*, pages 158–163. IEEE, 2012.

33. D. Lorenz, G. Georgakos, and U. Schlichtmann. Aging analysis of circuit timing considering NBTI and HCI. In *On-Line Testing Symposium, 2009, 15th IEEE International*, pages 3–8. IEEE, 2009.

34. X. Xu, W. Burleson, and D. E. Holcomb. Using statistical models to improve the reliability of delay-based PUFs. In *VLSI (ISVLSI), 2016 IEEE Computer Society Annual Symposium on*, pages 547–552. IEEE, 2016.

35. S. S. Avvaru, C. Zhou, S. Satapathy, Y. Lao, C. H. Kim, and K. K. Parhi. Estimating delay differences of arbiter PUFs using silicon data. In *2016 Design, Automation & Test in Europe Conference & Exhibition*, pages 543–546. IEEE, 2016.

36. H. Qin, Y. Cao, D. Markovic, A. Vladimirescu, and J. Rabaey. SRAM leakage suppression by minimizing standby supply voltage. In *5th International Symposium on Quality Electronic Design*, pages 55–60, 2004.

37. S. Li and C. D. Salthouse. Compact algorithmic time-to-digital converter. *Electronics Letters*, 51(3):213–215, 2015.

38. S. Li, X. Xu, and W. Burleson. CCATDC: A configurable compact algorithmic time-to-digital converter. In *VLSI (ISVLSI), 2017 IEEE Computer Society Annual Symposium on*, pages 501–506. IEEE, 2017.

39. D. Marioli, C. Narduzzi, C. Offelli, D. Petri, E. Sardini, and A. Taroni. Digital time-of-flight measurement for ultrasonic sensors. *IEEE Transactions on Instrumentation and Measurement*, pages 93–97, 1992.

40. S. Li and C. Salthouse. Digital-to-time converter for fluorescence lifetime imaging. In *Instrumentation and Measurement Technology Conference (I2MTC), 2013 IEEE International*, pages 894–897. IEEE, 2013.

41. N. Bar-Gill, L. M. Pham, A. Jarmola, D. Budker, and R. L. Walsworth. Solid-state electronic spin coherence time approaching one second. *Nature Communications*, 4:1743, 2013.

42. A. S. Yousif and J. W. Haslett. A fine resolution TDC architecture for next generation PET imaging. *IEEE Transactions on Nuclear Science*, pages 1574–1582, 2007.

43. J.-P. Jansson, A. Mantyniemi, and J. Kostamovaara. A CMOS time-to- digital converter with better than 10 ps single-shot precision. *IEEE Journal of Solid-State Circuits*, 41(6):1286–1296, 2006.

44. A. Mantyniemi, T. Rahkonen, and J. Kostamovaara. A CMOS time-to- digital converter (tdc) based on a cyclic time domain successive approximation interpolation method. *IEEE Journal of Solid-State Circuits*, 44(11):3067–3078, 2009.

45. W.-Z. Chen and P.-I. Kuo. A δσ TDC with sub-ps resolution for pll built-in phase noise measurement. In *European Solid-State Circuits Conference, 42nd ESSCIRC Conference, 2016*, pages 347–350. IEEE, 2016.

46. S. Tisa, A. Lotito, A. Giudice, and F. Zappa. Monolithic time-to-digital converter with 20ps resolution. In *Solid-State Circuits Conference, 2003, Proceedings of the 29th European*. IEEE, 2003.

47. R. B. Staszewski, S. Vemulapalli, P. Vallur, J. Wallberg, and P. T. Balsara. 1.3 v 20 ps time-to-digital converter for frequency synthesis in 90-nm CMOS. *IEEE Transactions on Circuits and Systems II: Express Briefs*, 53(3):220–224, 2006.

48. P. Dudek, S. Szczepanski, and J. V. Hatfield. A high-resolution CMOS time-to-digital converter utilizing a Vernier delay line. *IEEE Journal of Solid-State Circuits*, 35(2):240–247, 2000.

49. W. Zhao and Y. Cao. New generation of predictive technology model for sub-45 nm early design exploration. *IEEE Transactions on Electron Devices*, 53(11):2816–2823, 2006.

3

SiGe BiCMOS Technology and Devices

Edward Preisler

CONTENTS

3.1 Introduction

Over the past two decades SiGe BiCMOS has evolved into a dominant technology for the implementation of high-frequency integrated circuits. By providing performance, power consumption, and noise advantages over standard CMOS while leveraging the same manufacturing infrastructure, SiGe BiCMOS technologies can offer a cost-effective solution for challenging Radio Frequency (RF) and high-performance analog circuit applications. Today many cellular phones, wireless-LAN devices, GPS receivers, and optical networking transceivers employ SiGe BiCMOS circuitry for either receive or transmit functions because of these advantages. Recently, advanced-node CMOS devices have achieved performance levels that enable some of these applications to be realized in CMOS alone. But SiGe BiCMOS continues to provide advantages for many leading-edge products, often at a fraction of the cost of the leading-edge CMOS technologies. These existing markets, as well as emerging applications in the use of SiGe for millimeter-wave products, continue to drive SiGe technology development.

In this chapter, we will review SiGe BiCMOS technology and its most significant applications. First, we will provide a basic understanding of how

SiGe devices achieve a performance advantage over traditional bipolar and CMOS devices. Next, we review historical application drivers for SiGe technology and project a roadmap of SiGe applications well into the future. Then, we discuss RF performance metrics for SiGe HBT (heterostructure bipolar transistor) devices, followed by a discussion of how the devices can be optimized to maximize these performance metrics. Finally, we discuss some of the components built around SiGe devices that are part of modern SiGe BiCMOS technologies and make them suitable for advanced RF product design.

3.2 SiGe HBT Device Physics

SiGe heterostructure bipolar transistor (HBT) devices are bipolar junction transistors created using a thin epitaxial base incorporating roughly 8% to 30% atomic germanium content. These devices are usually fabricated alongside CMOS devices with the addition of four to seven masking layers relative to a core CMOS process. SiGe HBTs derive part of their performance benefits from heterojunction effects and part from their epitaxial-base architecture. Heterojunction effects were first described in the 1950s by Kroemer (eventually earning him a Nobel Prize) and more recently summarized by the same author [1]. These effects arise from the combination of different materials (in this case a $Si_{1-x}Ge_x$ alloy and Si) to create a variation in bandgap throughout the device that can be manipulated to improve performance.

Two common techniques for using heterojunction effects to improve performance are depicted in Figure 3.1 where typical doping and germanium

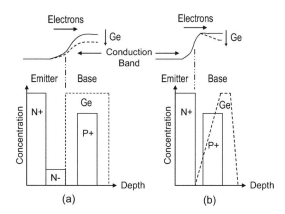

FIGURE 3.1
Common SiGe HBT doping and germanium profiles shown along with resulting band diagrams for (a) a box Ge profile and (b) a graded Ge profile.

profiles are shown along with the resulting conduction band energy profile. The first technique (c.f. Figure 3.1(a)) uses a box-shaped Ge profile. This creates an offset in the conduction band energy level at the emitter-base junction (due to the lower band gap of SiGe relative to Si), lowering the barrier for electron current flow into the base, and increasing the efficiency of electron injection into the base. The band offset in the valence band is relatively unchanged compared to a silicon homojunction and thus holes in the base are injected back into the emitter at roughly the same rate as they would be without the germanium. The achievement of greater electron injection efficiency without also increasing the efficiency of the back-injection of holes from the base results in higher current gain (collector current divided by base current, denoted as β for bipolar transistors). For a homojunction device, an increase in gain can be realized only by either thinning the metallurgical base width (increase in collector current) or increasing the doping in the emitter (decrease in base current). In the case of the SiGe HBT, however, the additional gain provided by the heterostructure effects can be traded off for increased base doping or lower emitter doping to improve base resistance and emitter-base capacitance, resulting in greater RF performance.

The second technique for utilizing heterojunction effects in an HBT (c.f. Figure 3.1(b)) employs a *graded* Ge profile to create a built-in (exists at zero bias) electric field in the base that accelerates electrons, reducing base transit time and improving high-frequency performance. This second technique somewhat offsets the effects of the first technique [2] since using the graded profile necessarily means a reduction in the Ge content at the emitter-base junction, thus reducing the conduction band-lowering effect discussed above. Further, a high Ge concentration on the *collector* side of the Ge base profile can lead to undesirable heterobarrier effects [3]. Thus careful design of the Ge profile throughout the device is a key factor in achieving optimal device performance. Today's SiGe bipolar HBTs make use of these two techniques to varying degrees to create a performance advantage over conventional bipolar devices.

Use of an epitaxially grown base rather than one formed by ion-implantation or diffused junctions is another reason SiGe HBTs exhibit better performance than conventional bipolar devices. The base of a conventional bipolar device is formed by implanting or diffusing base dopant into silicon, which results in a relatively broad base after subsequent thermal processing. Epitaxy allows one to "grow-in" the base doping profile through deposition of doped and undoped Si and SiGe layers controlled to nearly atomic dimensions. This allows the device designer to create an arbitrary as-deposited base profile. An implanted device is limited to skewed Gaussian dopant profiles whose width is a function of implantation energy. Usually, the epitaxy technique is used to distribute the same base dose in a narrower base width, improving transit time through the base and resulting in better high-frequency performance.

Despite the advantages introduced by the epitaxial growth of the base layer, the final dopant profile in the device is largely determined by the subsequent thermal processing of the wafers after the base growth. Due to the large diffusion coefficient of boron (typically used as the base dopant) in silicon, a narrow as-grown base profile might be diffused dramatically by the time the processing is completed. Germanium itself actually serves to arrest the diffusion of boron somewhat, but in modern SiGe HBTs another atomic species, carbon, is added in the epitaxial base of SiGe devices to further arrest the diffusion of boron [4]. A small amount of carbon is added (typically <1% atomic concentration of carbon is used) in the SiGe base during epitaxial deposition such that the electrical behavior is not significantly altered but the material properties are altered to reduce boron diffusion. The carbon helps to maintain a final boron profile closer to the as-deposited profile than it would be without the carbon. The electrical effect is a faster transit time due to a narrower base width, improving high-frequency performance. It should be noted that the introduction of carbon does reduce some of the beneficial band offsets introduced by germanium.

A final note about how the design of the epitaxial base growth affects HBT performance concerns strain. All SiGe HBTs are grown pseudomorphically on a silicon substrate, meaning that the SiGe (or in modern devices SiGe:C) is strained to take on the lattice constant of bulk silicon in the plane of the wafer. Any relaxation of the SiGe layers would generate dislocation-type defects that can short-circuit the emitter through the base to the collector of the device. Thus all SiGe base layers must necessarily be grown pseudomorphically. Since bulk (relaxed) SiGe has a lattice constant larger than that of silicon, the SiGe is always under compressive strain in the plane of the wafer and the lattice stretches out in a direction perpendicular to the surface of the wafer, according to the material's Poisson ratio. This strain can actually serve to enhance both the mobility of electrons traveling vertically through the device and of holes traveling horizontally from the extrinsic base to the intrinsic base. In bulk $Si_{1-x}Ge_x$ the mobility of electrons is actually *lower* than that of bulk silicon until the germanium concentration gets close to 100%. However, strained SiGe can have electron mobilities equal to or even superior to that of bulk silicon [5], thus enhancing the transit of the minority carrier electrons through the base. The introduction of carbon mitigates some of this strain by pushing the SiGe:C layer's lattice constant back closer to bulk silicon. So, again, trade-offs exist in the introduction of carbon in various locations of the epitaxial base growth.

In summary, modern SiGe HBT devices make use of the introduction of both germanium and carbon into the base of the transistor in order to manipulate both the electronic structure and metallurgical structure of the device to achieve performance not otherwise obtainable in bulk silicon devices.

3.3 Applications Driving SiGe Development

Several applications have driven advances in SiGe technology since the first high-speed SiGe bipolar devices were demonstrated in the late 1980s [6]. Initially, SiGe devices were conceived as a replacement to the Si bipolar device for emitter-coupled logic (ECL): high-speed digital ICs where SiGe transistors promised higher f_T, improving gate delay relative to their silicon bipolar or CMOS counterparts. However, advancements in the density, performance, and power consumption of CMOS technology quickly made it the logical choice for all but a few of these applications. So in the mid-1990s SiGe technology appeared to have a limited application base in only specialized, very high-speed digital functions.

However two analog applications have emerged which have subsequently driven the market and roadmap for SiGe BiCMOS technologies over the intervening two decades: commercial RF applications and high-speed wireline applications. These two markets and applications are discussed in detail in the succeeding sub-sections.

3.3.1 Commercial RF Applications

With the boom in wireless communications that began in the mid-1990s, a new application emerged as the primary driver for SiGe BiCMOS technology: the transceiver of a cellular phone. This application was tailor-made for SiGe BiCMOS because it required good high-frequency performance to support carrier frequencies in the 900 MHz to 2.4 GHz range, very low-noise operation (as very small signals must be received and amplified), and large dynamic range (as large output signals are required to drive the output RF signal). In the early 2000s, SiGe BiCMOS was used for a large percentage of the transceiver chips going into cellular phones, but the front-end module (FEM) components that sit between the antenna and the transceiver itself tended to be built using discrete III-V components.

As CMOS progressed to more advanced nodes, however, it became clear that the RF functionality of the transceiver could be realized with the same FET devices used in digital CMOS applications. Thus a low-cost CMOS-only transceiver could perform all of the high-frequency analog circuit functions previously carried out using SiGe HBTs while maintaining the same or better digital functionality as in a SiGe BiCMOS process. Thus, since the late 2000s almost all cellular transceivers have been built using CMOS-only processes.

But SiGe BiCMOS technologies have seen a dramatic resurgence in their utilization in cell phones over the past half-decade due to the proliferation of different RF bands in modern smart phones. While the base transceiver chip is still realized in basic CMOS as discussed above, the FEM chips alluded to previously have partially moved from traditional III-V based discrete devices to SiGe BiCMOS-based FEM-IC chips where multiple RF functions are realized on the same chip. Modern smart phones contain several FEMs within the phone: high- and low-band cellular, 2 and 5 GHz WiFi, Bluetooth, etc. These FEMs all have different requirements and many benefit from the advantages of a SiGe BiCMOS process. For example, the integration of a power amplifier with an RF switch device is very attractive for some of the lower-power FEMs and cannot be realized together monolithically on a III-V die. The SiGe HBT device is ideally suited for both the low-noise amplifier (LNA) on the receive side and the power amplifier (PA) on the transmit side. Thus, just as in the early 2000s when SiGe BiCMOS was the technology of choice for cell phone transceivers, SiGe BiCMOS-based devices are once again vital to the cell phone supply chain, now the largest end-market for semiconductors worldwide [7].

3.3.2 High-Speed Wireline Applications

In addition to wireless transceivers, high-speed fiber-optic and copper wire transceivers also provide a good application for SiGe transistors. This market can be divided by datarate, with the higher datarates requiring more advanced SiGe BiCMOS devices. The first intersection point of SiGe BiCMOS in this market was the 1–3 GBs node (meaning 1–3 billion bits of data per second). The market has since transitioned to 10 GBs and now 100 GBs, with 400 GBs the presumptive next node. The transceiver chips required for transmitting data at these speeds across long distances on copper wire (or now more commonly optical fibers) provide a strong market for SiGe devices, as many of the same characteristics required for wireless transceivers are important in these transceivers (high speed, low noise, large dynamic range). In fact it was this market, and the expected imminent transition to the 10 GBs and 100 GBs rates that pushed researchers in the late 1990s and early 2000s to invest in creating very high-speed SiGe transistors (with f_t and f_{max} of 200 GHz and above). However, with the dot-com bust, the ultra-high-speed telecommunications market was delayed for at least a decade. Nevertheless, the 10 GBs node eventually began to ramp up in the early 2010s with total shipments of 10 GBs ports now comparable to the 1–3 GBs ports [8]. For optical transceiver ports SiGe BiCMOS technology is commonly used on both the receive side to build trans-impedance amplifiers (TIAs) connecting to a photodiode, and on the transmit side to build laser or modulator drivers, connecting directly to a laser or to an electro-optic modulator. The data-conditioning chips that come behind the TIA in front of the laser driver have also been built with SiGe BiCMOS. Although it is now used less often for serializer/deserializer (SerDes) functions, SiGe BiCMOS is still a dominant technology for clock and data recovery circuits (CDRs) used at either end of almost every optical fiber data link.

3.3.3 Future Applications

Today, deep submicron CMOS is challenging SiGe for some of the applications discussed above for two reasons: the speed of CMOS is adequate for many applications (although SiGe maintains an advantage in noise and an even wider advantage in dynamic range), and the density of CMOS is now high enough to enable new architectures that rely more heavily on DSP rather than high-fidelity analog manipulation. In many cases, however, SiGe technology still offers a performance and power advantage and will continue to play a strong role in both the wireless FEM and wireline transceiver markets. It should also be noted that, at present, the cost of advanced SiGe BiCMOS wafers is significantly less than that of the RF-performance-equivalent CMOS node since the SiGe BiCMOS devices don't rely on nearly as advanced lithography nodes as their CMOS counterparts (see Figure 3.2). In addition to the incursion of CMOS devices in the traditional SiGe application space, current or even past-generation SiGe transistor performance is more than adequate to serve many of these traditional applications. Therefore these markets are becoming less important as drivers for future technology advancements.

Looking forward, however, new high-frequency and high-datarate applications are driving SiGe performance advancements. Millimeter-wave communications systems offer an excellent opportunity for the use of SiGe HBT devices. Millimeter-wave applications include, for example, proposed ~60 GHz WLAN standards [9], 77 GHz automotive collision avoidance systems, and 94 GHz and above "terahertz" passive imaging. And although not quite technically in the millimeter-wave range, next-generation cellular data standards, collectively termed "5G," will likely include a band at ~28 GHz

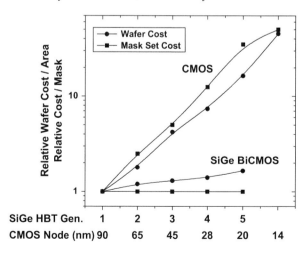

FIGURE 3.2
Costs of wafers (per unit area) and mask sets for different technology nodes. The CMOS and SiGe HBT nodes are lined up roughly in terms of equivalent RF performance.

which may end up being the main commercial driver for circuits pushing past the current communications bands which end at around 5 GHz. Most millimeter-wave radio systems require the use of phased array transmitters and receivers which allow directional communication between transmitter and receiver. The size of these phased array chips is dictated by the wavelength of the radiation and they are thus quite large with much empty space between the transmit or receive elements. Thus SiGe BiCMOS is a perfect technology for this application since the devices are capable of making low-noise LNA devices at the high frequencies and outputting high RF power levels at reasonable efficiency while maintaining low cost per unit area of silicon [11].

Next-generation wireline standards will also continue to push SiGe HBT performance, with the next standards not only doubling the clock speed from 28 to 56 GBaud, but transitioning from traditional non-return-to-zero (NRZ) modulation to the much more complicated PAM-4 modulation scheme [12]. The interfaces to electro-optical devices in these systems will likely continue to require SiGe HBTs even if CMOS will be used for DSP in the transceiver modules.

3.4 SiGe Performance Metrics

Two figures of merit are typically used to benchmark high-frequency device performance: (1) the cutoff frequency (F_t) which, for a bipolar device, is defined as the frequency at which the AC current gain is unity, and (2) the maximum frequency of oscillation (F_{max}) which, for a bipolar device, is defined as the frequency at which the power gain is unity (usually the unilateral power gain).

For a bipolar device, F_t and F_{max} are related to basic device parameters by the commonly used equations:

$$F_t = \frac{1}{2\pi \cdot \tau_F}, \tag{3.1}$$

$$\tau_F = \left(C_{BC} + C_{BE}\right) \cdot \left(R_E + \frac{kT}{qI_C}\right) + \frac{W_B^2}{2D_B} + \frac{W_C}{2v_S} + R_C \cdot C_{BC}, \tag{3.2}$$

$$F_{max} = \sqrt{\frac{F_t}{8\pi \cdot R_B \cdot C_{BC}}}, \tag{3.3}$$

where
  τ_F is the forward transit time
  C_{BE} is the emitter-base capacitance

C_{BC} is the base-collector capacitance
R_E is the emitter series resistance
I_C is the collector current
W_B is the vertical base width
D_B is the electron diffusion length in the base
v_S is the electron saturation velocity
W_C is the vertical collector-base depletion width
R_C is the collector resistance

At low collector current, F_t is dominated by the first term in Equation 3.2 where the junction capacitances combine with internal resistances to create an $R \times C$ time constant delay that is significantly longer than the other time constants in Equation 3.2 (see Figure 3.3). At high current the term W_B becomes a function of I_C. When the charge associated with the current through the collector-base depletion region becomes comparable to the intrinsic doping level on either side, the edges of the depletion region collapse and thus "push" the depletion region away from the base, effectively widening the base. Mathematically, this occurs when:

$$J_C \oplus q N_C v_{SAT} \tag{3.4}$$

Where N_C is the nominal doping in the collector and v_{SAT} is the electron saturation velocity in the collector. This base push-out, known as the Kirk effect [13], is responsible for F_t decreasing at high current rather than saturating, as would otherwise be predicted by Equation 3.2. So both F_t and F_{max} peak at a specific current density (see figure 3.3).

Obtaining ever-higher peak F_t and F_{max} are important because, while today's volume RF applications target modest operating frequencies relative to the peak f_T's shown in Figure 3.2, high peak F_t (and F_{max}), can be traded

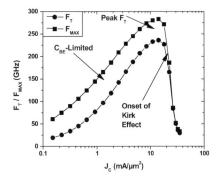
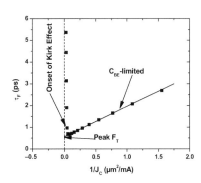

FIGURE 3.3
(a) Typical F_T vs. J_C plot for a state-of-the-art, volume-manufactured SiGe HBT showing the various regions indicated by the terms in Equation 3.2. (b) Typical τ_F $(1/(2\pi F_T))$ vs. $1/J_C$ plot indicating the various regions indicated by the terms in Equation 3.2.

off for other benefits, including: reduced power consumption, higher break-down voltage, and reduced noise.

Figure 3.4 shows an example of the power savings that can be achieved with higher f_T SiGe technology even when operating at relatively low frequencies. For instance, at 25 GHz or so there is a 3× improvement in current consumption going from the 0.3 μm node SiGe technology to the most advanced SiGe:C technology. At 50 GHz the advantage is almost an order of magnitude. Alternatively, when used as a gain stage in a multistage amplifier, one could operate the device closer to peak F_t and simply use fewer gain stages to achieve the same total circuit gain. For instance, in theory one could use any of the top three technologies shown in Figure 3.4 (1st–3rd generation SiGe:C HBT technologies) to achieve gain at 100 GHz. However, the most advanced technology would provide approximately 9 dB of gain at 100 GHz for peak F_t conditions, whereas the first-generation technology provides only 3.5 dB. Thus you could reduce the number of gain stages by more than half to achieve the same total gain, which provides an advantage in terms of power consumption, circuit area and the total noise added by the circuit.

The second advantage of higher F_t's, even in lower frequency applications, is in the RF noise figure. Minimum noise figure can be expressed by [14]:

$$NF_{min} = 1 + \frac{n}{\beta} + \frac{f}{F_t} \cdot \sqrt{\frac{2qI_C}{kT} \cdot (R_E + R_B) \cdot \left(1 + \frac{F_t^2}{\beta \cdot f^2}\right) + \frac{n^2 F_t^2}{\beta \cdot f^2}} \qquad (3.5)$$

where n is the collector current quality factor and β is the current gain. From Equation 3.5, it is seen that with a high β, as is typically seen in SiGe devices, the noise figure reduces to:

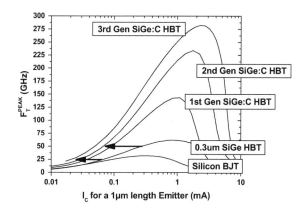

FIGURE 3.4
F_t for various TowerJazz BiCMOS technologies plotted as a function of I_c for a minimum width and unit length emitter. In addition to higher peak F_t, subsequent technology nodes lower power consumption even when biasing at low F_t as indicated by the arrows.

$$NF_{MIN} \approx \frac{f}{F_t} \cdot \sqrt{\frac{2qI_C}{kT} \cdot (R_E + R_B)}. \quad (3.6)$$

In this limit, a higher F_t and lower R_B result in a lower noise figure. Further, the ability to operate at a lower I_C and obtain the same f_T can lead to a reduction in the term under the radical in Equation 3.6. Also, when $F \ll F_t$ the n/β term in the front of Equation 3.5 becomes dominant and the ability to produce very high values of β discussed in Section 3.2 becomes almost as important as the RF characteristics (this becomes especially true at cryogenic temperatures [15]). Thus all of the advantages enabled by using SiGe in the base of an HBT are brought to bear when one is attempting to minimize noise figure: the high β enabled by putting SiGe at the emitter-base junction, the lower R_B for a given β enabled by increasing the base doping, and the higher F_t enabled by the germanium profile in the base.

Finally, F_t can be traded off for higher breakdown voltage by modulating the collector doping concentration through a collector implant mask such that multiple devices spanning a range of F_t and breakdown are made available on the same wafer. Figure 3.5 shows the family of devices realized by this technique across several generations of TowerJazz technology. Each subsequent generation supports devices with higher f_T but also improves the trade-off between F_t and breakdown voltage, improving large-signal performance for applications such as integrated drivers and power amplifiers. This is in contrast with CMOS where each new generation makes the integration of power devices more difficult due to the more brittle gate oxide forcing lower voltage ratings.

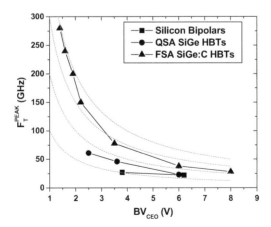

FIGURE 3.5
F_t vs. BV_{CEO} plotted for all devices available in several generations of bipolar technologies. The dashed lines are contours of constant $F_t \times BV_{CEO}$. All data is from TowerJazz electrical specifications.

3.5 Device Optimization and Roadmap

Higher F_t's are enabled by vertical scaling of the HBT device. The most fundamental device enhancement with each generation of higher f_T devices is scaling of the base width: the W_B term in Equation 3.2. The fundamental limit of scaling the base width occurs when the emitter-base and base-collector depletion regions touch and thus the device is "punched-through". It should be noted that the metallurgical base width in advanced SiGe HBTs is already many times narrower than all but the most aggressive CMOS channel lengths. The next most commonly adjusted vertical scaling parameter is the collector doping. An increase in collector doping offsets the Kirk effect as indicated by Equation 3.4 but, again, a fundamental limit is reached when the doping becomes so high that reverse bias leakage between the base and collector, either due to avalanche multiplication or tunneling, dominates the device behavior. Due to the reasons discussed above in Section 3.2, the additions of germanium and carbon into the base of the HBT serve to delay the point at which these fundamental limits are reached and thus allow further vertical scaling of the device than would be possible for a homojunction device. It should be noted that scaling down of the base width or scaling up of the collector doping both serve to reduce F_{max} by increasing R_B in the first case and by increasing C_{BC} in the latter (see Equation 3.3). Thus most techniques employed to enhance F_t trade off with a reduction in F_{max} and other techniques are needed to simultaneously improve F_{max}.

Higher F_{max}'s are enabled by lateral scaling of the devices. Smaller device dimensions serve to reduce the R_B and C_{BC} terms in Equation 3.3. In fact, most of the research involved in developing a new generation of SiGe HBT devices involves creating new ways to reduce these two parasitic parameters. At the heart of the scaling of SiGe HBT devices is the emitter width which, in turn, limits most of the other dimensions in the device as a whole. While the most advanced SiGe HBT devices constructed to date have emitter widths less than 100 nm [16], scaling of the emitter width is more than ten years behind the scaling of CMOS gate lengths. Figure 3.6 shows projection data from the ITRS roadmaps for CMOS and bipolar technologies [17], showing projected F_{max} vs. the minimum feature width in the given technology node. It shows that, on average, one can achieve the same F_{max} in a bipolar device with a minimum feature width roughly three times larger than CMOS. This difference in lithography requirement leads to a large difference in the cost of both wafers and mask sets for the given technology. Figure 3.2 shows the trend of wafer and mask set cost going to more advanced CMOS nodes vs. the trend for SiGe BiCMOS. The data shows an exponential increase in the cost of advanced CMOS nodes but a more gradual linear increase for the SiGe BiCMOS technologies.

Several device architectures have been developed in the past decade in order to allow scaling of SiGe HBT devices down to nano-scale dimensions.

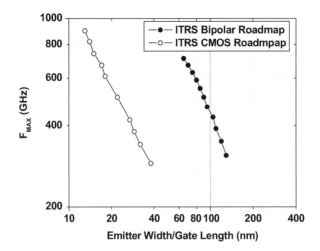

FIGURE 3.6

F_{max} vs. minimum feature size for bipolar vs. CMOS technologies. Data is from the ITRS roadmaps for CMOS and bipolar technologies [11].

Figure 3.7 shows a generic example of the device architecture used to construct most modern SiGe HBTs. The first large-scale manufactured SiGe HBT devices were built with a "quasi-self-aligned" architecture [18] where the extrinsic base is self-aligned by ion-implantation to the edges of the emitter poly but not the emitter itself (see the area labeled (a) in Figure 3.7). The next generation of devices split off into several different architectures which are "fully self-aligned", meaning that the extrinsic and intrinsic base alignment does not depend on mask alignment. One type of device uses a deposited polycrystalline extrinsic base followed by "selective epitaxy" of the intrinsic SiGe base [19]. A second method uses a sacrificial emitter post and spacer similar to the construction of a CMOS device [20]. Finally, various methods of growing a "raised extrinsic base" after the epitaxial growth of the intrinsic SiGe base have been developed [21, 22]. All of these modern techniques essentially serve to dope the region denoted (b) in Figure 3.7 at a higher p-type level than that of the SiGe epitaxy. This extra doping in the extrinsic base region serves to lower the total R_B of the device and thus improve F_{max}. While the techniques used to enhance F_t discussed above tend to reduce F_{max}, the scaling and architecture enhancements discussed here serve to improve F_{max} without any significant penalty to F_t. Thus these innovations have allowed continuous scaling of SiGe HBT devices akin to what is done in CMOS.

Figure 3.8 shows a compilation of F_t and F_{max} data from over 100 SiGe HBT publications overlaid with the 2010 ITRS roadmap for bipolar devices. The scatter plot shows the basic correlation of the progression of F_t and F_{max} despite the trade-offs mentioned above. The roadmap data predicts that the same lithography advancements responsible for the CMOS roadmap will enable improved SiGe performance for the foreseeable future.

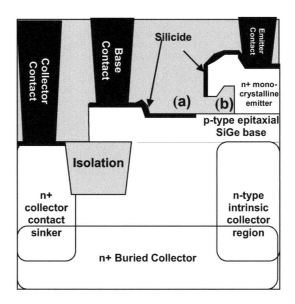

FIGURE 3.7
Example cross-section of a modern SiGe HBT: (a) denotes the silicided extrinsic base region and the implanted extrinsic base region in quasi-self-aligned devices, (b) denotes the "link" or "spacer" region which is doped (and thus implies "fully self-aligned") by the various techniques discussed in the text.

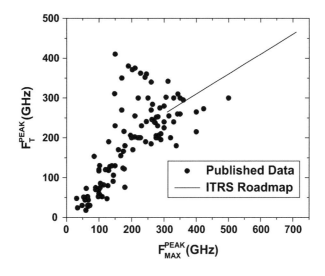

FIGURE 3.8
F_t vs. F_{max} scatter plot for published SiGe HBT data. The line is projection data from the ITRS bipolar roadmap [15].

To realize useful RF and analog circuits, however, more than just high-speed SiGe devices are necessary. In the next section, we will discuss modules integrated with SiGe transistors that help create a more complete modern platform for RF and analog IC design.

3.6 Modern SiGe BiCMOS RF Platform Components

Technology features integrated with SiGe transistors that make them useful for product design include active elements such as high-density CMOS, RF CMOS, high-voltage CMOS, high-performance PNPs as well as passive elements such as high-density MIM capacitors, high-quality inductors, and high-performance RF grounding features.

Today, most SiGe development is done in the context of a BiCMOS process in a CMOS node that typically trails the most advanced digital node by several generations. The critical hurdle to integrating advanced CMOS and SiGe devices is to marry their respective thermal budgets without degrading either device. The addition of carbon to SiGe layers as discussed in Section 3.2 has been used as a partial solution to this problem as it helps reduce boron diffusion allowing for a higher thermal budget after SiGe deposition. This, along with careful optimization of the integration scheme, has resulted in demonstrations of SiGe integration down to the 90 nm node and below [23, 24].

In many RF applications, including those discussed in Section 3.3A, an RF switch device is required to switch the RF signal between various nodes on the chip. RF switches are typically built using specially designed NFETs which are optimized to reduce losses in both on-state (R_{ON}) and off-state (C_{OFF}). The common figure of merit that combines these two is $R_{ON}C_{OFF}$, usually expressed in units of femtoseconds. While stand-alone RF switch technologies are usually built using thin-film SOI CMOS to reduce parasitic losses, excellent switch NFETs can also be integrated into a SiGe BiCMOS. Figure 3.9 shows the progression of improved RF switch NFET devices in both SOI CMOS processes and integrated into SiGe BiCMOS processes. While the dedicated RF SOI CMOS process devices always outpace the SiGe BiCMOS-integrated devices, the performance of the integrated devices are starting to come within range of the dedicated process devices; the best devices in production now show less than 300fs $R_{ON}C_{OFF}$.

Power management circuitry can be enabled with higher-voltage CMOS devices (typically requiring tolerance of 5 to 8 V). In smaller geometries that support only lower core voltage levels, these are enabled by introducing drain extensions to the CMOS devices that can enable higher drain bias than is supported in the native transistor. An example of such devices is shown

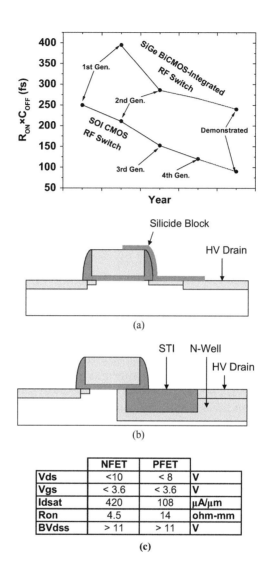

FIGURE 3.9
$R_{ON}C_{OFF}$ FOM for RF switch NFETs in RF SOI (dedicated RF switch process) and integrated in SiGe BiCMOS processes. Sketch of two types of commonly used extended drain devices: (a) silicide block extension and (b) STI extension as well as (c) a table showing characteristics of high-voltage devices available in a 0.18 µm SiGe BiCMOS technology using approach (b). Idsat is quoted for 3.3 V Vgs and 5 V Vds.

in Figure 3.9 and these are becoming common modules in SiGe technology offerings, often not costing additional masking layers to create.

A high-speed vertical PNP (VPNP) can form a complementary pair with the SiGe NPN and is important for certain high-speed analog applications such as fast data converters, push-pull amplifiers, and output stages for hard

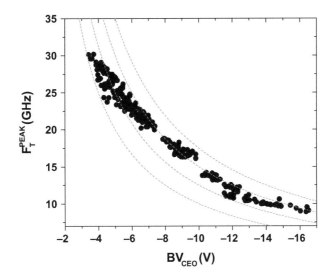

FIGURE 3.10
Performance (F_t) vs. breakdown (BV_{CEO}) trade-off of vertical silicon PNP devices integrated with SiGe NPNs to form a complementary pair. The dashed lines are countours of constant $F_t \times BV_{CEO}$. Data is from internal TowerJazz development wafers.

disk drive pre-amps. A VPNP can be made very fast by the use of a separate SiGe deposition step, and F_t's as high as 100 GHz have been reported [25]. But the cost associated with such a VPNP is prohibitive for most applications today. A more popular approach re-uses many of the steps needed to create the SiGe HBT and CMOS devices while adding specialized implants to optimize the performance of the VPNP. In this scheme devices with F_t's of up to 30 GHz can be achieved, as shown in Figure 3.10.

In addition to active components, high-quality passive components are necessary to enable advanced RF circuits. The most critical passive elements for RF design are capacitors and inductors as these can consume significant die area and, at times, limit performance of RF and analog circuits. Metal-insulator-metal (MIM) capacitors are available in most commercial SiGe BiCMOS and RF CMOS processes as they achieve excellent linearity and matching. The density of MIM capacitors has been steadily increasing over time, helping to shrink RF and analog die. Figure 3.11 shows a timeline of capacitance density for TowerJazz integrated MIM capacitors. An initial improvement in density from <1 fF/um² to 2 fF/um² was enabled by a move from oxide to nitride dielectrics [26]. Then, a move from 2 fF/um² to 4 fF/um² was enabled by the stacking of a 2 fF/um² capacitor on two consecutive metal layers. Finally, a further optimization of the nitride dielectric resulted in a density of 5.6 fF/um². Today, high-K dielectrics and various types of MIM trench capacitors are being investigated to enable even higher densities and it is conceivable that in the next few years densities of 10 to 20 fF/um² will be introduced.

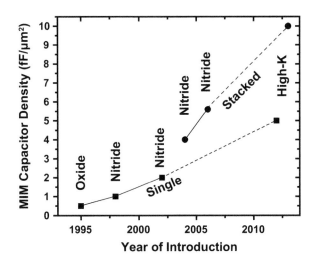

FIGURE 3.11
MIM capacitor density plotted as a function of year of first production (actual or planned) showing progression in dielectric technology (from oxide to nitride to high-K) and in integration (single to stacked capacitors).

Integrated inductor performance, measured as the quality factor (Q), is improved by the reduction of metal resistance made possible by thicker metal layers. Inductor Q can be traded off for reduced footprint such that a thicker metal layer can also help reduce chip area. This concept is demonstrated in Figure 3.12 where the area required to realize an inductor with

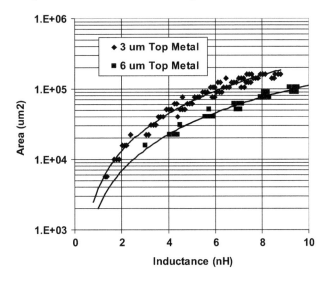

FIGURE 3.12
Inductor area as a function of inductance for a four-turn inductor with peak Q of 10 built in 3 um and 6 μm Al metal layers respectively.

Q of 10 is compared between use of a 6 um and 3 um top metal in a four-layer aluminum metal process. A 6 um metal inductor consumes half the die area of a 3 um metal inductor while achieving the same Q in this example.

For most RF circuits ground is provided by metal traces, eventually leading to a pad on the chip and then wire-bonded to a lead frame. This entire ground path has a significant amount of resistance and inductance, which leads to a very lossy and variable RF ground. While advancements in packaging techniques, such as copper pillars combined with "flip-chip" packaging, have given chip designers options to reduce this loss, built-in process methods for improving RF grounding are desired. Various types of "through-wafer-via" (TWV) approaches have been developed which connect the standard CMOS or BiCMOS metals to the back of the wafer, which is then epoxied or soldered to the lead frame. While TWVs are commonly available in III-V processes, they have only recently become available in mainstream silicon-based processes. However TWV processing usually requires complicated processing (thinning and backside metallization of the silicon wafer) once the wafer has left the CMOS factory. One low-cost approach to the TWV idea is a "deep silicon via" (DSV) [27], which offers the low inductance of TWV without the cost of the post-fab-processing of the TWV. Figure 3.13 shows a cross-sectional image of a wafer with active devices close to a DSV. DSVs use a highly doped silicon substrate capped with much lower doped silicon. The DSVs connect the first metal layer in the BiCMOS process to the highly doped substrate siting several microns below the active device region. The highly doped substrate becomes the effective ground plane and thus the path from device to "true" RF ground is very short. DSVs can provide <5pH of inductance per via and can be placed in close proximity to active BiCMOS devices to reduce the "loop inductance" of the path to ground.

Die scaling enabled by the advanced passive elements described in this section can often more than pay for the additional processing cost. An

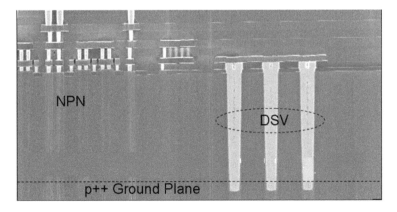

FIGURE 3.13
Cross-sectional SEM image of a deep silicon via.

optimized process can, in many cases, not only provide better performance than a digital CMOS process but also lower the die cost. Similarly, the integration of advanced active modules described in this section can help integrate more analog functionality on fewer die, reducing overall system-level costs.

3.7 Conclusions

In this chapter, we have reviewed SiGe BiCMOS technology and discussed how it has become important for many RF and high-performance analog applications by providing a performance advantage over stand-alone CMOS while sharing its manufacturing infrastructure to provide integration and cost advantages over III-V technology. In addition to higher speed, we have seen that an intrinsic advantage of SiGe over CMOS is its ability to maintain higher breakdown voltages and therefore support applications that require higher dynamic range. This gap will widen with more advanced generations of both CMOS and SiGe as each new generation of CMOS results in lower breakdown voltages while each new generation of SiGe results in a better trade-off between speed and breakdown. In addition, we have seen that performance of SiGe devices can be improved with advanced lithography in much the same way as with CMOS devices, such that a raw performance gap will continue to exist between SiGe and CMOS as more advanced nanometer nodes are created in the future. This will continue to enable a market for SiGe at the leading edge of performance, which today is translating into interest for SiGe in several millimeter-wave applications and very high-speed networks.

The biggest threat to SiGe advancements is the failure to identify high-speed, high-volume applications that take advantage of these benefits in the future but, much like Moore's law for CMOS has held true for decades and applications have taken full advantage, the imagination of the industry has never let us down before and is not likely to do so in this case.

Acknowledgments

The authors would like to acknowledge the help of current and former coworkers at TowerJazz including Volker Blaschke, Dieter Dornisch, David Howard, Chun Hu, Paul Hurwitz, Amol Kalburge, Arjun Karroy, Paul Kempf, Lynn Lao, Zachary Lee, Pingxi Ma, Greg U' Ren, Jie Zheng, and Bob Zwingman, as well as contributions from many of TowerJazz's customers and partners. A final special acknowledgment is given to Jay John of Freescale Semiconductor for compiling the large dataset of published f_T and f_{MAX} data used to generate Figure 3.8.

References

1. H. Kroemer, Heterostructure bipolar transistors and integrated circuits, *Proc. IEEE*, Vol. 70, pp. 13–25, 1982.
2. D. L. Harame, J. H. Comfort, J. D. Cressler, E. F. Crabbe, J. Y.-C. Sun, B. S. Meyerson, and T. Tice, Si/SiGe epitaxial-base transistors-part I: Materials, physics, and circuits, *IEEE Trans. Elec. Dev.*, pp. 455–482, 1995.
3. J. W. Slotboom, G. Streutker, A. Pruijmboom, D. J. Gravesteijn, Parasitic energy barriers in SiGe HBTs, *IEEE Elec. Dev. Lett.*, Vol. 12, No. 9, pp. 486–488, 1991.
4. L. D. Lanzerotti, J. C. Sturm, E. Stach, R. Hull, T. Buyuklimanli, and C. Magee, Suppression of boron outdiffusion in SiGe HBTs by carbon incorporation, *IEDM Digest*, pp. 249–252, 1996.
5. T. Manku and A. Nathan, Electron drift mobility model for devices based on unstrained and coherently strained $Si_{1-x}Ge_x$ grown on <001> silicon substrate, *IEEE Trans. Elec. Dev.*, Vol. 39, No. 9, pp. 2082–2089, 1992.
6. G. L. Patton, D. L. Harame, J. M. C. Stork, B. S. Meyerson, G. J. Scilla, and E. Ganin, Sige-base, poly-emitter heterojunction bipolar transistors, *Proc. Symp. on VLSI Technology*, pp. 35–36, 1989.
7. www.semiconductors.org, "End-Use Products that Drove Semiconductor Sales in 2016," 2017. http://blog.semiconductors.org/blog/what-end-use-products-drove-semiconductor-sales-in-2016.
8. "LightCounting June 2017 Quarterly Report." LightCounting Market Research, June 2017.
9. Task Group 3c. *Millimeter Wave Alternative PHY*, IEEE 802.15.3c, http://www.ieee802.org/15/pub/TG3c.html, 2018.
10. Wireless LAN Working Group. IEEE 802.11ad, https://standards.ieee.org/find-stds/standard/802.11ad-2012.html, 2012.
11. K. Kibaroglu, M. Sayginer, and G. Rebeiz, An ultra low-cost 32-element 28 GHz phased-array transceiver with 41 dBm EIRP and 1.0-1.6 Gbps 16-QAM link at 300 meters, *2017 RFIC Proc.*, pp 73–76, 2017.
12. S. Shahramian, J. Lee, J. Weiner, R. Aroca, Y. Baeyens, N. Kaneda, and Y.-K. Chen, A 112Gb/s 4-PAM transceiver chipset in 0.18µm SiGe BiCMOS technology for optical communication systems, *Proc. 2015 Compound Semiconductor Integrated Circuit Symposium*, 2015.
13. C. T. Kirk, A theory of transistor cutoff frequency (fT) falloff at high current densisites, *IRE Trans. Elec. Dev.*, Vol. 9, No. 2, p. 164, 1962.
14. S. P. Voinigescu, M. C. Maliepaard, J. L. Showell, B. E. Babcock, D. Marchesan, M. Schroter, P. Schvan, and D. Harame, A scalable high-frequency noise model for bipolar transistors with application to optimal transistor sizing for low-noise amplifier design, *Journal of Solid-State Circuits*, Vol. 32, No. 9, pp. 1430–1439, Sept. 1997.

15. S. Weinreb, J. C. Bardin, and H. Mani, Design of cryogenic SiGe low-noise amplifiers, *IEEE Trans. Microwave Theory and Techniques*, pp. 2306–2012, 2007.

16. T. Lacave, P. Chevalier, Y. Campidelli, M. Buczko, L. Depoyan, L. Berthier, G. Avenier, C. Gaquiere, and A. Chantre, Vertical profile optimization for +400GHz fMAX Si/SiGe:C HBTs, *2010 BCTM Proc.*, pp 49–52, 2010.

17. International Technology Roadmap for Semiconductors, International SEMATECH, Austin, Texas, 2011.

18. H. Miyakawa, M. Norishima, Y. Niitsu, H. Momose, K. Maeguchi, A 3.3V, 0.5um BiCMOS technology for BiNMOS and ECL gates, *1991 CICC Proc.*, pp. 18.3/1–18.3/4, 1991.

19. F. Sato, T. Hashimoto, T. Tashiro, T. Tatsumi, M. Hiroi, and T. Niino, A novel selective SiGe epitaxial growth technology for self-aligned HBTs, *Symp. on VLSI Tech. Digest*, pp. 62–63, 1992.

20. M. Racanelli, K. Schuegraf, A. Kalburge, A. Kar-Roy, B. Shen, C. Hu, D. Chapek, et al., Ultra high speed SiGe NPN for advanced BiCMOS technology, *IEDM Proc.*, pp. 15.3.1–15.3.4, 2001.

21. B. Jagannathan, M. Khater, F. Pagette, J.-S. Rieh, D. Angell, H. Chen, J. Florkey, et al., Self-aligned SiGe NPN transistors with 285 GHz fMAX and 207 GHz fT in a manufacturable technology, *IEEE Elec. Dev. Let.*, Vol. 23, No. 5, pp. 258–260, 2002.

22. H. Rucker, B. Heinemann, R. Barth, D. Bolze, J. Drews, U. Haak, W. Hoppner, et al., SiGe:C BiCMOS technology with 3.6ps gate delay, *BCTM Proc.*, pp. 121–124, 2003.

23. K. Kuhn, M. Agostinelli, S. Ahmed, S. Chanbers, S. Cea, S. Christensen, P. Fischer, et al., A 90nm communication technology featuring SiGe HBT transistors, RF CMOS, precision R-L-C RF elements and 1 μm^2 6-T SRAM cell, *2002 IEDM Tech. Dig.*, pp. 73–76, 2002.

24. P. Chevalier, G. Avenier, G. Ribes, A. Montagné, E. Canderle, D. Céli, N. Derrier, et al., A 55 nm triple gate oxide 9 metal layers SiGe BiCMOS technology featuring 320 GHz fT / 370 GHz fMAX HBT and high-Q millimeter-wave passives, *IEDM Proc. 2014*, pp. 3.9.1–3.9.3, 2014.

25. B. Heinemann, R. Barth, D. Bolze, J. Drews, P. Formanek, O. Fursenko, M. Glante, et al., A complementary BiCMOS technology with high speed npn and pnp SiGe:C HBTs, *IEDM Tech. Dig.*, pp. 117–120, 2003.

26. A. Kar-Roy, C. Hu, M. Racanelli, C. A. Compton, P. Kempf, G. Jolly, P. N. Sherman, J. Zheng, Z. Zhang, and A. Yin, High density metal insulator metal capacitors using PECVD nitride for mixed signal and RF circuits, *Interconnect Technology IEEE International Conference*, pp. 245–247, 1999.

27. V. Blaschke, T. Thibeault, L. Lanzerotti, C. Cureton, R. Zwingman, A. KarRoy, E. Preisler, D. Howard, and M. Racanelli, A deep silicon via (DSV) ground for SiGe power amplifiers, *Proc. 2010 Silicon Monolithic Integrated Circuits Meeting*, pp. 208–211, 2010.

4

Base Doping Profile Engineering for High-Performance SiGe PNP Heterojunction Bipolar Transistors

Guangrui (Maggie) Xia and Yiheng Lin

CONTENTS

4.1 Introduction

4.1.1 Industry Background and Motivations

Thanks to its many excellent material properties, low cost, and superior manufacturability, silicon-based products account for over 90% of the microelectronics market [1]. Silicon-germanium (SiGe) and SiGe:C alloys emerged and have been applied widely in the microelectronics industry for the past two to three decades due to their capability in strain and band gap engineering and their compatibility with mainstream Si processing. The high-speed SiGe heterojunction bipolar transistor (HBT) is one such example. Since the first sale in 1998, SiGe HBTs have seen increasing applications in wireless communications, hard disk drives (HDD), optical disk drives (ODD) for CD, DVD and Blu-ray disk, thin-film transistor (TFT) displays, high-resolution video, optical networking components and so on.

4.1.2 Requirement of Narrow Base Doping Profiles for HBTs

For HBTs in high-speed and high-frequency applications, there are two important figures of merit: the common-emitter short-circuit cut-off frequency f_T and the maximum frequency of oscillation f_{max}. Higher f_T and f_{max} represent better HBT device performance. HBTs were dominated by III-V compound semiconductors before the emergence of SiGe HBTs, which were enabled by the lattice-matched strained-SiGe epitaxy technology in the 1990s [2]. Reducing the base width is an effective way to reduce the base transit time and thus increase the f_T. Physically, higher base doping reduces electrical resistance, while narrower base doping profile reduces capacitance and carrier transit time in the base region. These all contribute to a higher f_T/f_{max} and better high-frequency performance. Heinemann *et al.* recently reported experimental results of f_T/f_{max} for up to 505/720 GHz which resulted from an aggressively scaled base doping profile and an advanced annealing technique [3].

PNP SiGe HBTs are much less explored due to the inherently slower minority carrier transport and thus lower frequency response. However, complementary SiGe HBTs (with both NPN and PNP HBTs) have many advantages over an NPN-only technology for numerous analog applications requiring high speeds, low noises, and large voltage swings [4, 5]. Furthermore, a complementary SiGe bipolar complementary metal-oxide-semiconductor (SiGe C-BiCMOS) technology platform containing both NPN and PNP devices with, ideally, equal performance, and CMOS can offer compelling advantages in many types of analog circuits. It can be used for lowering voltage supply rails, improving current sources, and implementing fast and low distortion driver amplifiers. To date, more and more effort is involved in the R&D of complementary SiGe BiCMOS technology (C-BiCMOS), which utilizes both NPN and PNP HBTs [6, 7]. The International Technology Roadmap for Semiconductors (ITRS) 2013 edition on radio frequency (RF) and analog/mixed-signal technologies for wireless communications has addressed the need for high-speed PNP HBTs (HS-PNP), which is driven by applications that require high-performance analog and mixed-signal ICs [8]. Aggressively scaled vertical doping profiles are essential in improving HS-PNP HBTs' performance. The state-of-the-art PNP HBTs for C-BiCMOS technologies use phosphorus (P) as the base layer dopant and strained SiGe as the base layer material. Therefore, to achieve narrow base doping, it is a must to *understand and predict P diffusion and segregation in strained SiGe*, which is the focus of this chapter.

4.1.3 Problem Definitions

To achieve this goal, we need to first introduce major approaches to suppress or limit P (P) diffusion in Si and SiGe. To date, there are two major approaches for this purpose:

1. Defect engineering methods such as carbon (C) incorporation or thermal nitridation
2. Advanced thermal processing techniques such as sub-millisecond annealing techniques with less thermal budget and therefore less diffusion.

This work is not about option 2, as rapid thermal processing (RTP) is the industry mainstream thermal processing technique. Existing studies of P diffusion in strained SiGe have been performed mostly using furnace annealing, mainly because of the availability of the tools, long anneal time and better temperature accuracy for diffusion study. Studies using RTP tools are in great need to match the state-of-the-art of industry practice.

Concerning option 1, studies have shown that C incorporation in Si and strained SiGe can cause the unsaturation of Si self-interstitials [9, 10], therefore interstitial-mediated dopants are suppressed, such as boron (B) [11, 12] and P diffusion [11, 13]. Karunaratne *et al.* reported the first quantitative comparison of B diffusivities in Si, $Si_{0.9}Ge_{0.1}$, $Si_{0.999}{:}C_{0.001}$ and $Si_{0.899}Ge_{0.1}{:}C_{0.001}$ to extract C retardation factors [14]. It has been a common industry practice to use C in SiGe to retard B diffusion for NPN SiGe HBTs, while there are only a handful of studies on P diffusion in SiGe:C for PNP HBTs. Tillack *et al.* studied C retardation effect in $Si_{0.8}Ge_{0.2}$ on P diffusion for two different C concentrations after annealing at 850°C [7]. However, this temperature is far from the temperature range where most diffusion happens in industry practice, which is above 1,000°C. *Systematic and quantitative study of C retardation effect on P diffusion in SiGe is still lacking*. This is highly desired in industry for accurate device and process design of HS-PNP HBTs and will be addressed as Topic 1 of Section 4.4, and will be discussed in Session 4. The results in

Topic 1 are also needed to serve as the base line in the study of thermal nitridation effect in Topic 2.

Thermal nitridation is also known to suppress B and P diffusion in Si via exposing bare Si surface in ammonia ambient, where vacancies are injected into Si substrates and retard interstitial diffusers' motion [15]. This effect was studied in the late 1980s and early 1990s, mainly for diffusion mechanism study [16, 17], and has no published record regarding its industry applications. R&D on SiGe PNP HBTs provides a good opportunity for the application of this method, as the dopants in PNP HBTs, B and P, are both interstitial diffusers, and can be retarded at the same time with the vacancy injection method. However, there is no study available on whether this method also works in strained-SiGe systems and to what extent, if any. It is also not clear whether this effect will work with other approaches (e.g. C incorporation) or to what extent. If this method is effective, it is a relatively low-cost and simple approach to be adopted by industry, as the only change is in the annealing ambient. These questions will be address as Topic 2 in Section 4.5.

The last topic is about P segregation across graded SiGe layers (common SiGe layer design for HBTs), which happens simultaneously with P diffusion. As P prefers to stay in Si rather than in SiGe, this effect is not desired for narrow P profiles, and needs to be taken into account in the mass transport modeling. While several studies on this effect are available, none of them are for graded SiGe layers and lack experimental verification and appropriate modeling. This topic will be discussed in Section 4.6.

4.2 Fundamentals of Dopant Diffusion in Semiconductors

4.2.1 Macroscopic Description of Diffusion

Adolf Fick proposed a mathematical framework to describe diffusion phenomena macroscopically [18]. It was postulated that the flux of matter j in the x direction is proportional to the pertaining gradient of concentration C:

$$j = -D\frac{\partial C}{\partial x} \tag{4.1}$$

This is what we call Fick's first law today. D is denoted as the diffusion coefficient, or diffusivity. Fick also derived a second law of diffusion

$$\frac{\partial C}{\partial t} = \frac{\partial}{\partial x}\left(D\frac{\partial C}{\partial x}\right) \tag{4.2}$$

If we look at Fick's original paper, one may notice that the above equations have been updated and optimized compared to his original expression

$$\frac{\delta y}{\delta t} = k \frac{\delta^2 y}{\delta x^2} \tag{4.3}$$

where y is the concentration (an extra minus sign is ignored here). There was an implicit assumption behind this equation that diffusivity is independent of concentration and of its gradient.

4.2.2 Microscopic Description of Diffusion

4.2.2.1 Atomic-Scale Description of Diffusion in Solids

Above zero Kelvin, every atom or ion in a solid material is vibrating very rapidly about its lattice position within the crystal at a frequency on the order of 10^{13} Hz with a typical vibration amplitude of about 10^{-12} m [19]. For simplicity, we will use "atoms" to stand for the elementary particles that constitute a solid diffusion medium material instead of using "atoms or ions", as here the diffusion media are Si and SiGe. A direct interchange mechanism is not favored in terms of energetic considerations [19–24]. On the other hand, the assistance of point defects in diffusion is believed to be of dominant importance in most crystalline materials, such as metals and semiconductors.

4.2.2.2 Point Defects and their Roles in Dopant Diffusion

The idea that dopant diffusion in silicon happens through interstitial-assisted and vacancy-assisted mechanisms has now been generally accepted, based on both experimental observations and theoretical calculations [23, 25]. This concept can be formulated by

$$D_A^{\text{eff}} = D_A^* \left(f_I \frac{[I]}{[I]^*} + f_V \frac{[V]}{[V]^*} \right) \tag{4.4}$$

where:

D_A^{eff} is the effective diffusivity of the dopant A measured under conditions where the point-defect populations are perturbed

D_A^* is the normal, equilibrium diffusivity under inert annealing conditions

$[I]$ and $[V]$ are the local interstitial and vacancy concentrations

$[I]^*$ and $[V]^*$ are the interstitial and vacancy concentrations under equilibrium conditions

f_I and f_V are the diffusion fractions mediated by interstitials and by vacancies respectively, and by definition

$$f_I + f_V = 1. \tag{4.5}$$

TABLE 4.1

Terminology Used for Diffusion Conditions

Terminology	Conditions
Intrinsic diffusion	Doping level $< n_i$
Extrinsic diffusion	Doping level $> n_i$
Unperturbed thermal equilibrium	$[I] = [I]^*, [V] = [V]^*, [I][V] = [I]^*[V]^*$
I and V recombine to a steady state	$[I] \neq [I]^*, [V] \neq [V]^*, [V] \neq [V]^*$
Non-steady state	$[I][V] \neq [I]^*[V]^*$

These formulae are capable of catching the diffusion behaviors for both equilibrium and non-equilibrium diffusion. Before further discussions, we need to clarify some terminologies to avoid any confusion. Table 4.1 shows the definitions of several diffusion conditions. n_i is the intrinsic carrier concentration, which is a material parameter and depends strongly on the temperature. In our study, the P concentrations are normally higher than n_i, so the P diffusion of this work is extrinsic diffusion.

Fahey *et al.* provided comprehensive analysis to derive the boundary values of f_I or f_V for P, Sb and As [26]. Ural *et al.* offered improvements by carrying out numerical calculations and using secondary ion mass spectrometry (SIMS) rather than spreading resistance analysis, which was used in earlier work [27]. Although there are debates on the values of f_I and f_V in some cases, such as in Si self-diffusion [28–30], this theoretical framework has been generally accepted. More details can be found in [23]. In the following sections, we will use the theoretical framework as discussed above.

The determination of f_I and f_V for common dopants in Si has always been a central issue, as summarized by Pichler [23]. One important fact that we should always keep in mind is that, although these parameters should have specific values under certain conditions (e.g. at a certain temperature and for a certain dopant), they cannot be measured directly from experiments. All the procedures that were used to indirectly obtain the f_I and f_V values involve experiments under perturbed conditions and subsequent analytical derivations. Therefore, the results obtained from various sources over the past several decades highly depend on the experiment conditions and the assumptions/approximations used. Without the capability to measure point-defect concentrations directly, the best criterion that we can utilize so far is how well these f_I and f_V values agree with experimental data not only for P, but also for other dopants.

4.2.2.3 Fermi Level Effect in Dopant Diffusion

Compared to diffusion in metals, there is a special effect in dopant diffusion in semiconductors, namely the Fermi effect, which comes from that fact that dopants exist as ionized impurities in semiconductors at room temperature and higher processing temperatures. The Fermi effect is related to the diffusion mechanisms, where diffusion is carried out by both neutral and charged defects [22]. For example, interstitials in semiconductors can be neutral interstitials (I^0), or single negatively charged interstitials (I^-), or double negatively charged interstitials ($I^=$), or single positively charged interstitials (I^+), or double positively charged interstitials (I^{++}), etc. The same is true for vacancies. The concentration of each charged defect is closely related to the doping level. The Fermi effect can be very strong, and introduce increases by factors of hundreds or thousands in the dopant diffusivity for heavily doped semiconductors. Experimentally, if we examine the actual concentration profile of many diffused dopants, we find that the diffusion appears to be faster in the higher-concentration regions. A commonly accepted mathematical form to describe the Fermi effect on the diffusivity is polynomial. For example, for an n-type doped semiconductor, the diffusivity is expressed as:

$$D_{eff} = D^0 + D^- \left(\frac{n}{n_i} \right) + D^= \left(\frac{n}{n_i} \right)^2, \tag{4.6}$$

where n is the local free electron concentration. For P-type dopant, the individual diffusivity terms that are significant are D^0, D^+ and D^{++} and so forth. Each of these individual diffusivities can be written in an Arrhenius form as

$$D = D_0 \exp \left(-\frac{E_A}{k_B T} \right) \tag{4.7}$$

with a preexponential factor D_0 and an activation energy E_A. Since this effect is thought to be related to the Fermi energy level that is set mainly by the doping level, it is often called the Fermi level effect. This effect has been well modeled and is included in TSUPREM-4™ [31]. In our modeling of coupled diffusion and segregation, we establish our own MATLAB® code, which includes this effect.

4.2.3 Methods of Dopant Profile Engineering

Dopant profile engineering during semiconductor fabrication can be categorized into two steps: one is the initial introduction of dopants such as predeposition, ion-implantation, in-situ doping, etc.; the other is the control of diffusion during high-temperature process steps. In the case of PNP SiGe HBT devices where the P profile in the base region is of central importance, the mission of the first step is to introduce the dopant with a high and narrow profile. For example, the atomic layer deposition (ALD) developed recently is a technique for this purpose [32]. In our work, in-situ doping of phosphorus profile was performed during the epitaxial growth, which is also the method used in industry. Our goal is to find out how to control P diffusion after the initial introduction step.

A method of suppressing dopant diffusion should satisfy two requirements simultaneously: it should reduce dopant diffusivity, and it should not bring significant negative changes to other material or device characteristics. The fact that dopant diffusion was mediated with the assistance of either vacancies of self-interstitial atoms gives us the opportunity to control diffusion. For an analogy, cars rely on bridges to cross a river, so by controlling the number of bridges it is possible to control the transport capacity. Since dopant diffusion relies on the help of point defects, we can control dopant diffusion by disturbing point-defect concentrations.

Point-defect concentrations can be disturbed externally or internally, as mentioned in the previous section. Excess vacancies or interstitials can be injected externally by gas ambient annealing (e.g. ammonia or oxygen [26]), or proton radiation [33]. However, proton radiation increases both interstitial and vacancy concentration, thus contradicting our technological mission to reduce P diffusion. Thermal nitridation is a possible method of suppressing P diffusion in SiGe since it reduces interstitial concentrations, and P diffusion in Si and low-Ge content SiGe alloys is mostly via the interstitial mechanism.

The incorporation of neutral alien impurities is another possible way of suppressing dopant diffusion. For example, C can suppress P and B diffusion in Si [34]. The current understanding is that C diffuses via the interstitial mechanism (or kick-out mechanism), thus reducing the concentration of self-interstitials [9]. We will discuss this method and our experiments in more detail in Section 4.4.

One thing to be noted is that, to our best knowledge, there has been no experimental study on the P diffusion mechanism in SiGe yet. The literature shows that P diffusion in Si is exclusively via interstitials while its diffusion in Ge is via vacancies. A natural logic may deduce that the P diffusion in SiGe should evolve from an interstitial mechanism to a vacancy mechanism with the increasing Ge molar fraction. In Sections 4.4 and 4.5, we see that this hypothesis is supported by our quantitative experimental evidence.

4.3 Experiment and Modeling Methodologies

4.3.1 Sample Structure Design and Growth

In a SiGe HBT, strained SiGe is used in the base region to create heterojunctions between Si on both sides. To preserve the crystallinity, strained SiGe layers are grown on top of Si in such a way that the lateral (in-plane) lattice spacing is maintained while the vertical (out-of-plane) lattice spacing can expand or contract freely. This can be done by epitaxial growth using a chemical vapor deposition (CVD) system, which is widely used in the semi-conductor industry.

Ge has a 4.2% larger lattice constant than Si, and the lattice constants for SiGe alloys obey Vegard's law. The lattice mismatch between Si and SiGe imposes a thickness limit, called the "critical thickness," below which the strain can be maintained. Beyond the critical thickness, defects and misfit dislocations form to relieve the strain, which degrades the device performance. A higher difference in Ge content results in a larger mismatch strain, thus a smaller critical thickness. This practical thickness limit depends on the epitaxial growth condition, the thickness of the Si capping layer on top and the subsequent thermal budget [35]. This metastable regime is the regime where most industry CMOS and HBT devices belong. In this work, the SiGe layers were designed to be in the metastable regime. Therefore, we needed to design the structures and the thermal annealing conditions such that the strain in the SiGe layers is fully maintained. We also needed to monitor the strain status closely in the diffusion experiment.

The design of SiGe layer thickness was based on two competing factors: it should be thick enough to accommodate sufficient P diffusion that SIMS can detect it, and the layers can't be so thick as to cause strain relaxation. A thicker layer can accommodate more diffusion, so it helps to obtain better accuracy in diffusivity extraction. C incorporation is helpful to increase the practical thickness limit as it helps to reduce the compressive strain due to the much smaller C size in a SiGe lattice. The equilibrium critical thickness and the practical thickness limit both decrease rapidly with the increasing Ge fraction. With 18% Ge and no carbon, sample Ge18C000 tends to relax the most as it has the highest Ge fraction in this study and no carbon. The Matthews-Blakeslee critical thickness model gives the equilibrium critical thickness, which is a very conservative thickness limit, below which the layers will not relax even with very high thermal budgets [36]. For 18% Ge, the Matthews-Blakeslee equilibrium critical thickness is less than 20 nm, which is too narrow to give sufficient room for dopant diffusion study in strained SiGe. In practice, strain relaxation happens above a much thicker limit than the Matthews-Blakeslee critical thickness limit, as layers can be under a metastable state before strain relaxation starts. Therefore, some trials were necessary to find the thickness and thermal budget range without strain relaxation. Test structures with different SiGe layer thicknesses were grown and annealed. The strain statuses before and after anneals were checked by X-ray diffraction (XRD). Based on these trial experiments, we found that 80 nm $Si_{0.82}Ge_{0.18}$ can stay fully strained after diffusion experiments using the thermal budgets of our interest.

There are two sets of samples studied in this work. One set of samples have uniform SiGe or SiGe:C regions where P diffusion happened. The other set of samples were used to study the segregation phenomenon across graded SiGe regions, thus the material composition in these regions were designed to be non-uniform. For the latter set of samples, the maximum Ge molar fraction is 0.18. The thickness and thermal budget requirements for the latter set of samples are less stringent, as the averaged Ge concentration is only half of the peak Ge concentration.

All the designs are shown in more detail in the corresponding sections. All the epitaxial layers were grown on 8-inch (100) Czochralski p-type Si wafers in an ASM Epsilon 2000 reduced-pressure chemical vapor deposition (RPCVD) reactor at Texas Instruments Deutschland GmbH in Germany. The epitaxial growth temperature of all structures was 550°C.

4.3.2 Annealing of Diffusion Samples

The selection of the annealing tool was based on the thermal budget design (temperature and time). For technological relevance, the industry main-stream annealing tools are rapid thermal annealing (RTA) tools, which have a fast ramp rate of about 100°C/sec. Soak RTA anneals normally last from one second to a few minutes at the peak temperature. However, if the annealing time at the peak temperature is too short, there will not be sufficient time for diffusivity extraction at this specific temperature. Therefore, all our samples were annealed for no less than 15 seconds. In terms of temperature choice, mainstream industry uses 1000°C–1100°C in the highest temperature anneals, so we used 1000°C–1050°C in the study of P diffusion in Sections 4.4 and 4.5.

In this work, three annealing tools were used.

1. In Section 4.4, inert anneals for C and Ge impact studies were per-formed in nitrogen ambient using an Applied Material Radiance rapid thermal processing tool at 1000°C for 15 to 120 seconds at Texas Instruments Deutschland GmbH in Germany.

2. In Section 4.5, ammonia anneals were performed using an AccuThermo AW610 atmospheric rapid thermal processing (RTP) system at Stanford Nanofabrication Facility (SNF), which has the rare capability of using NH_3 as an ambient and can anneal wafer pieces.

3. In Section 4.6, inert anneals were performed using a Linkam Scientific Instruments' high-temperature stage system TS1200 in nitrogen ambient at 900°C in our lab at UBC. The reason for using the heating stage was to compare our results with literature data in similar temperature ranges and longer anneal times ($\geq$10 mins), which the RTP tool at TI was not able to achieve.

4.3.3 Dopant Profile Characterization: Secondary Ion Mass Spectrometry

The central task of this thesis is to investigate the P diffusion in strained SiGe. The thin thickness of strained SiGe layer, tens of nanometers, poses a requirement for precise dopant profiling on the scale of angstroms.

Secondary ion mass spectrometry (SIMS) is the most appropriate analytical technique for this purpose, because it has the highest detection sensitivity (10^{17} cm^{-3}) for measuring elemental concentration, and is able to profile in the depth dimension with sub-nanometer precision. All the SIMS measurements were performed by Evans Analytical Group, which is the industry leader for commercial SIMS analysis. The samples were sputtered with 1 KeV Cs primary ion beam obliquely incident on the samples at 60° off the sample surface normal. The sputter rate was calibrated by a stylus profilometer measurement of the sputtered carter depth and corrected on a point-by-point basis. The measurement uncertainty for Ge fraction is ±0.5 at. %. ±1.5 % and ±5%. For P and C, the concentration uncertainty is ±3% and ±10%, respectively. The depth/thickness uncertainty is approximately 5%. Generally, the P dose variation for each structure is less than 10% among as-grown and annealed samples. It should be noted that C concentrations measured from SIMS are total C concentrations regardless of the C positions in the lattice.

4.3.4 Strain Characterization by X-ray Diffraction

As discussed earlier, what we studied was P diffusion in strained SiGe or SiGe:C. Long thermal anneals or thick strained layers may cause strain to relax. As strain impacts diffusion, it is important to monitor the strain status before and after diffusion closely. The most suitable characterization technique for strain measurements in our work is X-ray diffraction (XRD). In XRD, "Bragg condition" is

$$2d \sin \theta = n\lambda \tag{4.8}$$

where:
   d is the spacing between diffracting planes
   θ is the incident angle
   n is any integer
   λ is the wavelength of the incident X-ray beam.

High-resolution X-ray diffraction (HRXRD) was used in this work.

Reciprocal space mapping (RSM) is a special technique in XRD analysis. It generates a two-dimensional cut of the three-dimensional reciprocal space of the sample. In this way, the lattice strain status can be directly revealed in the result, which is usually an X-ray intensity contour. The principle of RSM is illustrated in Figure 4.1.

In a fully strained epitaxial layer, its in-plane lattice constant is equal to that of the substrate. Thus, their reciprocal lattice points will align vertically in the reciprocal space map. If they are not aligned, it means that their in-plane lattice constants are different and relaxation has occurred. In this work, XRD measurements were performed on the as-grown and the

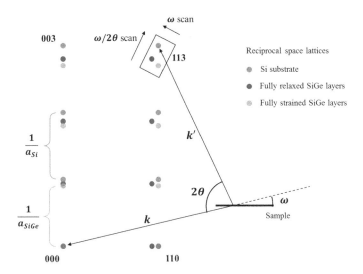

FIGURE 4.1

Two-dimensional reciprocal lattice and the Ewald sphere representation. In a reciprocal space map, the alignment of the diffraction spot from a SiGe layer with respect to the spot from the underlying Si substrate shows whether this layer is fully strained (i.e., with the same in-plane lattice constant) with the Si substrate.

annealed samples to confirm the strain status and substitutional C molar fraction. All the measurements were performed using a PANalytical X'Pert PRO MRD with a triple axis configuration, and (113) reciprocal space mapping was performed to further confirm the strain status for the sample with the highest thermal budget. $\omega - 2\theta$ rocking curves of (004) Bragg reflection were measured to determine the substitutional C molar fraction for each sample. The X-ray tube was operated at 45 kV and 40 mA in the line focus mode. The $Cu - K\alpha_1$ wavelength ($\lambda = 1.5406$ Å) was selected with a monochromator.

4.3.5 Computer Modeling and Simulations

In this study, we needed to include the concentration-dependent diffusivity (i.e. the Fermi effect), the impact of other impurities (C and germanium in this work), and dopant segregation. In such cases, numerical analysis to simulate a natural process becomes a must. On the other hand, computer simulations and fitting can also be part of the experiment analysis procedure, e.g. the extraction of diffusivity in our case.

4.3.5.1 Simulations with Process Simulator TSUPREM-4™

TSUPREM-4™ [31] is an industry mainstream two-dimensional technology computer-aided design (TCAD) tool to simulate processes used in the manufacturing of silicon integrated circuits. It has well-calibrated models and material parameters based on several decades of industrial practices, such as the defect and diffusion models and the diffusivities of common dopants in Si and SiGe. In Sections 4.4 and 4.5, the P diffusivity in each annealing condition was extracted by TSUPREM-4 fitting in terms of a ratio with regards to a reference diffusivity. TSUPREM-4 took the as-grown SIMS profiles as the pre-annealed profiles. The diffusivity used was the TSUPREM-4 default diffusivity of P in Si multiplied by a fitting parameter, i.e. the diffusivity ratio ($R_C^{A,Si_{1-x}Ge_x} \equiv \dfrac{D_A^{Si_{1-x}Ge_x:C_y}}{D_A^{Si_{1-x}Ge_x}}$ in Section 4.4 and $R_{NTD} \equiv \dfrac{D_P^{nitridation}}{D_P^{inert}}$ in Section 4.5). The fitting parameter value, i.e. $R_C^{A,Si_{1-x}Ge_x}$ or R_{NTD}, was determined until the best match of simulated diffusion profile to the post-annealed SIMS profile was achieved. In this way, we could study the impact of C or nitridation using the diffusivity ratios $R_C^{A,Si_{1-x}Ge_x}$ and R_{NTD}, respectively.

TSUPREM-4™ has the capability to accommodate some user-defined models. However, it was not able to take the segregation flux introduced by the Ge concentration gradient into account. Therefore, we had to find another method of simulating the coupled diffusion-segregation phenomenon. We used MATLAB to numerically solve the diffusion and segregation transport equations using the finite difference method.

4.3.5.2 Finite Difference Method

Equations describing mass transport phenomena are often partial differential equations, and we usually use the finite difference method for numerical analysis. We can obtain a numerical expression of Fick's law as

$$C_i^{t+\Delta t} = C_i^t + \frac{D\Delta t}{\Delta x^2}\left(C_{i-1} - 2C_i + C_{i+1}\right), \tag{4.9}$$

where the concentration of element i at time $t + \Delta t$ is explicitly expressed by the information at time t. By solving Equation 4.9 across the desired space and time range, we can obtain the numerically solved diffusion profile. The method described here is sometimes called the "FTCS explicit method": Forward in Time, Central in Space. The required condition for this method to be stable so that a convergent result can be obtained is:

$$\Delta t \leq \frac{1}{2}\frac{\Delta x^2}{D} \tag{4.10}$$

In this work, we used the finite difference method and implemented it in MATLAB to simulate the diffusion-segregation phenomenon in inhomogeneous materials.

4.4 The Effect of C on P Diffusion in SiGe

As mentioned previously, there are fewer studies available on P diffusion in SiGe:C. In Section 4.4, we present a systematic and quantitative study of C impact on P diffusion in strained SiGe through experiments and modeling. This topic is of great industry relevance, and can give guidance to the design of SiGe:C composition, structure and thermal processing.

4.4.1 Introduction

The fact that C can retard B and P diffusion in Si can be explained by two points: (1) C can reduce interstitial concentrations of Si and (2) B and P have very strong preferences for the interstitial mechanism, which can be represented by the value of f_I, i.e. $f_I^{B,Si} \approx 1$, $f_I^{P,Si} \approx 1$. For other common dopants, we have $f_I^{As,Si} \approx 0.4$ and $f_I^{Sb,Si} \approx 0.02$ [22]. The value of f_I is crucially important from a technological point of view because it indicates whether and how we can change the diffusivity of a specific dopant. The incorporation of an extra chemical element is a typical technological choice for the control, mostly the suppression, of dopant diffusion if that element can change the point-defect concentrations in a desired way. Besides the impact on diffusion, two extra properties are desired: that it is electrically inactive (neutral impurity) and that its impact on material properties is negligible at the given concentration. Based on these criteria, C became the ideal solution to suppress interstitial diffusers, such as B and P, as C can reduce the concentration of interstitials. Also, C is a neutral element for a Si/SiGe system so that it does not affect charge distribution. The size of the C atom is smaller than Si and Ge so that it slightly compensates the Ge-induced lattice mismatch strain and increases the stability and critical thickness of the strained SiGe layer [37].

It has been reported that C of 0.1% concentration can suppress B and P diffusion in Si by a factor of ten and even more with higher concentration [38]. When this method is extended from Si to SiGe, the situation has changed. The C incorporation method is still effective for B, but not for P. This drew us to question the underlying reason for their different responses to the same stimulus. To address this question, let's begin with literature review on the effect of C on dopant diffusion.

4.4.2 Literature Review: The Effect of C on Dopant Diffusion in Si and SiGe

4.4.2.1 C in Si and SiGe

C can be introduced either from the polysilicon used in crystal growth, or through contamination during this process [23]. When C is used for bandgap, stress or diffusion engineering, epitaxial deposition or solid-phase epitaxial regrowth of carbon-doped layers allow the incorporation of C up to 10^{21} cm^{-3} (2 at.%) in Si and SiGe [39, 40]. C has many formats of existence in a silicon lattice. It can exist as substitutional C (C_S) or interstitial C (C_I) in the silicon lattice, or it can form small or large complexes with other atoms or C clusters. Newman and Wakefield concluded that, in bulk monocrystalline silicon, C resides predominantly on substitutional sites based on the observations that the diffusion coefficient and the activation energy of C are similar to those of the dopants [41]. The substitutional character of C in silicon was experimentally demonstrated by Baker *et al.* [42] and Windisch and Becker [43], who found that the presence of C causes a reduction of the lattice parameter to an extent close to what can be expected from a linear interpolation between pure silicon and β-SiC. A second prominent atomic configuration is that C atoms reside in the interstitial sites instead of lattice sites [44]. This configuration usually results from the reaction of substitutional C with a silicon self-interstitial.

Because of the difference in covalent bond radius between Si and C atoms, C atoms in a Si lattice are surrounded by a strain field, with substitutional C atoms under tensile strain and interstitial ones under compressive strain fields [45]. The minimum formation energy of an interstitial C in split (100) configuration is 4.6 eV, which exceeds the formation energy of the substitutional C by 3 eV [46], which explains why C atoms are favored to occupy substitutional sites. In terms of epitaxially grown Si:C and SiGe:C films, the ratio of substitutional C concentration with respect to total C concentration (C_s / C_{total}) depends on the growth methods and conditions, and the value of this ratio was experimentally measured in this work.

4.4.2.2 Substitutional C Incorporation in Epitaxial Grown $Si_{1-y}C_y$ and $Si_{1-x-y}Ge_xC_y$

Unlike the Si-Ge system, silicon and carbon are only miscible to a very small degree under equilibrium conditions [23]. According to the Si-C phase diagram, stoichiometric SiC (silicon carbide) is the only stable compound [47]. Any alloy with a smaller C concentration is thermodynamically metastable. Such alloy layers can be obtained by kinetically dominated growth methods, such as molecular beam epitaxy (MBE) [48], solid-phase epitaxy [49], or CVD [50]. All of these methods generally work under far from thermodynamic equilibrium conditions, allowing a kinetic stabilization of metastable phases. Although the bulk solubility of C in silicon is small, 3×10^{17} cm^{-3}, at the melting point of silicon [51], epitaxial layers with more than 1%, 5×10^{20} cm^{-3}, C can be achieved. However, during an epitaxial growth step, it is not the equilibrium bulk solubility that is important but the "surface solubility" [52].

Substitutional C fraction or substitutionality (C_s / C_{total}) is strongly influenced by the growth conditions [40, 53]. The C substitutionality was determined by the cross-comparison between the XRD and SIMS measured C concentration. The XRD results represent substitutional C concentration while the SIMS results represent total C concentration. In this work, due to the presence of Ge, and the 1% SIMS error in Ge molar fraction measurement, we cannot use the cross-comparison between the XRD and SIMS measured C to determine the C substitutionality. Instead, we use the comparison between XRD of SiGe and SiGe:C and SIMS results to measure the C substitutionality, as discussed in 4.4.3.2.

The C substitutionality in epitaxial $Si_{1-x}Ge_x:C_y$ for different growth temperatures and C sources can be found in the literature, such as [53]. The general trend is similar to the case of $Si_{1-y}C_y$. In this work, all our wafers were grown at 550°C and the C concentration is up to 0.32%, which is well inside the region of complete substitutional incorporation. To distinguish these materials from the $Si_{1-y}C_y$ or $Si_{1-x-y}Ge_xC_y$ systems discussed above, we will use the same notation as for dopants, that is Si:C or SiGe:C. This notation also helps to convey our assumption that the C concentration is too low to significantly affect band alignment and strain of the Si:C or SiGe:C layers [52].

4.4.2.3 Mechanisms of C in Defect and Diffusion Engineering

The physical mechanisms of the C impact on dopant diffusion lie in the undersaturation of Si self-interstitials [9–11]. The diffusion of C is generally assumed to proceed via an interstitial-assisted mechanism [54]. A significant fraction of C diffusion via interstitial mechanism is also evidenced by the experiments of Kalejs *et al.* who observed enhanced C diffusion under conditions where the concentration of self-interstitials is enhanced by thermal oxidation or diffusion of P [55, 56]. Therefore, when a high and non-uniform concentration of C_s is grown in, the immobile C_s needs to change to mobile C_I for diffusion to happen [57]. The reaction of this step and its reverse reaction can be described as:

$$C_S + I \leftrightarrow C_I \qquad (4.11)$$

$$C_I + V \leftrightarrow C_S \qquad (4.12)$$

The first reaction describes that a C_s kicks a Si interstitial out from its site, and moves to the interstitial lattice site, also known as the kick-out mechanism. It consumes one Si interstitial per C_s atom. It is worth noting that the transition from C_s to C_I, and thus the consumption of Si interstitials, can happen without C diffusion.

Rücker *et al.* studied the impact of C on B diffusion where no supersaturation of interstitials due to ion-implantation or other processing steps was present [10, 11]. The diffusivity of B was found to be reduced by a factor of about twenty for temperatures between 750°C and 900°C due to the presence of 10^{20} cm^{-3} substitutional C. This effect has been attributed to a reduced density of Si self-interstitials [9, 58, 59]. These results show that C can be used as a diffusion suppressing agent if the target dopant is an interstitial diffuser. The use of C-rich layers to reduce B diffusion has been widely applied in SiGe HBTs [60, 61]. Transistors with improved static and dynamic performance can be fabricated with epitaxial SiGe:C layers [62].

4.4.2.4 C Impacts B and P Diffusion in Si Similarly

It has been generally agreed that B and P diffusion in Si is almost exclusively via the interstitial-assisted mechanism, i.e. the interstitial-assisted diffusion fraction $f_I \approx 1$ [22, 23]. It means that any method which can reduce the interstitial concentration will suppress B and P diffusion in Si. According to these studies, the degree of the C suppression effect on B and P diffusion in Si is very similar, presumably because the values of f_I for B and P are almost identical.

4.4.2.5 C Impacts B and P Diffusion in SiGe Differently

A handful of studies are available on B and P diffusion in low-Ge content SiGe. Zangenberg *et al.* and Christensen *et al.* both assumed that B and P were interstitial diffusers in SiGe [63, 64]. When the diffusion medium changes from Si to strained SiGe, the approach of using C to control diffusion has also worked quite well for B in SiGe bases for NPN HBTs [65]. It has been reported that C can retard B diffusion very similarly in strained SiGe with Ge molar fraction up to 20% [13]. However, the case for P is quite different. In $Si_{0.8}Ge_{0.2}$, the extent of the suppression effect on P diffusion is similar for 0.06% (3×10^{19} cm^{-3}) and 0.2% (1×10^{20} cm^{-3}) C molar fractions, while for B diffusion, the higher 0.2% C molar fraction clearly retards B stronger than the 0.06% C [7]. The above studies suggest that the diffusion mechanisms of B and P in Si may be similar, but they are different enough in strained SiGe to cause the difference of C retardation effectiveness. So far, there has been no systematic and quantitative study on the C retardation effect on P diffusion in SiGe, which is addressed in this work. Next, we present the experiment and modeling work on P diffusion in SiGe and SiGe:C.

4.4.3 Diffusion Experiments and Results

4.4.3.1 Sample Structures and Annealing Conditions Design

To understand the P diffusion behavior and mechanisms in $Si_{1-x}Ge_x$ and $Si_{1-x}Ge_x : C_y$, a wafer matrix with various Ge and C contents was designed and grown by epitaxy. Figure 4.2 shows the epitaxial structure design, where $Si_{1-x}Ge_x$ and $Si_{1-x}Ge_x : C_y$ are sandwiched between two Si layers. To be relevant to PNP SiGe HBT applications, the x value was chosen to be 0.18 and the y values ranged from 0 to 0.0032. The naming convention of the structures is to use the Ge and C percentages to name a $Si / Si_{1-x}Ge_x : C_y / Si$ structure. For example, Ge18C018 is used to name a $Si / Si_{0.82}Ge_{0.18} : C_{0.0018} / Si$

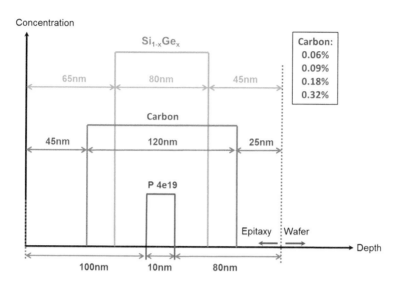

FIGURE 4.2

Schematic diagram of the structure design and splits in C mole fractions. The dotted line indicates the original wafer surface. The C concentration split is from 0.06 to 0.32 at.%, which corresponds to 3e19 to 1.6e20 cm^{-3}. The Ge and C profiles are designed to be constant with a certain thickness as shown above. (The first publication of this figure was in *Journal of Applied Physics* by AIP Publishing at https://doi.org/10.1063/1.4897498. Reprinted with permission.)

structure, which has 0.18% C as measured by SIMS. For each type of structure, several wafers were grown under identical growth conditions to be used as as-grown wafers and annealed wafers. Inert anneals were performed in nitrogen ambient using an Applied Material Radiance rapid thermal processing tool at 1000°C for 15 to 120 seconds. It should be noted that the concentration of substitutional C in epitaxial layers can exceed the solid solubility limit (e.g. 3.5×10^{17} cm^{-3} near the melting point) by several orders of magnitude [66]. Under optimized growth conditions, almost all C atoms can be incorporated on substitutional sites, which are the lowest energy configuration for isolated C atoms [40]. In this study, we confirmed this condition by XRD measurements for all as-grown and annealed samples.

The SiGe layers were grown pseudomorphically on Si substrates so that biaxial compressive strain was maintained. P was introduced in-situ by epitaxial growth with peak concentrations in the range of $2 \sim 4 \times 10^{19}$ cm^{-3}. As discussed in Section 4.3, by using trial experiments, we found that 80 nm $Si_{0.82}Ge_{0.18}$ can stay fully strained after diffusion experiments using the thermal budgets of our interest.

4.4.3.2 Strain and Depth Profile Analysis

We performed (113) reciprocal space mapping (RSM) on as-grown and annealed samples at room temperature. Figure 4.3 shows the RSMs of some

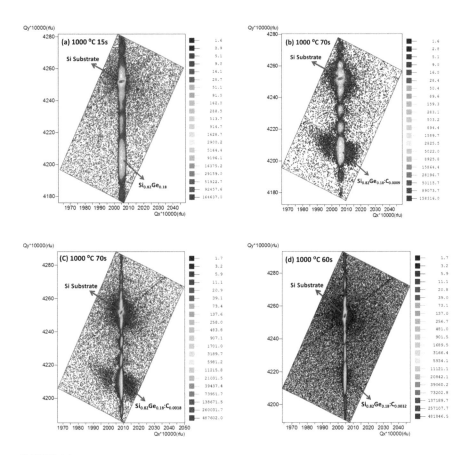

FIGURE 4.3
(113) reciprocal space mapping of annealed samples with the largest thermal budget for each type of structures: (a) Ge18C000; (b) Ge18C009; (c) Ge18C018; (d) Ge18C032. (The first publication of this figure was in *Journal of Applied Physics* by AIP Publishing at https://doi.org/10.1063/1.4897498. Reprinted with permission.)

annealed samples with the largest thermal budget for each type of structure. These results clearly show that the in-plane lattice constants of the $Si_{0.82}Ge_{0.18}$ and $Si_{0.82}Ge_{0.18} : C_y$ layer are the same as that of the Si substrate after annealing, which means that these two layers are fully strained to the substrate during diffusion. The (113) RSMs were done on all structures and annealing conditions, and no strain relaxation was observed for the structures after diffusion in this work. In the following discussion, SiGe stands for fully strained SiGe unless otherwise noted.

We extracted the as-grown C_S by comparing the XRD peak difference between an as-grown SiGe control sample and an as-grown SiGe:C sample, which were grown with the same epitaxy recipe except for C incorporation. By comparing the XRD simulations and SIMS data, we concluded that

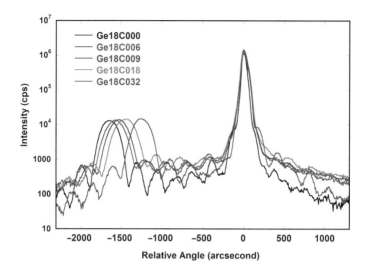

FIGURE 4.4

XRD (004) scan of SiGe and SiGe:C as-grown samples showing C molar fraction change. (The first publication of this figure was in *Journal of Applied Physics* by AIP Publishing at https://doi. org/10.1063/1.4897498. Reprinted with permission.)

C in as-grown samples is almost exclusively substitutional. Comparing the as-grown and annealed samples, we found that the substitutional C molar fraction is only reduced by less than 0.005% for Ge18C020 samples (see Figure 4.4). That means the substitutional C in annealed samples is more than 97% of that in as-grown samples. In fact, our growth condition has been optimized to achieve complete substitutional incorporation [52, 53]. Therefore, we approximate that the molar fraction of substitutional C is 100% in the modeling section, i.e. $C_s = C_y$.

SIMS measurements were performed for all as-grown and annealed samples. Figure 4.5 shows an example of SIMS profiles for as-grown and annealed wafers. Both the as-grown and annealed P peaks are within the constant Ge and C region. The shape of the diffused P peak is close to a Gaussian distribution, indicating that the concentration dependence of P diffusivity in SiGe is not strong. There is about 1% Ge concentration difference between the as-grown and annealed samples, which is within the SIMS measurement error. Ge profile also broadens slightly due to Si-Ge interdiffusion. C diffusion is seen to be much faster than P and Ge diffusion. The Ge fraction measured by XRD is 17.9% ± 0.1%. The Ge fraction measured by SIMS is consistent with that obtained from XRD within measurement errors.

4.4.3.3 P Diffusivity Ratio Extraction

The P diffusivity in each annealing condition was extracted by TSUPREM-4 fitting. TSUPREM-4 took the as-grown SIMS profiles as pre-annealed

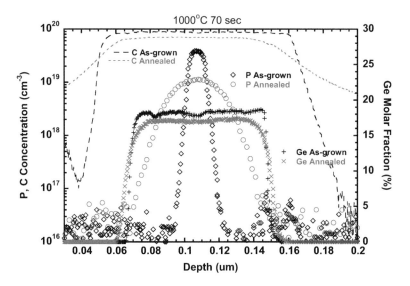

FIGURE 4.5

As-grown and annealed Ge, P and C SIMS profiles of Ge18C018 samples. (The first publication of this figure was in *Journal of Applied Physics* by AIP Publishing at https://doi.org/10.1063/1.4897498. Reprinted with permission.)

profiles. The diffusivity used was the TSUPREM-4 default diffusivity of P in Si, D_P^{Si}, multiplied by a fitting parameter. The fitting parameter value was modified until a good match to the post-annealed SIMS profile was achieved. At the annealing temperature of 1000°C, our SIMS and fitting results showed that P diffusivity in $Si_{0.82}Ge_{0.18}$, $D_P^{Si_{0.82}Ge_{0.18}}$, is very close to the diffusivity of P in Si in TSUPREM-4, D_P^{Si}, which is:

$$D_P^{Si_{0.82}Ge_{0.18}} \approx D_P^{Si} \tag{4.13}$$

The default P diffusivity in TSUPREM-4 has been calibrated based on a large amount of experimental data. We define $R_C^{A,Si_{1-x}Ge_x}$, the C impact factor, as the ratio of dopant element A diffusivity in $Si_{1-x}Ge_x{:}C_y$ over that in $Si_{1-x}Ge_x$:

$$R_C^{A,Si_{1-x}Ge_x} \equiv \frac{D_A^{Si_{1-x}Ge_x:C_y}}{D_A^{Si_{1-x}Ge_x}} \tag{4.14}$$

Figure 4.6 shows an example of the best fitting to a SIMS profile for R_C extraction. The extracted R_C values and the corresponding C molar fraction C_y are plotted in Figure 4.7 in reference to the available literature $R_C^{B,Si_{1-x}Ge_x}$ values($x \le 0.2$). It can be seen that $R_C^{P,Si_{0.82}Ge_{0.18}}$ are within the range of 0.43 to 0.85, which shows that C is still effective in retarding P diffusion in strained SiGe. However, the strongest retardation is in C molar fraction range of 0.05%

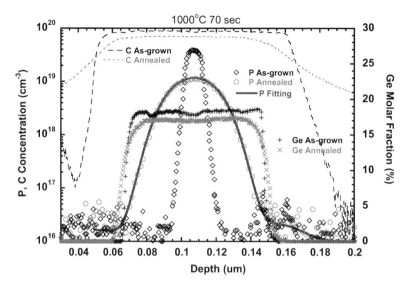

FIGURE 4.6
Example of TSUPREM-4™ fitting with SIMS profiles of as-grown and annealed samples Ge18C018. (The first publication of this figure was in *Journal of Applied Physics* by AIP Publishing at https://doi.org/10.1063/1.4897498. Reprinted with permission.)

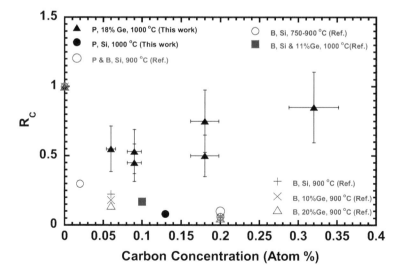

FIGURE 4.7
Summary of the C impact factor for P and B diffusivity in $Si_{1-x}Ge_x{:}C_y$. Literature data are from [10, 13, 14, 38]. (The first publication of this figure was in *Journal of Applied Physics* by AIP Publishing at https://doi.org/10.1063/1.4897498. Reprinted with permission.)

to 0.1%, above which the retardation is less effective. This trend is different from P in Si, B in Si and B in $Si_{1-x}Ge_x$ $(x \leq 0.2)$, where higher C gives more retardation. For C molar fraction at 0.20%, $R_C^{P,Si}$, $R_C^{B,Si}$ and $R_C^{B,Si_{1-x}Ge_x}$ $(x \leq 0.2)$ are all around 0.1 or below, which shows much stronger retardation than C effect on P in $Si_{0.82}Ge_{0.18}{:}C_y$. This difference between $R_C^{P,Si_{1-x}Ge_x}$ and $R_C^{B,Si_{1-x}Ge_x}$ clearly indicates that the diffusion mechanism for B and P in SiGe may be different.

In the following, the difference between B and P diffusion will be discussed based on point-defect-mediated diffusion mechanisms, i.e., interstitial-assisted and vacancy-assisted mechanisms. Considering the relatively low concentration of C, we will treat Si and SiGe as the diffusion media material, as the role of C is to change the point-defect concentrations and stress in SiGe.

4.4.4 Modeling of C Effect on P Diffusion

The effect of C on point defects is a dynamic process, and can be modeled by solving carbon-point defect reaction and transport equations [22,23]. In major process simulators such as TSUPREM-4 [31] and Sentaurus Process [67], C effect is modeled using multiple reaction equations, such as Equations 4.11 and 4.12. Higher-order carbon-defect complexes, typically involving more C and interstitial numbers, are sometimes used to model C effects [68]. However, since these parameters are vastly beyond experimental verification, it is very difficult to calibrate these equations and use them in a reliable manner. Considering that in the current HBT industry, the SiGe:C structures and the annealing conditions are in a certain range, it is then meaningful to establish empirical models to capture the C effect in simpler formats, which are easier to calibrate and more time-efficient for simulations. Therefore, our modeling will be based on the assumption that the local point-defect concentrations recombine to a steady-state condition, which is

$$[I][V] = [I]^*[V]^* \tag{4.15}$$

Furthermore, it is also assumed that

$$[AI]/[AI]^* = [I]/[I]^* \text{ and } [AV]/[AV]^* = [V]/[V]^* \tag{4.16}$$

where $[AI]$ and $[AV]$ are the dopant-interstitial pair and dopant-vacancy pair concentrations, and * denotes their counterparts under equilibrium condition.

Equation 4.16 shows the point-defect dynamics under perturbations: when the interstitial population of a material is decreased by X times, the vacancy population increases by X times. A similar approach has been applied to explain the effect of C on B diffusion in Si and SiGe where B was treated as an exclusive interstitial diffuser, i.e. $f_I^B = 1$ [69].

Several studies on the modeling of C effects have been reported. Rucker *et al.* considered the "kick-out" reaction involving C_S, I and C_I (Equation 4.11) and expressed the equilibrium self-interstitial concentrations with regard to substitutional C concentration using a simple analytical function [65]:

$$\frac{\widetilde{[I]}}{[I]^*} \approx \frac{1}{1+k_p[C_s]} \tag{4.17}$$

where k_p is the equilibrium constant for the reaction shown in Equation 4.11. Sibaja-Hernandez *et al.* modeled the C effect considering C clustering and the corresponding reaction dynamics [68]. This approach requires more than 15 fitting parameters to solve the model system. In this work, the focus is to obtain an empirical and reliable estimation of the extent of interstitial suppression caused by C incorporation. Based on the trend of $R_C^{B,Si}$ and $R_C^{B,Si_{1-x}Ge_x}$ data in Figure 4.7 and Equation 4.15, we established the C impact on the ratios of the time-averaged point-defect concentrations $\left(\widetilde{[I]}\text{ and }\widetilde{[V]}\right)$ over the corresponding equilibrium values $\left([I]^*\text{ and }[V]^*\right)$ as:

$$\frac{\widetilde{[I]}}{[I]^*} = \frac{1}{1+k[C_s]} \tag{4.18}$$

$$\frac{\widetilde{[V]}}{[V]^*} = 1+k[C_s], \tag{4.19}$$

where k is a fitting parameter.

In Equations 4.18 and 4.19, the constant k is in front of the C molar fraction and it shows the strength of C in reducing interstitial concentration, and should not depend on dopant species. It effectively shrinks or expands the C molar fraction axis in Figure 4.7. In the following discussions, C is in molar fractions so that k is dimensionless. Figure 4.7 shows the C impact on B diffusivity in the temperature range of 750~1000°C and up to $x_{Ge}=0.2$. It can be seen that the dependence of R_C^B on the temperature and Ge fraction is negligible within the given ranges. Therefore, in this work, we assume that k is constant for $x_{Ge} \leq 20\%$ within the range of 750~1000°C.

Now we can combine Equations 4.4, 4.5, 4.18 and 4.19 and describe the C effect on P diffusivity as

$$\frac{D_A^{\text{eff}}}{D_A^*} = \left\{ f_I^A \cdot \frac{1}{1+k[C_s]} + \left(1-f_I^A\right)\cdot\left(1+k[C_s]\right) \right\} \tag{4.20}$$

Since the majority of the C is located in substitutional sites according to XRD results and the post-annealed C profiles remain as box-like with wide plateaus, we can replace the substitutional C fraction $[C_s]$ by total C fraction $[C_y]$ measured by SIMS. Therefore Equation 4.20 is translated to

$$\frac{D_A^{Si_{1-x}Ge_x:C_y}}{D_A^{Si_{1-x}Ge_x}} \equiv R_C^{A,Si_{1-x}Ge_x} = f_I^A \cdot \frac{1}{1+k[C_y]} + \left(1-f_I^A\right)\cdot\left(1+k[C_y]\right) \qquad (4.21)$$

In this equation, both k and f_I determine the shape of $R_C(C_y)$. In particular, the C dependence is strongly influenced by f_I, when f_I is close to 1: see Figure 4.8. This model can explain the "U-shape" of R_C which we observed for P in SiGe. Figure 4.8 shows the relation between R_C and self-interstitial suppression $[I]/[I]^*$. When $f_I = 1$, C retards diffusion monotonically with the increase of C content. When $f_I = 0.99$, the C retardation effect will saturate at $[I]/[I]^* \approx 10$ because the vacancy supersaturation will become dominant beyond this point. This feature implies that C can be used as an effective tool to "magnify" the small vacancy contribution.

The above modeling only included the C impact on dopant diffusion in terms of changing the point-defect concentrations. Moreover, C also affects the diffusion media strain by compensating compressive strain in SiGe and thus affects the equilibrium diffusivities of a dopant. The highest C molar fraction in this study is 0.32%, which means that the strain compensation introduced by C is equivalent to 3.2% of Ge, or about 0.13% in strain.

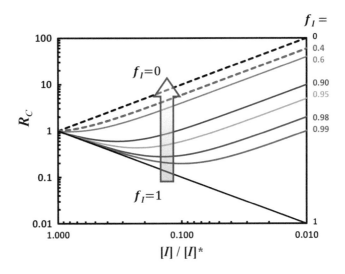

FIGURE 4.8
The relation between the interstitial suppression ratio ($[I]/[I]^*$) and dopant diffusivity suppression ratio ($R_C = D/D^*$) for different diffusion mechanism under the assumption of $[I][V] = [I]^*[V]^*$. (The first publication of this figure was in *Journal of Applied Physics* by AIP Publishing at https://doi.org/10.1063/1.4897498. Reprinted with permission.)

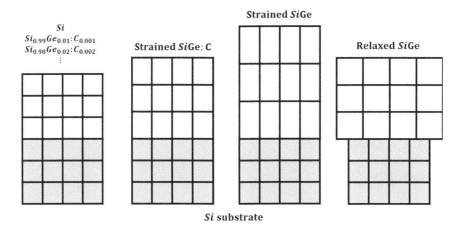

FIGURE 4.9
Simplified illustration of cross-section view of $Si_{1-x}Ge_x$ and $Si_{1-x}Ge_x:C_y$ lattice spacing under different strain status: (a) Si and fully strained $Si_{1-x}Ge_x:C_y$ with $x/y \approx 10$, and the out-of-place lattice constant is the same as the Si out-of-plane lattice constant, (b) fully strained $Si_{1-x}Ge_x:C_y$ with $x/y > 10$, (c) fully strained $Si_{1-x}Ge_x$ and (d) fully relaxed $Si_{1-x}Ge_x$. (The first publication of this figure was in *Journal of Applied Physics* by AIP Publishing at https://doi.org/10.1063/1.4897498. Reprinted with permission.)

However, these two types of C effects cannot be dissociated experimentally since they are always in effect simultaneously.

Figure 4.9 shows a simplified illustration of the lattice configuration for typical structures grown on a Si substrate under different strain statuses. It can be noticed that the strained SiGe:C lattice is different from the relaxed Si:C or the relaxed SiGe or the strained SiGe lattices, which were previously studied. Therefore we can only estimate the strain compensation effect, commonly expressed as a change in the diffusion activation energy, based on previous studies on strain impact on dopant diffusion in strained-SiGe and relaxed-SiGe structures [70].

Now we add an apparent strain factor to Equation 4.21 and revise the model to

$$\frac{D(P, SiGe:C)}{D(P, SiGe)} = \left\{ f_I \cdot \frac{1}{1+k[C_y]} + (1-f_I) \cdot (1+k[C_y]) \right\} \cdot \exp\left(-\frac{q' \cdot \varepsilon_c}{k_B T} \right), \qquad (4.22)$$

where:

ε_c is the change of strain due to substitutional C ($\varepsilon_c < 0$)
q' is the strain derivative with a unit of eV per unit strain
k_B is the Boltzmann constant.

Christensen *et al.* reported an apparent activation energy of $q' = -13$ eV per unit strain, assuming that the strain ε_c is proportional to the Ge content by $\varepsilon_c = -0.0418x$ [64]. Based on this value, 0.2% C will increase P diffusivity by about 10% due to the strain compensation effect. At 0.2% carbon, it was shown that the effective diffusivities of B and P were suppressed by a factor of more than ten, as shown in Figure 4.7. Therefore, in Equation 4.22, the term involving point defects in the curly brackets is dominant. We will neglect the strain compensation effect of C (the exponential term in Equation 4.22) in the following sections, and use Equation 4.21 in the rest of the work. After the model is established, we need to extract the parameters k and f_I in Equation 4.21.

4.4.5 Model Parameter Extraction

Let's consider the case of Si first. $f_I^{P,Si} \approx f_I^{B,Si} \approx 1$ has been generally agreed upon. Taking this assumption, we can extract k from the $R_C^{P,Si}$ and $R_C^{B,Si}$ data, which overlay within measurement errors. Therefore, k is independent of dopant species, as discussed in Section 4.4.2. Based on the fitting to this data (shown as the blue dotted line in Figure 4.10), we estimate that $k^{Si} \approx 5500$ for both B and P in Si. For SiGe, the literature data for B shows that the C impact factor R_C is very close to that in Si, i.e. $R_C^{B,Si_{1-x}Ge_x} \approx R_C^{B,Si}$ for $0 \leq x \leq 0.2$ with the C range up to 0.2%, which indicates that $f_I^{B,Si} \approx f_I^{B,SiGe}$ and $k^{Si} \approx k^{SiGe} \approx 5500$. It can also be deduced that $f_I^{B,Si_{1-x}Ge_x} \approx 1$ for $0 \leq x \leq 0.2$.

Next, we applied $k^{SiGe} = 5500$ and extracted the value of $f_I^{P,Si_{0.82}Ge_{0.18}}$ from Equation 4.21, fitting it to our experimental data of $R_C^{P,Si_{0.82}Ge_{0.18}}$. The best fit was determined by minimizing the root-mean-square error and it gives $f_I^{P,Si_{0.82}Ge_{0.18}} = 0.95$. Referring to Figure 4.8, it is clear that a small vacancy

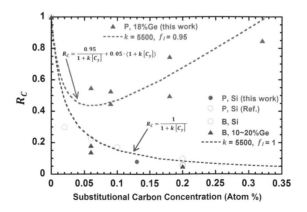

FIGURE 4.10

Modeling of C effect on P and B diffusivity in Si and strained SiGe. Literature data are from the same references as in Figure 4.7. (The first publication of this figure was in *Journal of Applied Physics* by AIP Publishing at https://doi.org/10.1063/1.4897498. Reprinted with permission.)

contribution ($f_V^{P,Si_{0.82}Ge_{0.18}} = 0.05$) is responsible for the U-shape of the $R_C^{P,Si_{0.82}Ge_{0.18}}$.

4.4.6 Comparison of C Impact on Different Dopants

Now that the C impact models on defects and dopant diffusivities have been established and applied to B and P in Si and SiGe, it is natural to ask whether these models apply to other dopants such as arsenic (As) and antimony (Sb). Only one literature work on As and Sb diffusion in Si:C was found [38], where the diffusion structures are 150 nm Si/150 nm Si:C/Si. One side of the As and Sb profiles are inside the Si:C, and the diffusion anneals were at 900°C for 2 to 6 hours. We compared the literature $R_C^{As,Si}$ and $R_C^{Sb,Si}$ with the calculations using our R_C model. We found that the experimental data agree with our model using $k^{Si} \approx 5500$, $f_I^{As,Si} = 0.4$ and $f_I^{Sb,Si} = 0.02$, the latter two of which are well accepted values derived from numerous literature data [22,23]. This comparison shows that our models can also apply to other dopants and to non-HBT-like structures and annealing conditions. Further experiments can be performed to investigate the dynamics of carbon-defect interactions with varied structure design and time-dependent annealing conditions.

Here we converted the C molar fraction into relative interstitial concentration ($[\tilde{I}]/[I]^*$), using $k^{Si} \approx 5500$ and Equation 4.18, and plotted it against the dopant diffusivity ratio D/D^* in a log-log scale so that the trend is more clear (see Figure 4.11). The choice of using $[\tilde{I}]/[I]^*$ instead of using C concentration is to include other defect engineering approaches such as thermal nitridation.

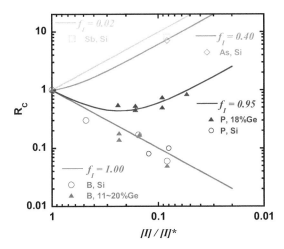

FIGURE 4.11
Experiment R_C data for Sb, As, P and B vs. model predictions. Literature data are from the same references as in Figure 4.7. (The first publication of this figure was in *Journal of Applied Physics* by AIP Publishing at https://doi.org/10.1063/1.4897498. Reprinted with permission.)

It shows that this model can effectively catch the C impact behavior for all these dopants, i.e. B, P, As, and Sb. The C impact on Sb in the original reference [38] was given as 8 ± 2 (not shown in this figure), which is close to the prediction of our model.

4.4.7 Section Conclusions

In this section, the C effects on P diffusion in SiGe:C and Si:C under rapid thermal anneal conditions were studied quantitatively. The Ge molar fractions studied are 0% and 18%, and the C molar fractions are up to 0.32%. The results showed that the C retardation effect is less effective for $Si_{0.82}Ge_{0.18}$ than Si. In $Si_{0.82}Ge_{0.18} : C$, there is an optimum C content at around 0.05% to 0.1%, beyond which more C incorporation does not add more retardation to P diffusion. These phenomena can be explained by the increased vacancy-mediated diffusion fraction f_V^P in strained SiGe as Ge content increases. Compared to B, P diffusion in strained $Si_{0.82}Ge_{0.18}$ has a bigger vacancy-mediated diffusion fraction by about 5%.

Empirical models were established to calculate the C impact on the time-averaged point-defect concentrations and effective diffusivities as a function of C content. The strain effect due to C incorporation was shown to be negligible. The models were shown to be consistent with experiments on antimony and arsenic diffusion in Si:C and B diffusion in SiGe:C. These empirical models are easy to calibrate and implement in process simulators. These results also indicate that C is very effective in perturbing point-defect concentrations, which in turn may provide a useful approach to investigate dopant diffusion and self-diffusion mechanisms.

4.5 The Effect of Thermal Nitridation on P Diffusion in SiGe and SiGe:C

R&D on SiGe PNP HBTs provides a good opportunity for the application of thermal nitridation, as the dopants in PNP HBTs, B and P, are both mainly interstitial diffusers. It is of technological interest to investigate whether this method is effective. If it is the case, it is a relatively low-cost and simple approach to be adopted by industry, as the only change is in the annealing ambient. These questions will be addressed in Section 4.5.

4.5.1 Literature Review

4.5.1.1 Physical Effects of Thermal Nitridation

The term "thermal nitridation" was historically used in several types of silicon wafer processing technologies which involve the chemical element nitrogen [71–73]. In this work, we refer to thermal nitridation or nitridation with regard to only "annealing silicon wafers in ammonia (NH$_3$) ambient". This process involves the reaction of ammonia gas with either SiO$_2$ (oxynitridation) or the bare silicon surface (direct nitridation).

The early studies of thermal nitridation mainly focused on how to produce high-quality thermal silicon nitride films [71, 74, 75]. Hayafuji *et al.* first investigated the physical phenomena during thermal nitridation of silicon, such as lattice defects [76]. They studied the shrinkage and growth of preexisting oxidation-induced stacking faults during thermal nitridation of silicon with and without an oxide film. They observed that stacking faults in silicon without an oxide film shrink linearly with nitridation time while stacking faults in silicon with an oxide film grew during nitridation. It was also found that the stacking faults' shrinkage and growth rates depend on the partial pressure of ammonia. Based on these results, a model was suggested that the shrinkage and growth of stacking faults were caused by the emission and absorption of silicon self-interstitials by the Frank partial dislocations [77]. It also implies that thermal nitridation with silicon causes the change in intrinsic point-defect concentrations. This theory was backed by Fahey *et al.*, who studied the impact of thermal nitridation on dopant diffusion [15].

4.5.1.2 Effect of Thermal Nitridation on Dopant Diffusion

It was found that direct nitridation of the silicon surface retarded B and P diffusion while oxynitridation enhanced diffusion of both impurities [15]. The responses of P, B, antimony and arsenic diffusion to identical external conditions were compared [16, 78]. The results were interesting in that some impurities exhibit the opposite behavior to other impurities under identical conditions. During direct nitridation where bare silicon wafers were annealed in ammonia, P and B diffusion was retarded while antimony diffusion was enhanced. In contrast, for oxynitridation, P and B diffusion was enhanced while antimony diffusion was retarded. These results revealed that direct nitridation and oxynitridation cause different changes to the silicon wafers and different impurities have various responses to the same change. Coupled with the theory proposed by Hayafuji, a full picture was drawn that direct nitridation and oxynitridation caused the point-defect concentrations to change in opposite ways, and subsequently each impurity's diffusivity was affected corresponding to its specific diffusion mechanism.

Although these arguments could not prove a one-to-one correspondence between macroscopic impurity profiles and microscopic diffusion mechanisms, it is still useful for us to understand the various experimental results within a consistent theoretical framework. It also provided convincing clues as to the dominant diffusion mechanism of each impurity, which is of great technological importance. A systematic review was given by Fahey *et al.* with a focus on the relation between intrinsic point defects and impurity diffusion [26]. The experimental results were explained by assuming Sb diffusion to be dominated by the vacancy mechanism, P and B diffusion to be dominated by the interstitial mechanism, and As to have comparable components of both types of mechanism.

The early studies in the 1980s were conducted mainly using spreading resistance analysis to obtain impurity profiles and using analytic arguments to derive results. Ural *et al.* offered improvements by implementing the SIMS technique and performing numerical simulations [27]. The results of impurity diffusion mechanisms were given with six sets of numerical simulations with various levels of approximations and assumptions on point-defect dynamics. These results generally agreed with previous studies and confirmed the hypotheses on defect injection conditions and impurity diffusion mechanisms.

Another issue of thermal nitridation experiments is the kinetics of injected point defects. Mogi *et al.* reported a time-dependent study using doped superlattices and demonstrated the transport capacity of injected point defects [17]. They used antimony diffusion profiles as the indicator to reveal the vacancy supersaturation during thermal nitridation as a function of time and temperature. They showed that vacancies diffused through the entire epilayer (650 nm) within 60 minutes at 860°C or 15 minutes at 910°C.

4.5.1.3 Thermal Nitridation on Si and SiGe

Several studies have been reported on thermal nitridation effects on both Si and SiGe. Karunaratne *et al.* studied the effect of thermal nitridation on B diffusion in Si, strained $Si_{0.89}Ge_{0.11}$ and $Si_{0.89}Ge_{0.11}$:C and the diffusion suppression factors in the different materials were compared [14, 79]. The results showed that the extent of B diffusion suppression in Si and $Si_{0.89}Ge_{0.11}$ is comparable between 940°C and 1050°C. The nitridation-induced suppression factor in the carbon-incorporated sample was weaker than in a carbon-free sample. This indicated that the effects of thermal nitridation and C incorporation can both suppress B diffusion and their impacts can be added to some extent, although the total effect is not simply the sum of the two effects. Bonar *et al.* studied the effect of thermal nitridation on Sb diffusion in Si and SiGe [80]. They showed that the Sb diffusion enhancement factor was smaller in $Si_{0.9}Ge_{0.1}$ than in Si.

Given the present understanding of point-defect injection conditions, it is interesting to explore its technological potential to control dopant diffusion. R&D on SiGe PNP HBTs provides a good opportunity for the application of this method, as the dopants in PNP HBTs, B and P, are both interstitial diffusers, and may be retarded at the same time with the vacancy injection method. However, there is no experimental study available on whether this method works for P in strained-SiGe systems and to what extent. It is also not clear whether this effect will work with other approaches (e.g. C incorporation) or to what extent. If this method is effective, it is a relatively low-cost and simple approach that can be adopted by industry, as the only change is in the annealing ambient. These questions are addressed in this section.

4.5.2 Diffusion Experiments

We designed P in SiGe and SiGe:C structures and annealed them in three different defect injection conditions, including the thermal nitridation (vacancy injection) condition. The effectiveness of the thermal nitridation method was studied based on these experiments. Apart from the industry applications, thermal nitridation also provides an important approach to reveal P diffusion mechanisms in SiGe and SiGe:C.

4.5.2.1 Sample Structures and Annealing Conditions Design

To understand the thermal nitridation effect on P diffusion behavior and mechanisms in $Si_{1-x}Ge_x$ and $Si_{1-x}Ge_x : C_y$, a wafer matrix with various Ge and C contents was designed and grown epitaxially. Figure 4.12 shows the epitaxial structure design, where $Si_{1-x}Ge_x$ and $Si_{1-x}Ge_x : C_y$ are sandwiched between two Si layers. To be relevant to PNP SiGe HBT applications, the x values were chosen to be 0.09 and 0.18 and the y values ranged from 0% to 0.09%. The naming convention of the structures is to use the Ge and C percentages to name a $Si/Si_{1-x}Ge_x : C_y/Si$ structure (Table 4.2).

As discussed in Section 4.3.1, all the wafers were grown on 8-inch (100) Czochralski p-type Si wafers in an ASM Epsilon 2000 RPCVD reactor. Ammonia anneals were performed using an AccuThermo AW610 atmospheric rapid thermal processing (RTP) system. The target annealing temperatures was 980–1000°C, and the annealing time ranged from 35 seconds to 6 minutes. The thermal annealing conditions were designed such that P diffusion happened inside the fully strained SiGe or fully strained SiGe:C layers.

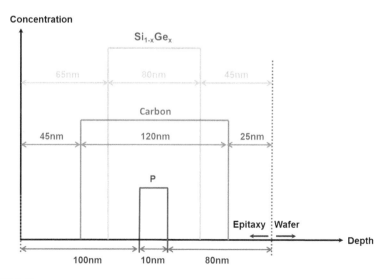

FIGURE 4.12
Schematic diagram of the structure design.

TABLE 4.2

Epitaxial Wafer Matrix Used in this Work

Structure	Wafer Name	Ge mole Fraction	C mole Fraction
$Si_{0.91}Ge_{0.09}$	Ge09C0	9%	0
$Si_{0.91}Ge_{0.09}$:C	Ge09C009	9%	0.09%
$Si_{0.82}Ge_{0.18}$	Ge18C0	18%	0
$Si_{0.82}Ge_{0.18}$:C	Ge18C006	18%	0.06%
$Si_{0.82}Ge_{0.18}$:C	Ge18C009	18%	0.09%

Source: The first publication of this table was in *Journal of Material Science* by Springer at https://doi.org/10.1007/s10853-015-9475-1. Reprinted with permission.

4.5.2.2 Defect Injection Using Masking Layers

To extract the thermal nitridation effect, control samples were needed to show diffusion under inert conditions (no defect injection). To avoid the run-to-run temperature variation, we needed to anneal the control samples together with those under thermal nitridation conditions in one RTA anneal. Therefore, for the control samples, we used SiO_2/SiN_x as masking layers, which has been proven to protect the underlying structure and create equivalently inert diffusion conditions, even when ammonia is present [57]. An oxide layer was also used without nitride to create interstitial injection conditions due to the oxynitridation reaction in ammonia. The oxide and nitride masking layers were deposited by plasma-enhanced chemical vapor deposition (PECVD). Figure 4.13a–b show the cross-sections of bare samples and masked control samples. By annealing the three types of samples in the same RTA run, we avoided run-to-run variation, and made the samples directly comparable to show the effect of vacancy injection and interstitial injection. The annealing conditions were designed carefully to avoid the strain relaxation in SiGe and SiGe:C layers. After annealing, the masking layers were etched away by 1:10

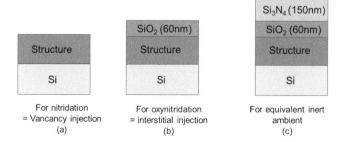

FIGURE 4.13

Structures used for different defect injection conditions during thermal nitridation experiments. (The first publication of this figure was in *Journal of Material Science* by Springer at https://doi.org/10.1007/s10853-015-9475-1. Reprinted with permission.)

hydrogen fluoride (HF) containing buffered oxide etch (BOE) solution, and SIMS was used to obtain P, Ge and C concentration profiles.

4.5.3 Experimental Results

4.5.3.1 Strain and Composition Analysis

We performed (113) reciprocal space mapping on as-grown and annealed samples at room temperature. Figure 4.14 shows the RSMs of some annealed samples, which were annealed with the largest thermal budget for the corresponding structure. These results clearly show that the in-plane lattice constants of the $Si_{1-x}Ge_x$ and $Si_{1-x}Ge_x : C_y$ layer are the same as that of the Si substrate after annealing. This means that these two layers are fully strained

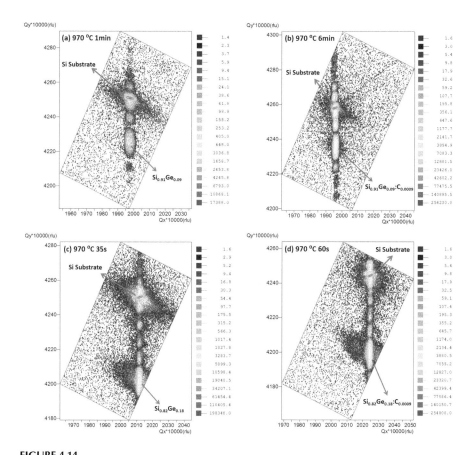

FIGURE 4.14

(113) reciprocal space mapping of annealed samples with the largest thermal budget for each type of structures: (a) Ge09C0; (b) Ge09C009; (c) Ge18C0; (d) Ge18C009. (The first publication of this figure was in *Journal of Material Science* by Springer at https://doi.org/10.1007/s10853-015-9475-1. Reprinted with permission.)

to the substrate during diffusion, and no strain relaxation was observed for the structures after diffusion. These results confirm that P diffusion happened in fully strained SiGe or fully strained SiGe:C. In the following discussion, SiGe stands for fully strained SiGe unless otherwise noted. $\omega - 2\theta$ rocking scans of (004) Bragg reflection were performed to confirm the substitutional C fraction of each structure. The measurement procedures are the same as described in Section 4.4.3.2.

4.5.3.2 Concentration Profiling and P Diffusivity Ratio Extraction

Figure 4.15 shows the SIMS profiles of as-grown and annealed samples for each structure. Both the as-grown and annealed P peaks are within the constant Ge and C region. The shape of the diffused P peak is close to a Gaussian

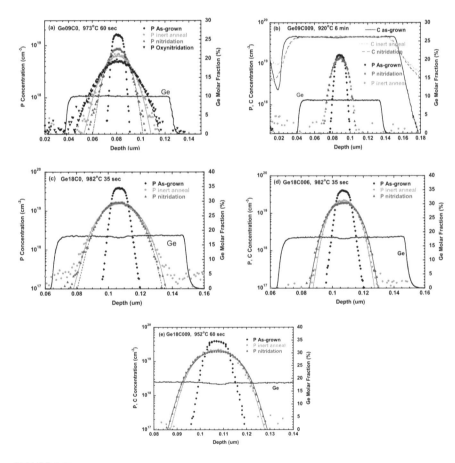

FIGURE 4.15
SIMS profiles of as-grown and annealed samples and TSUPREM-4™ fittings. The solid lines on top of the P profiles are the TSUPREM-4 fittings.

distribution, indicating that the concentration dependence of P diffusivity in SiGe is not strong. C diffusion is seen to be much faster than P and Ge diffusion.

Figure 4.15a shows that the P diffusion under the nitridation condition was retarded compared to the inert condition, and it was enhanced under the oxynitridation condition as expected, which showed that the defect engineering methods by capping layers were valid. As Si-Ge interdiffusion is much slower than P diffusion, annealed Ge profiles are very close to the as-grown Ge profiles; therefore, only Ge as-grown profiles are shown for clarity. In sample Ge09C009, shown in Figure 4.15b, there is no clear difference between P diffusion under the nitridation condition and that under the inert condition. However, C diffusion is retarded under the nitridation condition, as seen in the C rising edge at the depth below 40 nm and in the falling edge in the 140–180 nm depth range. As C is an interstitial diffuser in Si, this is a clear indication of the interstitial undersaturation caused by vacancy injection. The C profiles in all other diffusion structures are similar, and are not shown for clarity. It is also shown that this interstitial undersaturation penetrates to a depth of at least 200 nm. Figure 4.15c–e shows the results from samples with 18% Ge but different C contents. Interestingly, in Ge18C006 and Ge18C009, the P profile under the nitridation condition is not showing any retardation, but rather a slight enhancement in P diffusion is seen.

The nitridation impact factor on P diffusivity, $R_{\mathrm{NTD}} = \dfrac{D_P^{\mathrm{nitridation}}}{D_P^{\mathrm{inert}}}$, was extracted by fitting SIMS profiles and is listed in Table 4.3. As the temperature accuracy of the RTP system is estimated to be a few tens of degrees, some temperature adjustments were needed to fit the SIMS profiles. These adjustments do not affect the nitridation impact factor, as each set of samples for comparison were annealed in the same run. The error bar of $R_{\mathrm{NTD}} = \dfrac{D_P^{\mathrm{nitridation}}}{D_P^{\mathrm{inert}}}$ is estimated to be $\pm 20\%$ due to the SIMS error and TSUPREM-4 fitting errors. Figure 4.15 includes some examples of the best-fitting curves and the corresponding SIMS profiles for R_{NTD} extractions.

The nitridation impact factors, $R_{\mathrm{NTD}} = \dfrac{D_P^{\mathrm{nitridation}}}{D_P^{\mathrm{inert}}}$, are shown in Figure 4.16 as a function of C molar fraction. A value greater than one indicates a diffusivity enhancement, whereas smaller than one means retardation. In reality, during annealing, the point-defect concentrations, and therefore the diffusivity ratios, are functions of time. Therefore, the diffusivity values extracted in this work are time-averaged values. From Figure 4.16, we can see that thermal nitridation is still effective in retarding P diffusion in SiGe for up to 18% Ge, but when 0.06% or 0.09% C is present, this retardation is no longer effective. As a comparison, relevant literature data are also plotted in this figure ([14] and [27]).

Similar to the P case, R_{NTD} for B ($R_{\mathrm{NTD},B}$) increases with the C concentration for Si and $Si_{0.89}Ge_{0.11}$. However, $R_{\mathrm{NTD},B}$ is less than 1 with C, which means

TABLE 4.3

Nitridation Impact Factor $R_{NTD,P}$ for P under Non-equilibrium Point-Defect Conditions Caused by Nitridation

Wafer Name	Annealing Condition*	Nitridation Impact Factor $R_{NTD,P}$
Ge09C0	973°C 1 min	0.45 ± 0.09
Ge09C009	920°C 6 min	1.00 ± 0.20
Ge18C0	982°C 35 s	0.79 ± 0.16
Ge18C006	982°C 35 s	1.41 ± 0.28
Ge18C009	952°C 60 s	$1.19 + 0.24$

* The annealing temperatures quoted here are the temperature extracted using TSUPREM-4™ fitting to account for uncertainty.

Source: The first publication of this table was in *Journal of Material Science* by Springer at https://doi.org/10.1007/s10853-015-9475-1. Reprinted with permission.

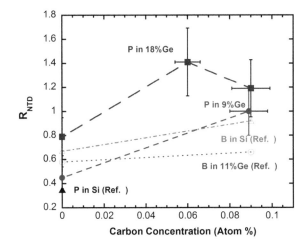

FIGURE 4.16

Nitridation impact factor $R_{NTD,P}$ as a function of C % at various Ge %. Error bars of the data without C are not shown for clarity. The literature data quoted in this figure are obtained from [27] for P in Si and from [14] for B in Si and SiGe. (The first publication of this figure was in *Journal of Material Science* by Springer at https://doi.org/10.1007/s10853-015-9475-1. Reprinted with permission.)

that even with C, the nitridation still retards B diffusion on top of the C retardation. While for $Si_{0.82}Ge_{0.18}$:C samples, $R_{NTD,P}$ can be larger than 1, showing that the thermal nitridation of Si increases P diffusion.

4.5.4 Discussions

4.5.4.1 Major Observations

Two major correlations can be observed from the experimental results shown in Figure 4.16. First, for a certain Ge molar fraction, R_{NTD} goes up with the increasing C content, i.e., the effectiveness of P diffusivity retardation under the vacancy injection condition decreases with the increasing C content. For the C molar fractions we used, 0.06% and 0.09%, no further retardation effect is seen from the thermal nitridation. Second, for a certain C molar fraction, R_{NTD} goes up with the increasing Ge fraction. These effects can be explained using the point-defect-assisted diffusion theory.

To analyze the thermal nitridation effect on P diffusion, applying the point-defect assisted diffusion theory, we define the impact of thermal nitridation as

$$R_{NTD,P}^{SiGe} \equiv \frac{D_{\mathrm{nitridation}}^{SiGe:C}}{D_{\mathrm{inert}}^{SiGe:C}} = f_I \frac{[I]}{[I]^*} + f_V \frac{[V]}{[V]^*}. \tag{4.23}$$

In SiGe:C samples, $R_{NTD,P}^{SiGe:C}$ shows only the thermal nitridation effect, not the overall C plus thermal nitridation effect.

4.5.4.2 Ge Fraction Dependence of Thermal Nitridation Impact Factor

On the Ge fraction dependence, the observation in Figure 4.16 is consistent with the results shown in Section 4.4. It was shown that when the matrix changes from Si to $Si_{0.82}Ge_{0.18}$, the vacancy-mediated diffusion fraction f_V increases from 0 to 0.05. Therefore, in Si, $f_V^{Si} = 0$, we have

$$R_{\mathrm{NTD},P}^{Si} = \frac{[I]}{[I]^*}. \tag{4.24}$$

In $Si_{0.82}Ge_{0.18}$, $f_V^{Si0.82Ge0.18} = 0.05$, and

$$R_{\mathrm{NTD},P}^{Si0.82Ge0.18} = 0.95 \frac{[I]}{[I]^*} + 0.05 \frac{[V]}{[V]^*}. \tag{4.25}$$

The experiment results in Figure 4.16 show that

$$R_{\mathrm{NTD},P}^{Si0.82Ge0.18} > R_{\mathrm{NTD},P}^{Si091Ge0.09} > R_{\mathrm{NTD},P}^{Si}. \tag{4.26}$$

Although we do not have direct evidence to show that $\dfrac{[I]}{[I]^*}$ values in $Si_{0.82}Ge_{0.18}$ and $Si_{0.91}Ge_{0.09}$ are close to that in Si under the same thermal nitridation condition, $f_V^{Si0.82Ge0.18} > f_V^{Si0.91Ge0.09} > f_V^{Si}$, and $\dfrac{[I]}{[I]^*}\bigg|_{Si0.82Ge0.18} \approx \dfrac{[I]}{[I]^*}\bigg|_{Si0.91Ge0.09} \approx \dfrac{[I]}{[I]^*}\bigg|_{Si}$ is one possible explanation of the Ge dependence observed.

When C is present, the C concentration is low and introduces interstitial undersaturation. We assume that C concentration studied does not change the P diffusion mechanisms, i.e., $f_V^{Si0.82Ge0.18} = f_V^{Si0.82Ge0.18:C}$, and

$f_V^{Si0.91Ge0.09} = f_V^{Si0.91Ge0.09:C}$. Therefore, it is expected that the Ge dependence doesn't change with the C addition, as seen in Figure 4.16:

$$R_{\text{NTD},P}^{Si0.82Ge0.18:C} > R_{\text{NTD},P}^{Si091Ge0.09:C} \tag{4.27}$$

4.5.4.3 C Fraction Dependence of Thermal Nitridation Impact Factor

On the C fraction dependence, the results indicate that when both are present, the nitridation retardation effect is weaker, i.e, $R_{\text{NTD},P}^{SiGe:C} > R_{\text{NTD},P}^{SiGe}$. For P in Si$_{0.82}Ge_{0.18}$:C, enhancement in P diffusion is seen instead of retardation, which may be a result of high vacancy supersaturation acting on the non-negligible f_V diffusion term. Therefore, the method of using thermal nitridation to control B and P diffusion only works for Si and very low C and Ge content SiGe:C, which is the area below the $R_{\text{NTD}} = 1$ line in Figure 4.16.

The trend of the increased $R_{\text{NTD},P}$ with the C presence agrees with the $R_{\text{NTD},B}$ results from [14]. However, $R_{\text{NTD},B} < 1$ still holds, where $R_{\text{NTD},P}$ can be larger than 1. This difference may be a result of $f_{V,P}^{SiGe} > f_{V,B}^{SiGe}$, as suggested from [81] and the above Ge dependence discussion.

As generally accepted, both the C incorporation and the thermal nitridation of Si act on dopant diffusion by reducing the concentrations of interstitials [9, 10, 59, 76]. Historically, the influence of thermal nitridation on point defects was observed indirectly by the growth or shrinkage of stacking faults and dopant diffusion [76], as the point-defect concentrations are too low to be measured directly. When both methods are applied simultaneously, the nitridation impact is less, i.e., $R_{\text{NTD},P}^{SiGe:C} > R_{\text{NTD},P}^{SiGe}$. As P is not a pure interstitial diffuser in SiGe for the Ge fractions studied, while B is, using B diffusion as an indirect measurement of the point-defect concentration change when both C method and nitridation method are combined, although outside the scope of this work, will be valuable to give a clearer defect concentration picture.

4.5.4.4 Thermal Nitridation and Equivalent $[I]/[I]^*$ in Si$_{0.82}$Ge$_{0.18}$

In Section 4.4, we systematically studied the C impact on defect engineering in Si and SiGe, and expressed the C impact factor $R_{C,P}^{SiGe:C}$ as a function of the C content and the interstitial undersaturation ratio $[I]/[I]^*$, which is less than 1. A natural thought is to compare $R_{C,P}$ and $R_{\text{NTD},P}$, and relate the latter with an equivalent C content and $[I]/[I]^*$. The relation between the interstitial undersaturation ratio $[I]^*/[I]$ in Si:C and Si$_{0.82}$Ge$_{0.18}$:C and the C molar fraction $[C]$ was established by (equivalent to Equation 4.18)

$$\frac{[I]^*}{[I]} = 1 + 5500[C] \tag{4.28}$$

By definition,

$$R_{C,P}^{SiGe:C} \equiv \frac{D_{inert}^{SiGe:C}}{D_{inert}^{SiGe}}. \tag{4.29}$$

Also, the steady-state condition holds:

$$R_{C,P}^{Si0.82Ge0.18:C} = 0.95\frac{[I]}{[I]^*} + 0.05\frac{[V]}{[V]^*} = 0.95\frac{[I]}{[I]^*} + 0.05\frac{[I]^*}{[I]} \tag{4.30}$$

By definition,

$$R_{NTD,P}^{SiGe:C} \equiv \frac{D_{nitridation}^{SiGe:C}}{D_{inert}^{SiGe:C}} = \frac{D_{nitridation}^{SiGe:C}}{D_{inert}^{SiGe} * R_{C,P}} \tag{4.31}$$

Therefore,

$$\frac{D_{nitridation}^{SiGe:C}}{D_{inert}^{SiGe}} = R_{NTD,P}^{SiGe:C} * R_{C,P}^{SiGe:C} \tag{4.32}$$

$R_{C,P}^{SiGe:C}$ has been measured in $Si_{0.82}Ge_{0.18}$:C in [81], and the annealing in that work was performed in nitrogen ambient at 1000°C for 15 to 120 seconds, close to the conditions in this work. Therefore, it is reasonable to assume $R_{C,P}^{SiGe:C}$ obtained from [81] can be applied to this work. Combining Equations 4.30 and 4.32, we can calculate $[I]/[I]^*$ for SiGe:C under thermal nitridation.

P diffusivity under a defect engineering condition divided by the equilibrium P diffusivity, i.e., $\frac{D_{P,defect-engineered}}{D_P^*}$, is plotted as a function of the interstitial undersaturation ratio $[I]/[I]^*$ in $Si_{0.82}Ge_{0.18}$ in Figure 4.17. Depending on which defect engineering method(s) is (are) used, $\frac{D_{P,defect-engineered}}{D_P^*}$ can be $\frac{D_{nitridation}^{SiGe:C}}{D_{inert}^{SiGe}}$ or $\frac{D_{nitridation}^{SiGe}}{D_{inert}^{SiGe}}$ or $\frac{D_{inert}^{SiGe:C}}{D_{inert}^{SiGe}}$.

From Table 4.4 we can see that thermal nitridation further injected vacancies and suppressed self-interstitial concentrations by 31% to 53%. These results are close to the observation of vacancy injection impact on B diffusion in strained SiGe:C with 11% Ge from [14], which provided an extra 8% to 42% interstitial suppression on top of the C impact for the temperature range of 940–1000°C.

To further investigate the effects of thermal nitridation on point defects and dopant diffusion in strained SiGe, although beyond the scope of this work, one might consider using the comprehensive procedure previously used for the study of dopant diffusion in Si [27]. This procedure combines the nitridation and oxidation processes, and uses identical conditions for multiple dopants, such as B and antimony, where a parallel analysis of different dopants

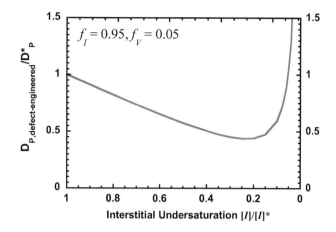

FIGURE 4.17
P diffusivity under a defect engineering condition divided by the P diffusivity ratio as a function of the interstitial undersaturation ratio $[I]/[I]^*$ in $Si_{0.82}Ge_{0.18}$. It is based on local equilibrium assumption of $[I][V] = [I]^*[V]^*$. This figure was adapted from [81].

TABLE 4.4

Summary of the impact factor of carbon, thermal nitridation, and both methods on P diffusion in $Si_{0.82}Ge_{0.18}$ and the corresponding interstitial undersaturation ratio $[I]/[I]^*$

C Concentration	$R_{C,P}^{Si_{0.82}Ge_{0.18}:C}$	$\dfrac{[I]}{[I]^*}$ (C effect only)	$R_{NTD,P}^{Si_{0.82}Ge_{0.18}:C}$	$\dfrac{D_{nitridation}^{SiGe:C}}{D_{inert}^{SiGe}}$ for $Si_{0.82}Ge_{0.18}$	$\dfrac{[I]}{[I]^*}$ (C+nitridation effect)
0	1	1	0.79 ± 0.16	0.79 ± 0.16	0.77
0.06%	0.44	0.200	1.41 ± 0.28	0.62 ± 0.12	0.094
0.09%	0.46	0.165	1.19 ± 0.24	0.55 ± 0.11	0.114

Source: The impact of C is adopted from [81]. (The first publication of this table was in *Journal of Material Science* by Springer at https://doi.org/10.1007/s10853-015-9475-1. Reprinted with permission.)

can be performed in a more rigorous way to reveal key parameters such as $[I]/[I]^*$ and $[V]/[V]^*$.

4.5.5 Section Conclusions

To conclude, we reported an experimental study of the thermal nitridation effects on P diffusion in strained $Si_{1-x}Ge_x$ and strained $Si_{1-x}Ge_x:C_y$ with up to 18 at.% Ge and 0.09 at.% carbon. P diffusivities under thermal nitridation (vacancy injection) and effective inert conditions were compared. The results show that thermal nitridation can retard P diffusion in SiGe, but the effectiveness of this retardation method decreases with increasing Ge and C content. When 0.06% or 0.09% C is present in $Si_{0.82}Ge_{0.18}$, thermal nitridation slightly increases P diffusivity compared to the inert conditions. The Ge dependence can be explained by the increasing contribution from the vacancy-assisted mechanism for P diffusion in strained SiGe with the increasing Ge content. In terms of the interstitial undersaturation ratio $[I]/[I]^*$ in $Si_{0.82}Ge_{0.18}$, thermal nitridation can further decrease $[I]/[I]^*$ by 31% to 53% on top of the C effect.

4.6 Segregation of P in SiGe

After the discussion of P diffusion in uniform Si and SiGe with a focus on diffusivities and diffusion mechanisms, let us turn to another issue: the dopant distribution across graded SiGe layers. Although SiGe layers with a constant Ge molar fraction are essential for diffusion studies, graded SiGe layers are more commonly used in SiGe HBTs. This involves another universal phenomenon regarding the redistribution of a chemical element in an inhomogeneous solid material system besides diffusion, which is called segregation. In strained-SiGe systems, P tends to segregate from higher Ge regions into lower Ge regions, while B has exactly the opposite tendency. This is especially a problem for thin SiGe layers in PNP SiGe HBTs because segregation causes broadened P profiles, which contradicts the goal to achieve high and narrow P profiles for high-speed PNP HBTs. The segregation phenomenon cannot be described by a simple modification in the diffusivity using Fick's law alone. As the segregation happens simultaneously with diffusion, we have to account for the segregation-related transport together with the diffusion transport. Without knowledge and models of segregation, one cannot predict dopant profiles accurately.

4.6.1 Introduction

Let us clarify some terms before further discussions. Here, we restrict the term "diffusion" to the net mass transport of an impurity in *homogeneous* solids where the material composition can be considered to be uniform. In contrast, the term "segregation" is used to describe the impurity distribution in *inhomogeneous* solids where the material composition is not constant. The $Si_{1-x}Ge_x : P$ material systems provide good examples of the "coupled diffusion and segregation" phenomenon. By using a colon in $Si_{1-x}Ge_x : P$, we indicate that only the concentrations of Si and Ge affect the material property of SiGe, as a binary alloy, and the impurity P has negligible impact except for providing charge carriers as a dopant. That being said, only when Ge molar fraction x changes do we refer to this material as *inhomogeneous*.

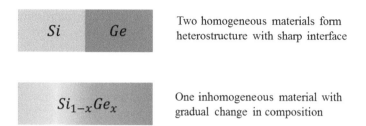

FIGURE 4.18
Illustration of sharp and graded interfaces.

In many practical structures, segregation and diffusion are decoupled in space, where segregation happens and is modeled only at interfaces. For Si and SiGe material systems, most dopant segregation studies have been limited to sharp interfaces between two different materials, such as Si/SiO_2 and Si/Si_3N_4. Segregation is typically modeled using a flux balance equation with a segregation coefficient in many commercial simulation tools such as TSUPREM-4™. It should be noted that when there is not a sharp interface, but a gradual change in the solid-state solvent, such as a gradual change in Ge concentration in a SiGe alloy (see Figure 4.18), segregation also happens, and it is coupled with dopant diffusion. In the Sentaurus Process™ [68], the segregation model for a graded solid solvent is available, but the calibration was based only on two sets of experiments using sharp SiGe interfaces under equilibrium conditions, which are not relevant to typical industry structures or annealing conditions.

SiGe HBTs are good examples for investigating coupled diffusion and segregation. The epitaxial strained heterostructures provide ideal testbeds. A typical SiGe HBT base layer has an epitaxial SiGe layer with a graded Ge slope on one or both sides, and the layer is doped with either B (B) for a NPN-type HBT or with P (P) for a PNP-type HBT. During fabrication steps with high temperatures, dopant diffusion and segregation happen simultaneously in the same space. Segregation is beneficial for NPN HBTs as B prefers to stay in SiGe base layers with higher Ge content, while in PNP HBTs, P segregates out of SiGe base layers to Si layers and makes the vertical doping profile hard to control. In this case, coupled dopant diffusion and segregation models are required to describe the dopant profile evolution across the Ge concentration slopes.

There are only a handful of experimental studies on dopant segregation in SiGe, which were typically performed on sharp Si/SiGe interfaces using long annealing times [13, 64, 82]. Due to the low spatial resolution of the concentration profiles, these data can hardly be used for model calibration purposes. Therefore, more accurate and industry-relevant experimental data for segregation model calibration are in great need.

In this study, we re-derived a diffusion-segregation model clearly showing the diffusion and segregation fluxes, and performed P segregation experiments at SiGe slopes to verify the model. Compared to segregation experiments using sharp SiGe/Si interfaces, where one segregation coefficient can be measured for one SiGe composition, segregation experiments using graded SiGe provide a new approach to calibrating and/or extracting the segregation coefficients for a range of SiGe compositions, and the experimental structures are directly relevant to the SiGe HBT industry. Our coupled diffusion-segregation model was then calibrated using these, which enables more accurate prediction of dopant profile evolution during diffusion. The theoretical derivations of the coupled diffusion and segregation model of this work is generic to all inhomogenous solid solutions, and the experiments and model calibration are specifically relevant to PNP SiGe HBTs.

4.6.2 Literature Review

Early studies on dopant segregation in SiGe material were mainly focused on how to introduce certain amounts of doping into elemental semiconductors for the development of fabrication technology [83–85]. This research focused mainly on the segregation coefficients (called "distribution coefficients" at that time) for various situations, together with chemical reactions during crystal growth, often from a thermodynamic equilibrium perspective [83, 86, 87]. Segregation, as a part of mass transport phenomena during high-temperature annealing after the structure fabrication, has been investigated along with the development of SiGe epitaxy techniques. Hu *et al.* performed segregation experiments where dopants diffused from a polycrystalline silicon layer on top of the Si/SiGe/Si epitaxial layers [88]. It was found that B tends to segregate into the SiGe layer while P and arsenic tend to segregate away from it. More experimental studies were reported with drive-in diffusion ([82, 89]) and in-situ doping ([13, 90]) conditions.

From a theoretical perspective, Hu laid the groundwork by analyzing the chemical potential of a dopant in inhomogeneous media [91, 92]. In an inhomogeneous material, the driving force for the diffusion of a dopant is no longer simply given by its concentration gradient, but by the gradient of its chemical potential. The chemical potential of a dopant, considering the atomic system and electronic subsystem, is given by [93]:

$$\mu_1 = kT \ln\left(\frac{N_1^2}{N_L\left(N_C N_V\right)^{1/2}}\right) - Z_1 \Delta W_i\left(N_2\right) - \Delta E_{b1}\left(N_2\right) + \frac{1}{2} E_g\left(N_2\right)$$

$$+ \theta\beta_1\beta_2 N_2 + \theta\beta_1^2 N_1 \tag{4.33}$$

where N_1, N_2 and N_L are the concentrations of the dopant, Ge and lattice sites, respectively. The bonding energy ΔE_{b1}, the intrinsic work function W_i, the bandgap E_g, the lattice contraction/expansion coefficient β_2, the conduction- and valence-band density of states N_C and N_V are all functions of the germanium concentration N_2. Theoretical calculations suggested that B segregation is dominated by stress effects, while P and As are dominated by changes in electronic band structure [93]. With approximation of low doping concentration for $N_1 < 1 \times 10^{20} \, \text{cm}^{-3}$, the dopant flux J_1 was given by the gradient of chemical potential as:

$$J_1 = -\frac{D_1 N_1}{kT} \frac{\partial \mu_1}{\partial x}$$

$$= -D_1 \frac{\partial N_1}{\partial x} + \frac{D_1 N_1}{2} \left[\frac{1}{N_C} \frac{\partial N_C}{\partial N_2} + \frac{1}{N_V} \frac{\partial N_V}{\partial N_2} + \frac{2}{kT} \left(Z_1 \frac{\partial W_i}{\partial N_2} + \frac{\partial E_{b1}}{\partial N_2} - \frac{1}{2} \frac{\partial E_g}{\partial N_2} - \theta \beta_1 \beta_2 \right) \right] \frac{\partial N_2}{\partial x}.$$

(4.34)

With the gradients $\partial N_1 / \partial x$ and $\partial N_2 / \partial x$, the diffusion and segregation flux can be calculated simultaneously. However, this model requires many parameters, such as the effective density of states (N_C and N_V), lattice contraction coefficients (β_1 and β_2) and several local energy terms, which are difficult to obtain from theory or experiments.

On the other hand, You *et al.* derived a simpler equation based on thermodynamic principles (parameter labels are modified here for consistency) [94]:

$$J_1 = -D_1 \left(\frac{\partial N_1}{\partial x} - \frac{N_1}{N_1^{\text{eq}}} \frac{\partial N_1^{\text{eq}}}{\partial x} \right),$$

(4.35)

where N_1^{eq} was defined as "normalized thermal equilibrium concentration". It was given by [95]:

$$\frac{N_1^{\text{eq}}}{N_L} = \exp\left(-\frac{g^f}{k_B T} \right),$$

(4.36)

where N_L was lattice site concentration and g^f was the formation Gibbs free energy of the diffusing species. However, there is no clear explanation on the precise meaning of N_1^{eq}. For example, N_1^{eq} was explained as solubility and was used to define segregation coefficient k_{seg} as [96], [65, 66]:

$$k_{\text{seg}} \equiv \frac{C_\beta^{A1}}{C_\beta^{A2}} = \frac{C_\beta^{A1} \, \text{solubility}}{C_\beta^{A2} \, \text{solubility}}.$$

(4.37)

By using solid solubility ratios, this definition implies that the segregation coefficient is a constant, independent of dopant concentration, which is theoretically inappropriate [92]. On the other hand, the definition of "solubility" is also ambiguous since it has at least two different meanings: (1) as described by Henry's law, the solubility of a gas in a liquid is directly proportional to the partial pressure of the gas above the liquid; (2) the maximum possible concentration of an element which can be dissolved into a solid-state material without creating a second phase. Therefore, the parameters' definitions were confusing and it was not clear on how to use Equation 4.35 for modeling of the coupled diffusion-segregation phenomenon.

From these references, one can see that a gap between experimental results and theoretical modeling still exists. In the following sections, we will address this issue through detailed derivation and definitions and specially designed experimental verifications.

4.6.3 Derivation of Coupled of Diffusion-Segregation Equation

In some classic diffusion textbooks, such as the one by Shewmon [97], we can find some discussions on ternary alloys, where the mass flux of one element is not only driven by its own concentration gradient $\partial N_1 / \partial x$, the normal diffusion term, but is also driven by the gradient of the second element $\partial N_2 / \partial x$, which is the segregation term. SiGe:P can be considered as a ternary alloy. Let's at first review the derivations of the mass fluxes, which are based on classic thermodynamics and previous work [23]. We use a more convenient definition of k_{seg}, and separate the diffusion flux and segregation flux in the data fitting and simulations in Section 4.6.4.

4.6.3.1 The Driving Force for Diffusion in Solid-State Materials

In solid-state materials, diffusion and segregation are statistical results of random jumps of atoms. This phenomenon can be described thermodynamically *as if* the atoms are driven by the chemical potential gradient of the diffusing species. The mass flux can be expressed as

$$J = N \cdot \vec{V} = N \cdot \left(M \cdot \vec{F} \right) \tag{4.38}$$

where N, $\vec{V}$, M and $\vec{F}$ denote concentration, velocity, mobility and the driving force, respectively [98]. For a one-dimension case along the z axis, the driving force is opposite to the chemical potential gradient,

$$\vec{F} \cdot e_z = -\frac{\partial \mu}{\partial z}, \tag{4.39}$$

where:

    μ is the chemical potential
    e_z is the unit vector of the z axis.

As pointed out by Atkins [99], such a force should be considered as a *"thermodynamic force"*. It is not necessarily a real *force* pushing the particles down the slope of the chemical potential. Considering the fact that solid-state diffusion is microscopically the random jump of atoms, this *force* may represent the spontaneous tendency of the atoms to disperse as a consequence of the second law of thermodynamics and the hunt for maximum entropy. Usually, this force is called the *driving force* for diffusion.

4.6.3.2 Derivation of Coupled Diffusion-Segregation Equation

Now let us consider a three-element system, e.g. $Si_{1-x}Ge_x : P$. Let N_1 be the number concentration of P, the solute, and N_2 be the number concentration of Ge, which represent the material composition. Here, N_1 is small enough that its impact on material property is negligible and N_2 is large enough to cause material property change. To differentiate their different roles, we put a colon ":" between $Si_{1-x}Ge_x$ and P. For the purpose of generality, we will treat P as a solute β and $Si_{1-x}Ge_x$ as the solution A, respectively.

Then the chemical potential gradient of solute β in solution A can be expressed as

$$\frac{\partial \mu_\beta{}^A}{\partial z} = \frac{\partial \mu_\beta{}^A}{\partial N_1}\frac{\partial N_1}{\partial z} + \frac{\partial \mu_\beta{}^A}{\partial N_2}\frac{\partial N_2}{\partial z}, \tag{4.40}$$

where the first term on the right-hand side is the traditional diffusion driving force term, as in the Fick's law, and the second term is the driving force of segregation.

As discussed in Section 4.6.1, here, "diffusion" is used in its narrower sense, which means the net transport of substance due to its chemical potential gradient in a given homogeneous diffusion medium. "Segregation" is used to describe the net transport due to inhomogeneous diffusion media. By separating the segregation term, users of a simulation tool can evaluate the magnitude of the segregation term to see if it is significant enough to be included for the price of longer computation time.

In the case of P in biaxially fully strained SiGe, we treat N_2 ($N_2 = x_{Ge} \cdot N_{total}$) as the sole representative of the material property. It can be approximated that $\partial N_{total} / \partial N_2 \approx 0$ when the Ge concentration is low enough so that its impact on lattice concentration is negligible. This approximation generally holds in this study. Combining Equations 4.38 through 4.40, we obtain

$$J = -N_1 \cdot M \cdot \left(\frac{\partial \mu_\beta{}^A}{\partial N_1}\frac{\partial N_1}{\partial z} + \frac{\partial \mu_\beta{}^A}{\partial N_2}\frac{\partial N_2}{\partial z} \right). \tag{4.41}$$

In a homogeneous structure, the material is uniform, i.e. $\partial N_2 / \partial z = 0$, and there is no segregation. When $\partial N_2 / \partial z \neq 0$, we need to find out the expression of $\partial \mu_\beta{}^A / \partial N_2$, which represents the non-uniform nature of the inhomogeneous structure.

Based on classic thermodynamics, at a fixed pressure and temperature, the chemical potential of solute β in solution A, such as P in P-doped SiGe, is

$$\mu_\beta^{A\circ} = \mu_\beta^{A,\Theta^\circ} + RT\ln\left(a_\beta^A\right), \quad \text{and} \quad a_\beta^A = x_\beta^A\gamma_\beta^A, \quad x_\beta^A = \frac{N_1}{N_{\text{total}}}, \quad (4.42)$$

where $\mu_\beta^{A,\Theta}$, R, T, a_β^A, x_β^A, γ_β^A, N_1 and N_{total} denote the standard chemical potential of β in solution A, ideal gas constant, the absolute temperature, the activity, the molar fraction, the chemical activity coefficient of β in A, the number of β atoms, and the total number of atoms in solution A. The chemical potential of β as a function of Ge concentration can now be expressed as

$$\frac{\partial\mu_\beta^A}{\partial N_2} = \frac{\partial\mu_\beta^{A,\Theta^\circ}(N_2)}{\partial N_2} + RT\frac{\partial\ln\left(x_\beta^A\cdot\gamma_\beta^A(N_2)\right)}{\partial N_2}. \quad (4.43)$$

To find out the expression of $\partial\mu_\beta^A/\partial N_2$, let us consider the situation when solute β is distributed within two adjacent and different solutions (or phases) A_1 and A_2. The concept of thermal equilibrium in a system requires that the chemical potential of each species involved is constant in the system, i.e. equal in adjacent phases. This condition can be expressed as

$$\mu_\beta^{A1}\left(x_\beta^{A1}\right) = \mu_\beta^{A2}\left(x_\beta^{A2}\right). \quad (4.44)$$

Using the relation of Equations 4.42, it can be further expressed as

$$\mu_\beta^{A1,\Theta^\circ} + RT\ln\left(\gamma_\beta^{A1}\right) + RT\ln\left(x_\beta^{A1}\right) = \mu_\beta^{A2,\Theta^\circ} + RT\ln\left(\gamma_\beta^{A2}\right) + RT\ln\left(x_\beta^{A2}\right). \quad (4.45)$$

Then the segregation coefficient k_{seg} is defined as

$$k_{\text{seg}}\left(\frac{N_2}{N_{\text{total}}}\right) \equiv \frac{x_\beta^{A2}}{x_\beta^{A1}} = \frac{\gamma_\beta^{A1}}{\gamma_\beta^{A2}}\cdot\exp\left(\frac{\mu_\beta^{A1,\Theta^\circ} - \mu_\beta^{A2,\Theta^\circ}}{RT}\right). \quad (4.46)$$

Since the Si, SiGe and Ge discussed here are all in solid-state phase, the standard chemical potential of β in these materials is identical [23] ($\mu_\beta^{A1,\Theta^\circ} - \mu_\beta^{A2,\Theta^\circ} = 0$). Therefore the segregation coefficient corresponds to the inverse ratio of activity coefficient

$$k_{\text{seg}} \equiv \frac{N_1^{A2}}{N_1^{A1}} = \frac{\gamma_\beta^{A1}}{\gamma_\beta^{A2}}. \quad (4.47)$$

Also, the concentration of β can be considered as independent of N_2 (i.e. $\dfrac{\partial\ln\left(x_\beta^A\right)}{\partial N_2} = 0$). Then Equation 4.43 can be expressed as

$$\frac{\partial\mu_\beta^A}{\partial N_2} = RT\frac{\partial\ln\left(\gamma_\beta^A(N_2)\right)}{\partial N_2} = -RT\frac{\partial\ln\left(k_{\text{seg}}(N_2)\right)}{\partial N_2}. \quad (4.48)$$

Combining Equations 4.40, 4.42 and 4.48, the chemical potential gradient of β can be expressed as

$$\frac{\partial \mu_\beta{}^A}{\partial z} = \frac{RT}{N_1} \cdot \frac{\partial N_1}{\partial z} - RT \frac{\partial \ln(k_{seg})}{\partial N_2} \frac{\partial N_2}{\partial z}. \tag{4.49}$$

Applying this equation to Equation 4.41, we get the mass flux expression as

$$J = -N_1 \cdot M \left[\frac{RT}{N_1} \cdot \frac{\partial N_1}{\partial z} - RT \frac{\partial \ln(k_{seg})}{\partial N_2} \frac{\partial N_2}{\partial z} \right] = -D_1 \frac{\partial N_1}{\partial z} + D_1 N_1 \frac{\partial \ln(k_{seg})}{\partial N_2} \frac{\partial N_2}{\partial z}. \tag{4.50}$$

On the right-hand side, the first term is the traditional diffusion flux term, as in the Fick's law, and the second term is the segregation flux. With a chosen reference material, if k_{seg} changes with N_2 (such as Ge fraction in SiGe alloys), then segregation happens in an inhomogeneous media with N_2 being non-uniform. Mathematically, Equation 4.50 can be written as

$$J = -D_1 \cdot \frac{1}{k_{seg}} \cdot \frac{\partial(N_1 k_{seg})}{\partial z}, \tag{4.51}$$

where both diffusion and segregation terms are integrated into one spatial derivative. This is the format of segregation model suggested in the Sentaurus Process Advanced Calibration Kit, where the segregation effect was included in the "stress model", as stress also causes segregation. It is, however, useful for a simulation tool to have these two terms separated, as in Equation 4.50. In many process simulation tools, diffusion flux is well modeled, where segregation flux is not commonly included except at material interfaces. By separating these two terms, one can do a rough estimation of the relative magnitude of both terms and decide whether the segregation flux is significant enough to be included, at the cost of longer simulation time.

During the above derivations, P in SiGe is used as an example, and Equation 4.50 is generic to all coupled diffusion and segregation phenomena in alloys due to alloy composition change only. Stress can also change the chemical potential of a species and thus drive segregation, which should add another term in Equation 4.50. As many semiconductor alloy systems are pseudomorphic, stress is a function of the alloy composition. Therefore, the stress effect can be included in k_{seg}. In other words, if k_{seg} in Equation 4.50 is the segregation coefficient for fully strained material, i.e. $k_{seg}(N_2) = \dfrac{N_1{}^{A2,\text{fully strained}}}{N_1{}^{A1}}$, then

Equation 4.50 can be used for pseudomorphic alloy systems. In SiGe HBTs, SiGe layers are designed to be fully strained, where Equation 4.50 applies. In this work, the segregation coefficients studied are for Si/fully strained SiGe.

4.6.4 Experimental Verification

4.6.4.1 Sample Structure Design and Experiments

Now that our model had been derived, experiments were designed to validate it. To observe the coupled diffusion-segregation phenomena, we designed structures to let dopants diffuse and segregate across the graded SiGe regions. The peak concentrations of Ge used were 14% and 18% in this work (see Figure 4.19). The P peak was designed to be inside the triangle SiGe region so that we could observe P segregation out from the Ge triangle to Si [88].

As discussed in Section 4.3.2.2, all the wafers were grown on 8-inch (100) Czochralski p-type Si wafers in an ASM Epsilon 2000 RPCVD reactor. Inert anneals at 900°C were performed in nitrogen ambient using an enclosed Linkam TS1200 high-temperature heating stage. The thermal budgets of the annealing were selected to avoid strain relaxation. At 900°C, the ramp-up rate was about 60°C/minute and ramp-down rate was about 150°C/minute. The temperature ramp rates were fast enough that we could neglect the thermal budget in the ramp-up and ramp-down stages. The temperature uncertainty of the heating stage is less than ±10°C, which is about ±50% in terms of P diffusivity at the annealing temperature. This uncertainty was taken into account in our simulations. Depth profiles of P and Ge were measured by SIMS for as-grown and annealed samples. Ion yield and sputter rate were calibrated to accommodate the matrix composition change.

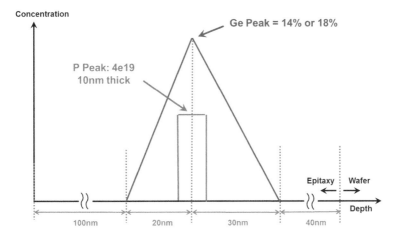

FIGURE 4.19

Schematic diagram of the structure design. The Ge peak concentrations used were 14% and 18%. (The first publication of this figure was in *Journal of Applied Physics* by AIP Publishing at https://doi.org/10.1063/1.4921798. Reprinted with permission.)

4.6.4.2 Parameters for Modeling

As discussed in Section 4.3, our first choice was to use for modeling and simulations. As it does not allow the incorporation of the second gradient term in its flux calculations, we needed to write our own numerical calculation code. The computer coding for the new algorithm was done in MATLAB. To implement the model according to Equation 4.50, a finite difference approach was used to numerically calculate P diffusion and segregation fluxes simultaneously and the P profile evolution with time. For this purpose, we needed to determine two parameters, the P diffusivity D_1 and the segregation coefficient as a function of Ge concentration, $k_{seg}(N_2)$.

Since these two parameters are both temperature dependent, we chose a fixed temperature of 900°C for our annealing experiments as a starting point. The P diffusivity D_1 is based on the calibrated values for Si including the Fermi effect [22]. The Ge fraction (x_{Ge}) dependence of D_1 was also taken into account according to the data from [64]. For P diffusivity in SiGe up to 18% Ge, P diffusivity increases monotonically with the Ge fraction, and the difference between P in Si and P in $Si_{0.82}Ge_{0.18}$ is about 40%, which is not big. Therefore, we approximated the P diffusivity in SiGe with a simple linear function of Ge molar fraction. A small factor D_{scale} is applied to accommodate the experimental temperature uncertainty. Then the P diffusivity D_1 is expressed as:

$$D_1 = D_{scale} \times \left(1 + x_{Ge}\right) \times \left[D^0 + D^- \left(\frac{n}{n_i}\right) + D^= \left(\frac{n}{n_i}\right)^2 \right], \qquad (4.52)$$

where $x_{Ge} = N_2 / N_{total}$. This expression includes the concentration-dependent diffusivity, as described in Section 4.4.2. The parameters such as D^0, D^-, $D^=$ and n_i are a set of experimentally calibrated values adopted by TSUPREM-4 (see Table 4.5). The value of n is calculated according to local charge neutrality condition.

To determine the values of k_{seg} as a function of x_{Ge}, we used the available literature data as a starting point. Figure 4.20 is a summary of experimental

TABLE 4.5

List of Major Parameters Used in TSUPREM-4™ Simulations

Parameter	Prefactor	Activation Energy (eV)
D^0	2.31×10^{10}	3.66
D^-	2.66×10^{10}	4.0
$D^=$	2.652×10^{11}	4.37
n_i	3.87×10^{16}	0.605

The diffusivities are in $\mu m^2/min$. concentrations are in cm^{-3}.

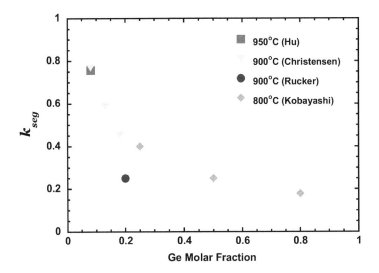

FIGURE 4.20
Experimental data of P segregation coefficient in SiGe. The data points in diamonds are from [82], which were obtained from SiGe films thicker than 0.2 micron where no strain analysis was available. The remaining data are extracted from [13, 64, 88]. (The first publication of this figure was in *Journal of Applied Physics* by AIP Publishing at https://doi.org/10.1063/1.4921798. Reprinted with permission.)

data where Ge concentration is expressed in at.%. It shows that $k_{seg}(x_{Ge})$ is not sensitive to temperature within the range of 800°C to 950°C at the low Ge end. This may be due to two possible reasons: (1) the temperature dependence is insignificant compared to the experimental accuracy; and (2) segregation coefficients were extracted as a ratio of P concentration on SiGe side over that on Si side from "on-going" diffusion-segregation profiles in these data. From the time evolution of P profiles (as shown later in this chapter), we can see that this ratio is not a constant. This data should be considered as showing a "trend of the segregation phenomenon" rather than the determination of the value of k_{seg}. Meanwhile, the experiment condition of some experiments may not be consistent, e.g. SiGe layers may be too thick to be dislocation-free [82]. In principle, the P segregation coefficient can be expressed by an Arrhenius relation where the activation energy is a function of Ge content [91, 100]:

$$k_{seg} = \exp\left(\frac{-x_{Ge} \cdot E_{seg}}{kT}\right). \tag{4.53}$$

Here, E_{seg} is the activation energy, which indicates how strongly the Ge affects P segregation. In the following simulations, the values of E_{seg} were determined by fitting to the experimental data.

4.6.4.3 SIMS Data and Data Fitting

The SIMS profiles of four different structures and/or annealing conditions and the corresponding fitting results are plotted in Figure 4.21a–d. As-grown SIMS profiles of P and Ge were taken as pre-annealed profiles and post-annealing profiles were numerically calculated. As mentioned, any annealing temperature uncertainty will influence the diffusivity, which was treated with a fitting parameter D_{scale}. The fitting parameters, D_{scale} and E_{seg}, were determined when the best match to the post-annealing SIMS profiles was achieved. For each diffused profile, D_{scale} determines the amount of diffusion, seen as the broadening of the annealed profiles, while E_{seg} determines the shape of the diffused peak, i.e. the dip inside the SiGe and the peak in the surrounding Si layers. These two effects are independent, so the best fitting D_{scale} and E_{seg} can be obtained for each diffused profile. Figure 4.21a–d compares the as-grown and annealed Ge and P SIMS with the best-fitting calculation for the four structure and condition combinations. It can be seen that our model can accurately describe the coupled diffusion-segregation behavior of P across the gradient $Si_{1-x}Ge_x$ region. In comparison, the diffusion-only profiles cannot catch the segregation phenomena. The best-fitting E_{seg} and the corresponding segregation coefficient (k_{seg}) for each set of data and segregation coefficients from literature data are summarized in Figure 4.22. From this comparison, within the temperature range of 800°C to 950°C and experimental errors, we can see that $E_{seg} = 0.5$ eV is a good compromise for all the available studies.

4.6.4.4 Model Prediction and P Profile Time Evolution

With the model and the calibrated parameter E_{seg}, we can simulate the time evolution of the P profile during a segregation-diffusion process, as seen in Figure 4.23. The pre-annealed P and Ge profiles are the as-grown Ge and P profiles in the 14% Ge structure. We can see that the segregation effect is more and more obvious as time increases, which is a dynamic process. It should be noted that in previous segregation coefficient studies [64, 82], the ratio of P concentration in SiGe over that in Si was measured and quoted as the segregation coefficient. As we can see from Figure 4.23b,c, this ratio also depends on time. Under the assumption that the SiGe layer stays fully strained during the annealing, it takes time for P segregation to reach a stable value, and only after that can one use the concentration ratio as the segregation value. The difficulty in doing that for a Si/SiGe system is that when the time is too long, SiGe layers may start to relax or interdiffuse with Si, and the segregation is then influenced by the stress relaxation and interdiffusion. Therefore, it makes more sense to use the coupled segregation and diffusion model to extract the segregation coefficients from experiments with annealing time short enough to avoid stress relaxation.

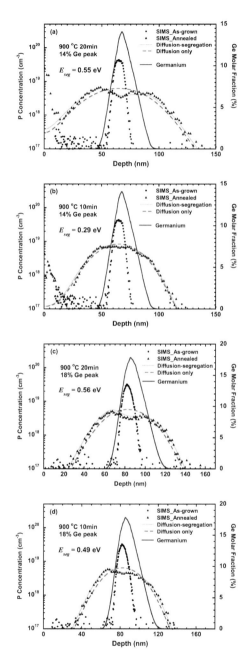

FIGURE 4.21
Ge and P SIMS profiles and best-fitting curves using the coupled diffusion-segregation model in Equation 4.50. Simulations with diffusion-only model were plotted for comparison. (The first publication of this figure was in *Journal of Applied Physics* by AIP Publishing at https://doi.org/10.1063/1.4921798. Reprinted with permission.)

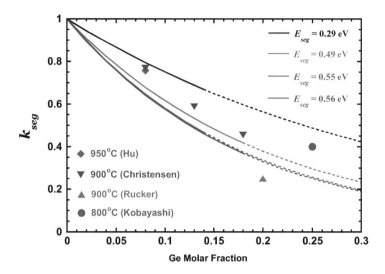

FIGURE 4.22
The best-fitting E_{seg} and the corresponding segregation coefficients extracted from coupled diffusion-segregation experiments. The dashed parts are extrapolation using the E_{seg} values. Literature data are from the same reference as in Figure 4.20. (The first publication of this figure was in *Journal of Applied Physics* by AIP Publishing at https://doi.org/10.1063/1.4921798. Reprinted with permission.)

The above experiments and analysis demonstrated a new approach in measuring segregation coefficients. As the coupled diffusion-segregation model calculates dopant fluxes, it is capable of calculating dopant profile evolution during short anneals before the concentration ratio reaches a stable value. The segregation coefficient extraction is done by fitting the entire dopant profiles instead of by calculating the concentration ratios at a sharp interface. The use of this model eliminates the need for long-time annealing and thick layers to extract segregation coefficients, which are not practical for many heterostructures with lattice mismatch strains. Future studies are necessary to fine-tune our model for P diffusion in SiGe and to apply our method to different material systems.

4.6.5 Impact of P Segregation on HBT Device Performance

At this point, one may wonder if the segregation effect is significant enough to impact the device performance of a PNP HBT. To address this question, we performed one-dimensional device simulations with a mainstream device simulation tool Medici™ [101] using their default models for PNP HBTs. The initial P, C and Ge profile design was modified from the NPN HBT design in [102]. We chose a 3 nm thick box-like P as-grown profile and calculated the annealed profiles using Equation 4.50. Two annealing conditions (5 s and 10 s at 1000°C) were simulated. Two

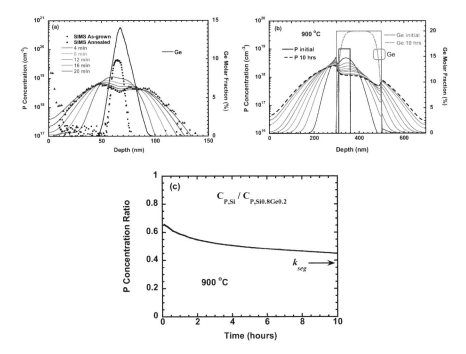

FIGURE 4.23

(a) Simulated time evolution of P profiles during annealing in comparison with SIMS data. The peak Ge fraction is 14%, (b) simulated time evolution of P profiles across sharp Si/SiGe interfaces during annealing and (c) extracted P concentration ratio at either side of the left Si/SiGe interface as a function of time (from 1 to 10 hours). The initial P and Ge have box-like profiles. The k_{seg} for Si/Si$_{0.8}$Ge$_{0.2}$ (fully strained) is shown for comparison. (The first publication of this figure was in *Journal of Applied Physics* by AIP Publishing at https://doi.org/10.1063/1.4921798. Reprinted with permission.)

types of P base doping profiles were loaded in Medici for comparison: one was simulated with the segregation-diffusion model and the other was simulated with the diffusion model only, as shown in Figure 4.24. The electrical characteristics were simulated using an industry-proven drift-diffusion model.

For HBT device performance, depending on the applications, two figures of merits are important. For HBTs in low-speed applications such as the low-speed input stages to amplifier cells, the common-emitter current gain (β) is very important, while the common-emitter short-circuit cut-off frequency f_T is less important. In radio frequency (RF) applications, f_T is a very important figure of merit. It is desired to achieve the fastest transit times possible in HBTs, therefore they are operated at large current densities to minimize impacts of junction capacitances.

The impact of the segregation effect on β of this PNP HBT is shown in Figure 4.25, where the collector voltage was set to 2 V. It shows that the segregation effect reduces the current gain over a wide range of collector

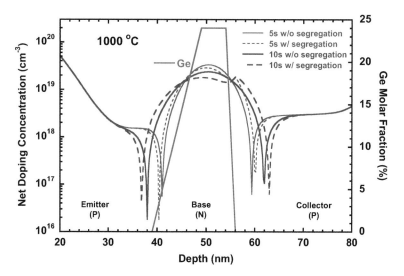

FIGURE 4.24
Doping and Ge profiles used for one-dimensional HBT device simulation.

current. The impact of segregation effect on f_T is shown in Figure 4.26. It shows that in the high collector current I_C window, where the RF applications usually operate, the segregation effect reduces the cut-off frequency f_T greatly.

These results show that the segregation effect has a strong impact on PNP HBT performance. It is therefore important to use the coupled diffusion-segregation model in the base doping and Ge profile design.

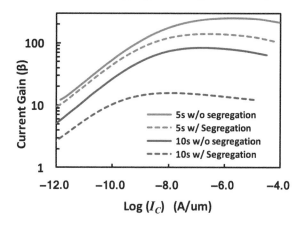

FIGURE 4.25
One-dimensional device simulation results showing the impact of the P segregation effect on PNP HBT current gain (β). The P base profiles are simulated using the diffusion-segregation model and the diffusion-only model for comparison. The annealing conditions used are 5 s and 10 s at 1000°C.

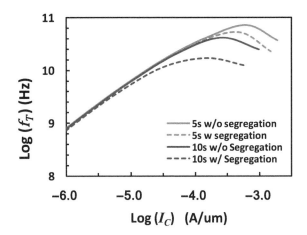

FIGURE 4.26
One-dimensional device simulation results showing the impact of the P segregation effect on PNP HBT cut-off frequency (f_T). The P base profiles are simulated using the diffusion-segregation model and the diffusion-only model for comparison. The annealing conditions used are 5 s and 10 s at 1000°C.

4.6.6 Section Conclusions

In this section, a coupled diffusion and segregation model was re-derived, where the contributions from diffusion and segregation to dopant flux are explicitly shown. The model is generic to coupled diffusion and segregation in inhomogeneous alloys, and provides a new approach in segregation coefficient extraction. This model is especially helpful for heterostructures with lattice mismatch strains. Experiments of coupled P diffusion and segregation were performed with graded SiGe layers for Ge molar fractions up to 0.18, which are relevant to PNP SiGe HBTs. The model was shown to catch both diffusion and segregation behavior well. The diffusion-segregation model for P in SiGe alloys was calibrated and $E_{seg} = 0.5\,eV$ is suggested for the temperature range from 800°C to 950°C. The above experiments and analysis demonstrates a new approach in measuring segregation coefficients, which eliminates the needs of long-time annealing and thick layers, which are not practical for many heterostructures with lattice mismatch strains.

4.7 Acknowledgments

The authors would like to acknowledge Texas Instruments for funding the research. We are grateful for our collaborators at Texas Instruments (TI): Hiroshi Yasuda and Dr. Stanley Philips for their constant support, helpful discussions and industry insights, Mr. Manfred Schiekofer and Mr. Bernhard Benna at TI Germany for growing all the wafers and performing the rapid thermal annealing experiments.

Bibliography

1. D. J. Paul, Si/SiGe heterostructures: From material and physics to devices and circuits, *Semiconductor Science and Technology*, vol. 19, no. 10, pp. R75–R108, Oct. 2004.

2. P. Ashburn, *SiGe Heterojunction Bipolar Transistors*. Chichester: Wiley, 2003.

3. B. Heinemann, H. Rücker, R. Barth, F. Bärwolf, J. Drews, G. G. Fischer, A. Fox, O. Fursenko, T. Grabolla, F. Herzel, J. Katzer, J. Korn, A. Krüger, P. Kulse, T. Lenke, M. Lisker, S. Marschmeyer, A. Scheit, D. Schmidt, J. Schmidt, M. A. Schubert, A. Trusch, C. Wipf, and D. Wolansky, SiGe HBT with fx/fmax of 505 GHz/720 GHz, in *Electron Devices Meeting, 2017. IEDM'17. International*, pp. 51–54.

4. J. D. Cressler, Issues and opportunities for complementary SiGe HBT technology, *ECS Transactions*, vol. 3, no. 7, pp. 893–911, 2006.

5. D. M. Monticelli, The future of complementary bipolar, in *Bipolar/BiCMOS Circuits and Technology, 2004. Proceedings of the 2004 Meeting*, 2004, pp. 21–25.

6. D. Knoll, B. Heinemann, K. E. Ehwald, A. Fox, H. Rucker, R. Barth, D. Bolze, T. Grabolla, U. Haak, J. Drews, B. Kuck, S. Marschmeyer, H. H. Richter, M. Chaimanee, O. Fursenko, P. Schley, B. Tillack, K. Kopke, Y. Yamamoto, H. E. Wulf, and D. Wolansky, A low-cost, high-performance, high-voltage complementary BiCMOS process, in *Electron Devices Meeting, 2006. IEDM'06. International*, 2006, pp. 1–4.

7. B. Tillack, B. Heinemann, D. Knoll, H. Rücker, and Y. Yamamoto, Base doping and dopant profile control of SiGe npn and pnp HBTs, *Applied Physics Letters*, vol. 254, no. 19, pp. 6013–6016, Jul. 2008.

8. "The International Technology Roadmap for Semiconductors." 2013. http://www.itrs2.net/.

9. R. Scholz, U. Gösele, J.-Y. Huh, and T. Y. Tan, Carbon-induced undersaturation of silicon self-interstitials, *Applied Physics Letters*, vol. 72, no. 2, pp. 200–202, Jan. 1998.

10. H. Rücker, B. Heinemann, W. Röpke, R. Kurps, D. Krüger, G. Lippert, and H. J. Osten, Suppressed diffusion of B and C in carbon-rich silicon, *Applied Physics Letters*, vol. 73, no. 12, pp. 1682–1684, Sep. 1998.

11. H. Rucker, B. Heinemann, D. Bolze, D. Knoll, D. Kruger, R. Kurps, H. J. Osten, P. Schley, B. Tillack, and P. Zaumseil, Dopant diffusion in C-doped Si and SiGe: Physical model and experimental verification, in *IEEE Int. Electron Devices Meeting Technical Digest*, 1999, pp. 345–348.

12. H. Rücker and B. Heinemann, Tailoring dopant diffusion for advanced SiGe:C heterojunction bipolar transistors, *Solid-State Electronics*, vol. 44, no. 5, pp. 783–789, May 2000.

13. H. Rucker, B. Heinemann, R. Kurps, and Y. Yamamoto, Dopant diffusion in SiGeC alloys, *ECS Transactions*, vol. 3, no. 7, pp. 1069–1075, Oct. 2006.

14. M. S. A. Karunaratne, A. F. W. Willoughby, J. M. Bonar, J. Zhang, and P. Ashburn, Effect of point defect injection on diffusion of B in silicon and silicon–germanium in the presence of carbon, *Journal of Applied Physics*, vol. 97, no. 11, pp. 113531–113531–7, Jun. 2005.

15. P. Fahey, R. W. Dutton, and M. Moslehi, Effect of thermal nitridation processes on B and P diffusion in <100> silicon, *Applied Physics Letters*, vol. 43, no. 7, pp. 683–685, Oct. 1983.

16. P. Fahey, G. Barbuscia, M. Moslehi, and R. W. Dutton, Kinetics of thermal nitridation processes in the study of dopant diffusion mechanisms in silicon, *Applied Physics Letters*, vol. 46, no. 8, pp. 784–786, Apr. 1985.

17. T. K. Mogi, M. O. Thompson, H.-J. Gossmann, J. M. Poate, and H. S. Luftman, Thermal nitridation enhanced diffusion of Sb and Si(100) doping superlattices, *Applied Physics Letters*, vol. 69, no. 9, pp. 1273–1275, Aug. 1996.

18. A. Fick, V. On liquid diffusion, *Philosophical Magazine Series 4*, vol. 10, no. 63, pp. 30–39, Jul. 1855.

19. V. V. Levitin, *Atom Vibrations in Solids: Amplitudes and Frequencies*. Cambridge: Cambridge Scientific Publishers, 2004.

20. A. Ural, Atomic-scale mechanisms of dopant and self-diffusion in silicon, Ph.D., Stanford University, CA, United States, 2001.

21. N. E. B. Cowern, S. Simdyankin, C. Ahn, N. S. Bennett, J. P. Goss, J.-M. Hartmann, A. Pakfar, S. Hamm, J. Valentin, E. Napolitani, D. De Salvador, E. Bruno, and S. Mirabella, Extended point defects in crystalline materials: Ge and Si, *Physical Review Letters*, vol. 110, no. 15, p. 155501, Apr. 2013.

22. J. D. Plummer, M. D. Deal, and P. B. Griffin, *Silicon VLSI Technology: Fundamentals, Practice and Modeling*. Upper Saddle River: Prentice Hall, 2000.

23. P. Pichler, *Intrinsic Point Defects, Impurities, and Their Diffusion in Silicon*. Vienna: Springer, 2004.

24. K. C. Pandey, Diffusion without vacancies or interstitials: A new concerted exchange mechanism, *Physical Review Letters*, vol. 57, no. 18, pp. 2287–2290, Nov. 1986.

25. H. Mehrer, *Diffusion in Solids: Fundamentals, Methods, Materials, Diffusion-Controlled Processes*. Berlin/Heidelberg: Springer, 2007.

26. P. M. Fahey, P. B. Griffin, and J. D. Plummer, Point defects and dopant diffusion in silicon, *Reviews of Modern Physics*, vol. 61, no. 2, pp. 289–384, Apr. 1989.

27. A. Ural, P. B. Griffin, and J. D. Plummer, Fractional contributions of microscopic diffusion mechanisms for common dopants and self-diffusion in silicon, *Journal of Applied Physics*, vol. 85, no. 9, pp. 6440–6446, May 1999.

28. A. Ural, P. B. Griffin, and J. D. Plummer, Self-diffusion in silicon: Similarity between the properties of native point defects, *Physical Review Letters*, vol. 83, no. 17, pp. 3454–3457, Oct. 1999.

29. H. Bracht and E. E. Haller, Comment on 'Self-diffusion in silicon: Similarity between the properties of native point defects', *Physical Review Letters*, vol. 85, no. 22, p. 4835, Nov. 2000.

30. A. Ural, P. B. Griffin, and J. D. Plummer, Ural, Griffin, and Plummer reply, *Physical Review Letters*, vol. 85, no. 22, pp. 4836, Nov. 2000.

31. TSUPREM-4™, version 2007.12 (Synopsys, Inc., Mountain View, CA, 2007).

32. B. Tillack and Y. Yamamoto, Atomic layer doping for future Si based devices, *ECS Transactions*, vol. 22, no. 1, pp. 121–131, 2009.

33. H. Bracht, J. F. Pedersen, N. Zangenberg, A. N. Larsen, E. E. Haller, G. Lulli, and M. Posselt, Radiation enhanced silicon self-diffusion and the silicon vacancy at high temperatures, *Physical Review Letters*, vol. 91, no. 24, p. 245502, Dec. 2003.

34. H. Osten, D. Knoll, and H. Rücker, Dopant diffusion control by adding C into Si and SiGe: Principles and device application, *Materials Science and Engineering: B*, vol. 87, no. 3, pp. 262–270, Dec. 2001.

35. A. Fischer, H.-J. Osten, and H. Richter, An equilibrium model for buried SiGe strained layers, *Solid-State Electronics*, vol. 44, no. 5, pp. 869–873, May 2000.

36. J. W. Matthews and A. E. Blakeslee, Defects in epitaxial multilayers: I. Misfit dislocations, *Journal of Crystal Growth*, vol. 27, pp. 118–125, Dec. 1974.

37. H. Nitta, J. Tanabe, M. Sakuraba, and J. Murota, C effect on strain compensation in $Si_{1-x-y}Ge_xC_y$ films epitaxially grown on Si(100), *Thin Solid Films*, vol. 508, nos. 1–2, pp. 140–142, Jun. 2006.

38. H. Rücker, B. Heinemann, and R. Kurps, Nonequilibrium point defects and dopant diffusion in carbon-rich silicon, *Physical Review B*, vol. 64, no. 7, p. 073202, Jul. 2001.

39. J. P. Liu and H. J. Osten, Substitutional C incorporation during $Si_{1-x-y}Ge_xC_y$ growth on Si(100) by molecular-beam epitaxy: Dependence on germanium and carbon, *Applied Physics Letters*, vol. 76, no. 24, pp. 3546–3548, Jun. 2000.

40. H. J. Osten, M. Kim, K. Pressel, and P. Zaumseil, Substitutional versus interstitial C incorporation during pseudomorphic growth of $Si_{1-y}C_y$ on Si(001), *Journal of Applied Physics*, vol. 80, no. 12, pp. 6711–6715, Dec. 1996.

41. R. C. Newman and J. Wakefield, The diffusivity of C in silicon, *Journal of Physics and Chemistry of Solids*, vol. 19, nos. 3–4, pp. 230–234, May 1961.

42. J. A. Baker, T. N. Tucker, N. E. Moyer, and R. C. Buschert, Effect of C on the lattice parameter of silicon, *Journal of Applied Physics*, vol. 39, no. 9, pp. 4365–4368, Aug. 1968.

43. D. Windisch and P. Becker, Lattice distortions induced by C in silicon, *Philosophical Magazine A*, vol. 58, no. 2, pp. 435–443, Aug. 1988.

44. G. D. Watkins and K. L. Brower, EPR observation of the isolated interstitial C atom in silicon, *Physical Review Letters*, vol. 36, no. 22, pp. 1329–1332, May 1976.

45. Sokrates T. Pantelides and Stefan Zollner, *Silicon-Germanium C Alloys: Growth, Properties and Applications*, New York: CRC Press, 2002.

46. J. Tersoff, Enhanced solubility of impurities and enhanced diffusion near crystal surfaces, *Physical Review Letters*, vol. 74, no. 25, pp. 5080–5083, Jun. 1995.

47. R. W. Olesinski and G. J. Abbaschian, The C–Si (Carbon-Silicon) system, *Bulletin of Alloy Phase Diagrams*, vol. 5, no. 5, pp. 486–489, Oct. 1984.

48. K. Eberl, S. S. Iyer, S. Zollner, J. C. Tsang, and F. K. LeGoues, Growth and strain compensation effects in the ternary $Si_{1-x-y}Ge_xC_y$ alloy system, *Applied Physics Letters*, vol. 60, no. 24, pp. 3033–3035, Jun. 1992.

49. J. W. Strane, H. J. Stein, S. R. Lee, B. L. Doyle, S. T. Picraux, and J. W. Mayer, Metastable SiGeC formation by solid phase epitaxy, *Applied Physics Letters*, vol. 63, no. 20, pp. 2786–2788, Nov. 1993.

50. T. O. Mitchell, J. L. Hoyt, and J. F. Gibbons, Substitutional C incorporation in epitaxial $Si_{1-y}C_y$ layers grown by chemical vapor deposition, *Applied Physics Letters*, vol. 71, no. 12, pp. 1688–1690, Sep. 1997.

51. R. I. Scace and G. A. Slack, Solubility of C in silicon and germanium, *The Journal of Chemical Physics*, vol. 30, no. 6, pp. 1551–1555, Jun. 1959.

52. J. D. Cressler, *SiGe and Si Strained-Layer Epitaxy for Silicon Heterostructure Devices*. Boca Raton: CRC Press, 2010.
53. J. L. Hoyt, T. O. Mitchell, K. Rim, D. V. Singh, and J. F. Gibbons, Comparison of $Si/Si_{1-x-y}Ge_xC_y$ and $Si/Si_{1-y}C_y$ heterojunctions grown by rapid thermal chemical vapor deposition, *Thin Solid Films*, vol. 321, no. 1–2, pp. 41–46, May 1998.
54. V. D. Akhmetov and V. V. Bolotov, The effect of C and B on the accumulation of vacancy-oxygen complexes in silicon, *Radiation Effects*, vol. 52, nos. 3–4, pp. 149–152, Jan. 1980.
55. J. P. Kalejs, L. A. Ladd, and U. Gösele, Self-interstitial enhanced C diffusion in silicon, *Applied Physics Letters*, vol. 45, no. 3, pp. 268–269, Aug. 1984.
56. L. A. Ladd and J. P. Kalejs, Self-interstitial injection effects on C diffusion in silicon at high temperatures, in *Symposium K – Oxygen, Carbon, Hydrogen and Nitrogen in Crystalline Silicon*, 1985, vol. 59.
57. U. Goesele, P. Laveant, R. Scholz, N. Engler, and P. Werner, Diffusion engineering by C in silicon, *MRS Online Proceedings Library*, vol. 610, https://doi.org/10.1557/PROC-610-B7.11, 2000.
58. R. F. Scholz, P. Werner, U. Gösele, and T. Y. Tan, The contribution of vacancies to C out-diffusion in silicon, *Applied Physics Letters*, vol. 74, no. 3, pp. 392–394, Jan. 1999.
59. P. Werner, U. Gösele, H.-J. Gossmann, and D. C. Jacobson, C diffusion in silicon, *Applied Physics Letters*, vol. 73, no. 17, pp. 2465–2467, Oct. 1998.
60. L. D. Lanzerotti, J. C. Sturm, E. Stach, R. Hull, T. Buyuklimanli, and C. Magee, Suppression of B outdiffusion in SiGe HBTs by C incorporation, in *Electron Devices Meeting, 1996. IEDM'96, International*, 1996, pp. 249–252.
61. H. J. Osten, G. Lippert, D. Knoll, R. Barth, B. Heinemann, H. Rucker, and P. Schley, The effect of C incorporation on SiGe heterobipolar transistor performance and process margin, in *Electron Devices Meeting, 1997. IEDM'97. Technical Digest, International*, 1997, pp. 803–806.
62. D. Knoll, B. Heinemann, H. J. Osten, B. Ehwald, B. Tillack, P. Schley, R. Barth, M. Matthes, K. S. Park, Y. Kim, and W. Winkler, Si/SiGe:C heterojunction bipolar transistors in an epi-free well, single-polysilicon technology, in *Electron Devices Meeting, 1998. IEDM'98. Technical Digest, International*, 1998, pp. 703–706.
63. N. R. Zangenberg, J. Fage-Pedersen, J. L. Hansen, and A. N. Larsen, B and P diffusion in strained and relaxed Si and SiGe, *Journal of Applied Physics*, vol. 94, no. 6, pp. 3883–3890, Sep. 2003.
64. J. S. Christensen, H. H. Radamson, A. Y. Kuznetsov, and B. G. Svensson, Diffusion of P in relaxed $Si_{1-x}Ge_x$ films and strained $Si/Si_{1-x}Ge_x$ heterostructures, *Journal of Applied Physics*, vol. 94, no. 10, pp. 6533–6540, Nov. 2003.
65. H. Rucker, B. Heinemann, and A. Fox, Half-Terahertz SiGe BiCMOS technology, in *Silicon Monolithic Integrated Circuits in RF Systems (SiRF), 2012 IEEE 12th Topical Meeting on*, 2012, pp. 133–136.
66. S. C. Jain, H. J. Osten, B. Dietrich, and H. Rucker, Growth and properties of strained $Si_{1-x-y}Ge_xC_y$ layers, *Semiconductor Science and Technology*, vol. 10, no. 10, p. 1289, Oct. 1995.
67. Sentaurus Process™, version 2013.12 (Synopsys, Inc., Mountain View, CA, 2013).
68. A. Sibaja-Hernandez, S. Decoutere, and H. Maes, A comprehensive study of B and C diffusion models in SiGeC heterojunction bipolar transistors, *Journal of Applied Physics*, vol. 98, no. 6, p. 063530, Sep. 2005.

69. H. Rucker, B. Heinemann, W. Ropke, G. Fischer, G. Lippert, H.-J. Osten, and R. Kurps, Modeling the effect of C on B diffusion, in, *1997 International Conference on Simulation of Semiconductor Processes and Devices, SISPAD'97*, 1997, pp. 281–284.

70. N. E. B. Cowern, P. C. Zalm, P. van der Sluis, D. J. Gravesteijn, and W. B. de Boer, Diffusion in strained Si(Ge), *Physical Review Letters*, vol. 72, no. 16, pp. 2585–2588, Apr. 1994.

71. T. Ito, I. Kato, T. Nozaki, T. Nakamura, and H. Ishikawa, Plasma-enhanced thermal nitridation of silicon, *Applied Physics Letters*, vol. 38, no. 5, pp. 370–372, Mar. 1981.

72. S. T. Ahn, H. W. Kennel, J. D. Plummer, and W. A. Tiller, Film stress-related vacancy supersaturation in silicon under low-pressure chemical vapor deposited silicon nitride films, *Journal of Applied Physics*, vol. 64, no. 10, pp. 4914–4919, Nov. 1988.

73. M. M. Moslehi and K. C. Saraswat, Thermal nitridation of Si and SiO_2 for VLSI, *Electron Devices, IEEE Transactions on*, vol. 32, no. 2, pp. 106–123, Feb. 1985.

74. T. Ito, T. Nozaki, H. Arakawa, and M. Shinoda, Thermally grown silicon nitride films for high-performance MNS devices, *Applied Physics Letters*, vol. 32, no. 5, pp. 330–331, Mar. 1978.

75. S. P. Murarka, C. C. Chang, and A. C. Adams, Thermal nitridation of silicon in ammonia gas: Composition and oxidation resistance of the resulting films, *Journal of the Electrochemical Society*, vol. 126, no. 6, pp. 996–1003, Jun. 1979.

76. Y. Hayafuji, K. Kajiwara, and S. Usui, Shrinkage and growth of oxidation stacking faults during thermal nitridation of silicon and oxidized silicon, *Journal of Applied Physics*, vol. 53, no. 12, pp. 8639–8646, Dec. 1982.

77. J. M. Silcock and W. J. Tunstall, Partial dislocations associated with NbC precipitation in austenitic stainless steels, *Philosophical Magazine*, vol. 10, no. 105, pp. 361–389, Sep. 1964.

78. S. Mizuo, T. Kusaka, A. Shintani, M. Nanba, and H. Higuchi, Effect of Si and SiO_2 thermal nitridation on impurity diffusion and oxidation induced stacking fault size in Si, *Journal of Applied Physics*, vol. 54, no. 7, pp. 3860–3866, Jul. 1983.

79. M. Karunaratne, J. Bonar, P. Ashburn, and A. Willoughby, Suppression of B diffusion due to C during rapid thermal annealing of SiGe based device materials—some comments, *Journal of Materials Science*, vol. 41, no. 3, pp. 1013–1016, 2006.

80. J. M. Bonar, A. F. W. Willoughby, A. H. Dan, B. M. McGregor, W. Lerch, D. Loeffelmacher, G. A. Cooke, and M. G. Dowsett, Antimony and B diffusion in SiGe and Si under the influence of injected point defects, *Journal of Materials Science: Materials in Electronics*, vol. 12, no. 4, pp. 219–221, 2001.

81. Y. Lin, H. Yasuda, M. Schiekofer, B. Benna, R. Wise, and G. (Maggie) Xia, Effects of C on P diffusion in SiGe:C and the implications on P diffusion mechanisms, *Journal of Applied Physics*, vol. 116, no. 14, p. 144904, Oct. 2014.

82. S. Kobayashi, M. Iizuka, T. Aoki, N. Mikoshiba, M. Sakuraba, T. Matsuura, and J. Murota, Segregation and diffusion of P from doped Si1–xGex films into silicon, *Journal of Applied Physics*, vol. 86, no. 10, pp. 5480–5483, Nov. 1999.

83. K. Weiser, Theoretical calculation of distribution coefficients of impurities in germanium and silicon, heats of solid solution, *Journal of Physics and Chemistry of Solids*, vol. 7, nos. 2–3, pp. 118–126, Nov. 1958.

84. P. Rai-Choudhury and E. I. Salkovitz, Doping of epitaxial silicon: Equilibrium gas phase and doping mechanism, *Journal of Crystal Growth*, vol. 7, no. 3, pp. 353–360, Nov. 1970.

85. D. T. J. Hurle, R. M. Logan, and R. F. C. Farrow, Doping of epitaxial silicon, *Journal of Crystal Growth*, vol. 12, no. 1, pp. 73–75, Jan. 1972.

86. W. H. Shepherd, Doping of epitaxial silicon films, *Journal of the Electrochemical Society*, vol. 115, no. 5, pp. 541–545, May 1968.

87. C. D. Thurmond and J. D. Struthers, Equilibrium thermochemistry of solid and liquid alloys of germanium and of silicon. II. The retrograde solid solubilities of Sb in Ge, Cu in Ge and Cu in Si, *Journal of Physical Chemistry*, vol. 57, no. 8, pp. 831–835, Aug. 1953.

88. S. M. Hu, D. C. Ahlgren, P. A. Ronsheim, and J. O. Chu, Experimental study of diffusion and segregation in a Si-(Ge_xSi_{1-x}) heterostructure, *Physical Review Letters*, vol. 67, no. 11, pp. 1450–1453, Sep. 1991.

89. S. Kobayashi, T. Aoki, N. Mikoshiba, M. Sakuraba, T. Matsuura, and J. Murota, Segregation and diffusion of impurities from doped Si1–xGex films into silicon, *Thin Solid Films*, vol. 369, no. 1–2, pp. 222–225, Jul. 2000.

90. R. F. Lever, J. M. Bonar, and A. F. W. Willoughby, Boron diffusion across silicon–silicon germanium boundaries, *Journal of Applied Physics*, vol. 83, no. 4, pp. 1988–1994, Feb. 1998.

91. S. M. Hu, General theory of impurity diffusion in semiconductors via the vacancy mechanism, *Physical Review*, vol. 180, no. 3, pp. 773–784, Apr. 1969.

92. S. M. Hu, Diffusion and segregation in inhomogeneous media and the Ge_{x} Si_{1-x} heterostructure, *Physical Review Letters*, vol. 63, no. 22, pp. 2492–2495, Nov. 1989.

93. C. Ahn and S. T. Dunham, Calculation of dopant segregation ratios at semiconductor interfaces, *Physical Review B*, vol. 78, no. 19, p. 195303, Nov. 2008.

94. H.-M. You, U. M. Gösele, and T. Y. Tan, Simulation of the transient indiffusion-segregation process of triply negatively charged Ga vacancies in GaAs and AlAs/GaAs superlattices, *Journal of Applied Physics*, vol. 74, no. 4, p. 2461, Aug. 1993.

95. T. Y. Tan, Mass transport equations unifying descriptions of isothermal diffusion, thermomigration, segregation, and position-dependent diffusivity, *Applied Physics Letters*, vol. 73, no. 18, p. 2678, Nov. 1998.

96. T. Y. Tan, R. Gafiteanu, H.-M. You, and U. Gösele, Diffusion-segregation equation for modeling in heterostructures, *AIP Conference Proceedings*, vol. 306, no. 1, p. 478, Jun. 1994.

97. Paul G. Shewmon, Problem 4–8 in Session 4, in *Diffusion in Solids*, Switzerland: Springer International Publishers, p. 136, 2016.

98. M. E. Glicksman, *Diffusion in Solids: Field Theory, Solid-State Principles, and Applications.* New York: Wiley, 1999.

99. P. W. Atkins, *Physical Chemistry.* W. H. Freeman, 1994.

100. L. J. Giling and J. Bloem, The incorporation of P in silicon; The temperature dependence of the segregation coefficient, *Journal of Crystal Growth*, vol. 31, pp. 317–322, Dec. 1975.

101. Medici™, version 2010.03 (Synopsys, Inc., Mountain View, CA, 2010).

102. D. Bolze, B. Heinemann, J. Gelpey, S. McCoy, and W. Lerch, Millisecond annealing of high-performance SiGe HBTs, in *17th International Conference on Advanced Thermal Processing of Semiconductors, 2009. RTP'09*, 2009, pp. 1–11.

List of Abbreviations

ALD	atomic layer deposition
BJT	bipolar junction transistor
CVD	chemical vapor deposition
CMOS	complementary metal-oxide-semiconductor
C-BiCMOS	complementary BJT-CMOS
FEA	finite element analysis
HBT	heterojunction bipolar transistor
HDD	hard disk drive
ITRS	International Technology Roadmap for Semiconductors
ODD	optical disk drive
RTA	rapid thermal annealing
RTP	rapid thermal processing
RSM	reciprocal space mapping
SiGe	silicon-germanium
SIMS	secondary ion mass spectrometry
TFT	thin-film transistor
TCAD	technology computer-aided design
XRD	X-ray diffraction

5

Performance Evaluation of AsF5-intercalated Top-Contact Multilayer Graphene Nanoribbons for Deeply Scaled Interconnects

Rohit Sharma and Atul Kumar Nishad

CONTENTS

5.1 Introduction

For current and future technology nodes, interconnect delay dominates over the transistor delay and is considered to be a major performance bottleneck. In the last few decades, extensive research has been done on improvement of interconnect performance [1–4]. There is a general consensus on replacing the industry-preferred interconnect material, i.e. copper (*Cu*). This is due to excessive increase in *Cu* resistivity on account of various width-dependent scattering mechanisms. In fact, *Cu* resistivity may extend up to 16 μΩ.cm for local/intermediate and 12.22 μΩ.cm for global interconnects toward the end of the International Technology Roadmap for Semiconductors (ITRS) [2]. In addition, *Cu* exhibits electromigration that comes into the picture due to reduced feature sizes. Of late, graphene nanoribbons (GNRs) have attracted a lot of attention for both interconnect and device applications [5–8]. This is due to their extraordinary properties, such as their larger mean free path (MFP) [9], higher thermal conductivity [10] and current carrying capability [11]. Single-layer GNRs offer very high resistance [12], prompting researchers to focus on multilayer GNRs (MLGNRs) for interconnect applications. MLGNRs can be further subdivided into top-contact MLGNRs (TC-MLGNRs) and side contact MLGNRs (SC-MLGNRs) based on their connection with surrounding devices and interconnects [13]. In the case of TC-MLGNRs, only the topmost GNR layer is connected with metal contacts, which makes its fabrication easier. On the other hand, in the case of SC-MLGNRs, all GNR layers are physically connected with the contacts. Based on the available literature [14, 15], the performance of SC-MLGNRs is superior to TC-MLGNRs but the former suffers from a complex fabrication process. Thus, there is a clear trade-off as far as interconnect application of MLGNRs for current and future technology nodes is concerned. Due to scaling of interconnect widths and/or edge-induced roughness, the MFP and the number of conduction channels in each GNR layer reduce dramatically [16]. This results in inferior performance of MLGNRs as compared to *Cu* interconnects. Generally, roughness at GNR edges is introduced during fabrication when wider graphene sheets are patterned into narrow GNR interconnects. However, there are promising chemical techniques available that can fabricate GNRs with smoother edges.

It has been reported in the available literature that the in-plane and interplane conductivity of MLGNRs can be increased by several tens of magnitude using intercalation doping [17, 18]. It is not therefore surprising that the effect of intercalation on the electrical properties of MLGNRs has attracted the greatest amount of our attention in recent times. However, previously reported works [19–21], have mostly analyzed the effect of intercalation on SC-MLGNRs.

Given the fabrication complexity involved in SC-MLGNRs, virtually all the experiments use top contacts [22–24]. Therefore, it is imperative to study the effect of intercalation doping on TC-MLGNRs if *Cu* interconnects are to be replaced for current and future technology nodes. In addition, the effect of edge roughness on the performance of AsF5-intercalated TC-MLGNRs needs to be investigated thoroughly. This happens to be the central theme of our present work.

In [21], the authors claim to have achieved higher MFP and Fermi level per GNR layer using AsF5 intercalation. The large increase in conductivity in AsF5-intercalated MLGNRs results from charge transfer from the intercalant layer (where the carriers have a lower mobility) to the GNR layers (where the mobility is usually higher). This leads to an upward shift in Fermi level. Moreover, intercalation increases interlayer spacing that leads to higher MFP due to lower scattering.

In this work, for the first time, we present the effect of AsF5 intercalation on TC-MLGNRs at a deeply scaled technology node (6 nm). Various interconnect performance metrics, such as propagation delay, energy-delay-product (EDP), bandwidth density (BWD) for local, intermediate and global levels, are presented considering both smooth and rough edges in easily fabricated TC-MLGNRs. The performance metrics of AsF5-intercalated TC-MLGNRs are compared with neutral MLGNRs and *Cu* interconnects. In our analysis, both fabrication-related defects and quantum effects on the contact resistance have been considered. The thickness of the GNR stack is optimized to obtain the lowest delay and EDP and the highest BWD for varying interconnect lengths. Our results show that TC-MLGNRs' performance closely matches to that of SC-MLGNRs considering optimized interconnect thickness, thereby eliminating the need for complex fabrication steps. It is found that optimized AsF5-intercalated TC-MLGNRs with smooth edges exhibit 58%, 60% and 50% less EDP as compared to Cu wires for local, intermediate and global interconnects, respectively. Similarly, bandwidth density improves by 1.62×, 2.4× and 1.86× as compared to Cu wires for local, intermediate and global interconnects, respectively. Our analysis proves that AsF5-intercalated TC-MLGNRs with optimized thickness can surpass the performance of Cu for local, intermediate and global interconnects and also overcome the inherent limitations posed by neutral MLGNRs. To the best of our knowledge, thickness optimization, performance prediction and analysis of AsF5-intercalated TC-MLGNR interconnects has not been reported so far, which establishes the novelty of our proposed work.

The remainder of the article is organized as follows: Section 5.2 presents the methodology to analyze the various performance parameters for MLGNR interconnects. In Section 5.3, where we have established the motivation behind the present work, a comparison of resistance between neutral TC- and SC-MLGNR interconnects with that of Cu interconnects for varying interconnect thickness is presented. The effect of intercalation on MLGNRs' effective resistance is described here. In this section, the delay, EDP and BWD of AsF5-intercalated TC-MLGNRs are analyzed and compared with Cu and neutral SC- and TC-MLGNR interconnects. Section 5.4 concludes this chapter.

5.2 Methodology to Analyze Various Performance Parameters of MLGNR Interconnects

In this section, the methodology to analyze the various performance parameters for MLGNR interconnect is described. The distributed circuit parameters such as resistance (r), kinetic inductance (l_K), quantum capacitance (c_Q) are analyzed based on analytical expressions while magnetic inductance (l_M) and electrostatic capacitance (c_E) are obtained by using Synopsys Raphael [25]. In this work, we have considered a three-line driver-interconnect-load (DIL) system. Mutual inductance and coupling capacitance are obtained by using Synopsys Raphael. After obtaining all circuit parameters, the complete DIL system is simulated using Synopsys HSPICE to analyze various performance metrics. Note that technology year 2026 (6 nm technology node) has been considered for obtaining delay, EDP and BWD in local, intermediate and global interconnects.

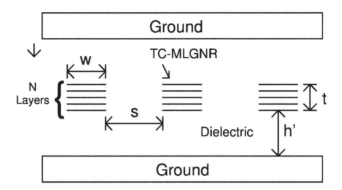

FIGURE 5.1
Schematic of a three-line MLGNR interconnect system embedded in a dielectric.

Figure 5.1 shows the schematic of a three-line MLGNR interconnect system. All interconnect dimensions are taken from ITRS [26]. Here, w is the interconnect width, t is the interconnect thickness, h' is the height of the substrate, δ represents the average interlayer distance between GNR layers and ε_r is the dielectric constant of the substrate. Figure 5.2 gives the driver-interconnect-load system in which R_S and C_S represent the source resistance and source capacitance, respectively. R_{c1} and R_{c2} are assumed as additional contact resistances that account for fabricated related defects. C_L is the load capacitance while r, l_K, l_M, c_Q and c_E are the p.u.l. resistance, kinetic inductance, magnetic inductance, quantum capacitance and electrostatic capacitance of the MLGNR interconnect.

In [14], the authors have proposed an accurate mathematical model for the calculation of the effective resistances of both TC- and SC-MLGNR interconnects, which is given by

$$R_{\text{eff}} = \sum_{m=1}^{m=M}\sum_{m=1}^{m=M}[A_{11}]\left(\frac{[A_{11}]}{R_b} - [A_{21}]\right)^{-1} \tag{5.1}$$

where $[A_{11}]$, $[A_{21}]$ are matrices of size $M \times M$, and depends on R_a and R_b. M represents the number of partitions along the interconnect length. The in-plane resistance, R_b, depends on the number of conduction channel, N_{ch},

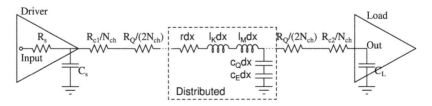

FIGURE 5.2
Driver-interconnect-load system used to evaluate the performance of MLGNR interconnects.

and the effective mean free path, λ_{eff}, of the GNR layer, while the perpendicular resistance, R_a, between the GNR layers depends on interlayer resistivity, ρ_c, and the interlayer spacing, δ. Here, δ for neutral GNRs is assumed to be 0.34 nm. R_a and R_b can be calculated as

$$R_a = \frac{\rho_c \delta}{w \Delta z} \tag{5.2}$$

$$R_b = \frac{R_Q \Delta z}{N_{\text{ch}} \lambda_{\text{eff}}} \tag{5.3}$$

where R_Q is the quantum resistance. Note that the inter-plane resistivity ρ_c becomes zero in case of SC-MLGNRs while it is 30 Ω.cm for TC-MLGNRs [14]. N_{ch} is the number of conduction channels in GNRs and is a function of Fermi energy, E_F, and interconnect width, w [27, 28]. Here, E_F for neutral GNRs is taken as 0.2 eV [13, 27]. The electron MFP, λ_{eff}, in GNRs is limited due to scattering from the edges and from substrate-induced scattering, and can be obtained by using the simple formulation reported in [14, 16]. The substrate-limited MFP, λ_{sub}, is taken as 300 nm. The electron MFP depends on interconnect width, w, subband index, m, and the edge-scattering coefficient in GNR, p. The value of p is between 0.2 and 1.0, as reported in [29, 30]. The *p.u.l.l*$_K$, c_Q of the MLGNR interconnect can now be computed using standard expressions given by authors in [21]:

$$l_K = \frac{h / 4e^2 v_F}{N N_{\text{ch}}} \tag{5.4}$$

$$c_Q = \frac{4e^2}{h v_F} N N_{\text{ch}} \tag{5.5}$$

where e is electronic charge, h is Plank's constant, N represents the number of GNR layers, and $v_F = \left(8 \times 10^5 \, m/s\right)$ is the Fermi velocity. The *p.u.l.* electrostatic capacitance c_E and magnetic inductance, l_M, can be obtained by using *Synopsys Raphael* [25]. However, during analysis, we have found that l_M has negligible effect on the performance metrics that are of interest to us. The equivalent capacitance, ' $\hat{c}$ ', in terms of the quantum and electrostatic capacitances, is given by [13, 31] as

$$\hat{c} = \left(c_Q^{-1} + c_E^{-1}\right)^{-1} \tag{5.6}$$

Quantum contact resistance is assumed to be 12.9 kΩ/conduction channel [14]. Additional contact resistances, R_{c1} and R_{c2}, introduced due to imperfect contacts, are assumed to be 4.3 kΩ/conduction channel. Therefore, the total contact resistance per conduction channel is taken as 17.2 kΩ, assuming a 75% electron transmission coefficient from contacts into the graphene channel. Based on ITRS projections, driver resistance, R_S, and capacitance, C_S, are taken as 41 kΩ and 3.5 aF, respectively [26]. Based on the above explanation, we are now in a position to simulate the three-line MLGNR interconnect system given in Figure 5.1.

The 50% propagation delay for the complete interconnect system including driver, load and contacts has been obtained using Synopsys HSPICE. The energy dissipation of the driver-interconnect-load system, shown in Figure 5.2, is given by [14] as

$$E = \frac{1}{2}\left(C_S + \hat{c}.L + C_L\right)V_{DD}^2 \tag{5.7}$$

where L is the interconnect length.

BWD is a key performance metric that highlights the trade-off between aggregate bandwidth and available routing width. BWD is defined as the ratio of data rate (DR) and pitch (p) and gives the throughput of interconnect through a unit cross-section, where DR can be calculated using rise time obtained from Synopsys HSPICE [32]. Higher BWD results in efficient interconnects while lower EDP improves their latency.

5.3 Results and Analysis

In this section, performance comparison of AsF5-intercalated MLGNRs (both top and side contacts) with neutral MLGNRs and *Cu* interconnects is presented. Note that ITRS technology year 2026, which represents the end of the roadmap, has been considered throughout this article, where interconnect width, thickness and substrate height are assumed to be 6 nm, 13.2 nm and 13.2 nm for local/intermediate, and 9 nm, 21.6 nm and 21.6 nm for global interconnects, respectively. The power supply, V_{DD}, and substrate dielectric constant, ε_r, are taken as 0.57 V and 1.87, respectively. Since the fabrication technology for *Cu* interconnects is mature, all types of contact resistances have been ignored in case of *Cu* interconnects.

5.3.1 Effect of AsF5 Intercalation on Effective Resistance

Due to AsF5 intercalation, c-axis (inter-plane) conductivity improves to 0.24 Ω^{-1}cm^{-1}, average layer spacing increases to 0.575 nm [17] and E_F shifts upward to 0.6 eV [21]. Increased interlayer spacing leads to higher MFP due to lower scattering. The Fermi level in AsF5-intercalated MLGNRs improves by 3× as compared to neutral MLGNRs, resulting in a higher number of conduction channels. This is due to an increase in carrier concentration in the GNR layers. An increase in MFP and N_{ch} due to AsF5 intercalation can lead to significant reduction in the effective resistance.

Based on the above data, the effective resistance *p.u.l.* of AsF5-intercalated MLGNRs for both top and side contacts is calculated using Equation 5.1 for local interconnects, as plotted in Figure 5.3. Here, thickness ratio (TR) is the ratio of MLGNR thickness (t_{MLGNR}) to *Cu* thickness (t_{ITRS}). At ITRS technology year 2026, *Cu* thickness (t_{ITRS}) is taken as 13.2 nm for local interconnect.

As clearly seen in the plots, the effective resistance decreases as the number of layers in the MLGNR stack increases for side contacts. However, its value saturates after few GNR layers in case of top contact due to the presence of non-zero interlayer resistance.

Here, resistance is plotted for two values of edge-scattering coefficient ($p=0$ and 0.2). However, resistance *p.u.l.* in case of TC-MLGNRs with perfectly

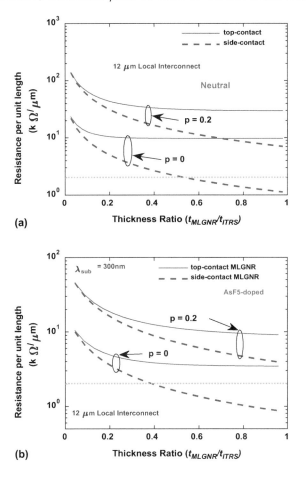

(a)

(b)

FIGURE 5.3
Variation of resistance per unit length (kΩ/μm) for (a) neutral, (b) AsF5-intercalated TC- and SC-MLGNR versus thickness ratio (t_{MLGNR}/t_{ITRS}). Green line represents the resistance *p.u.l.* of copper at 13.2 nm thickness. The analysis is done considering perfectly smooth edges ($p=0$) and an edge-scattering probability of 20% ($p=0.2$).

smooth edges ($p=0$) saturates earlier than that in TC-MLGNRs with rough edges ($p=0.2$). This happens due to a lower in-plane to inter-plane resistance ratio that leads to lesser penetration of current in the bottom layers. Further, it can be also be inferred that even with fewer GNR layers, AsF5 intercalation leads to lower resistance as compared to neutral TC-MLGNRs. As explained in the later part of this article, a smaller number of GNR layers will directly lower the coupling capacitance, thereby reducing delay and EDP.

5.3.2 Local Interconnects

A minimum-sized driver has been considered for analysis of local interconnects. For the ITRS technology year 2026 (1/2-M1 pitch of 6 nm), the *Cu* resistivity will reach 16 $\mu\Omega$.cm under the effect of various width-dependent scattering mechanisms [26]. *Cu* thickness (t_{ITRS}) is taken as 13.2 nm for local interconnects.

The delay variation in neutral and AsF5-intercalated TC-MLGNR with respect to its thickness at 12 μm interconnect length is shown in Figure 5.4. It can be noted that there exists a minima in the delay variation curve. This is due to the fact that effective resistance of MLGNRs decreases while its capacitance increases with thickness.

The variation of delay in neutral and AsF5-intercalated TC-MLGNRs for various interconnect lengths and different values of edge scattering is presented in Figure 5.5a,b. Optimum thickness has been considered for each interconnect length. In the case of perfectly smooth edges ($p=0$), optimized AsF5-intercalated TC-MLGNRs exhibit lower delay as compared to *Cu* for all interconnect lengths. For example, AsF5-intercalated TC-MLGNRs have 14% improvement in delay as compared to *Cu* at 12 μm interconnect lengths. However, due to edge roughness, delay in AsF5-intercalated TC-MLGNRs is higher than that of *Cu* for all lengths. The neutral TC-MLGNRs show higher delay for all interconnect lengths.

Figure 5.5c,d give the EDP for neutral and AsF5-intercalated TC-MLGNRs, respectively. It is compared with *Cu* for varying interconnect lengths. Similar to delay, the EDP of AsF5-intercalated TC-MLGNRs with respect to TR has a minima point and this is considered as optimized EDP for a given interconnect length. The EDP for optimized AsF5-intercalated TC-MLGNR with smooth edges ($p=0$) is lower than that of *Cu* for all interconnect lengths. In the case of a non-zero edge scattering coefficient, AsF5-intercalated TC-MLGNRs exhibit lower EDP only up to 9 μm interconnect length. The BWD of neutral and AsF5-intercalated MLGNRs have been compared with *Cu* for various interconnect lengths in Figure 5.5e,f, respectively. AsF5-intercalated TC-MLGNRs with smooth edges have higher BWD for all interconnect lengths and exhibit 1.62× improvement when compared to *Cu* at 12 μm interconnect length. However, due to edge roughness, AsF5-intercalated TC-MLGNRs exhibit higher BWD up to 3 μm lengths only.

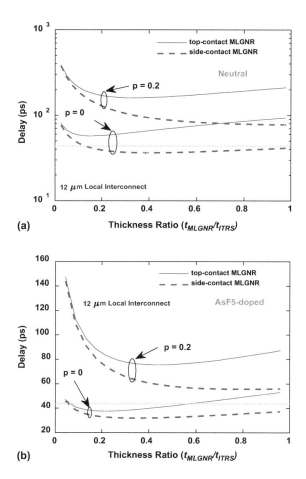

FIGURE 5.4

Delay variation in (a) neutral, (b) AsF5-intercalated MLGNRs versus thickness ratio (t_{MLGNR}/t_{ITRS}). The green line represents the delay of copper at 13.2 nm thickness. The analysis is done considering perfectly smooth edges ($p=0$) and an edge-scattering probability of 20% ($p=0.2$).

Summarizing the above discussion, AsF5-intercalated TC-MLGNRs with perfectly smooth edges provide improved delay, EDP and BWD as compared to *Cu*. However, for a more practical scenario with rough edges, they exhibit inferior performance in terms of delay, EDP and BWD. This happens due to higher TR with larger *rc* product that leads to higher propagation delay and rise time, which lowers BWD and increases the EDP of the interconnect.

5.3.3 Intermediate Interconnects

Although the lengths of intermediate-level interconnects are often shorter than those of global interconnects, repeaters are still required. In our

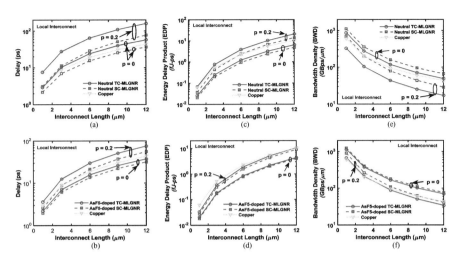

FIGURE 5.5
Variation in delay, EDP and BWD of (a,c,e) neutral, (b,d,f) AsF5-intercalated MLGNR for local interconnects. For MLGNR interconnects, two cases are considered: (1) circles: $t = t_{opt}$ with top contacts; and (2) squares: $t = t_{opt}$ with ideal contacts. The analysis is done considering perfectly smooth edges ($p = 0$) and an edge-scattering probability of 20% ($p = 0.2$).

analysis, to accommodate the effect of repeater insertion, the size of driver is increased by 50 times. For ITRS technology year 2026, Cu thickness (t_{ITRS}) is taken as 13.2 nm for intermediate interconnects. Typical interconnect length lies between 20 μm and 500 μm.

The delay comparison between neutral, AsF5-intercalated TC-MLGNRs and Cu interconnects for varying interconnect lengths is shown in Figure 5.6a,b. It can be inferred that optimized AsF5-intercalated TC-MLGNRs with perfectly smooth edges exhibit lower delay as compared to Cu. However, with the existence of rough edges ($p = 0.2$), AsF5-intercalated TC-MLGNRs exhibit higher delay as compared to Cu. At 500 μm interconnect length, AsF5-intercalated TC-MLGNRs with ($p = 0$) exhibit nearly 57% less delay as compared to Cu. This improvement has been achieved due to a higher Fermi level in intercalated interconnects. With optimized thickness at each interconnect length, the EDP of neutral, AsF5-intercalated TC-MLGNRs has been compared with Cu in Figure 5.6c,d. At 500 μm interconnect lengths, EDP improves by 60% in the case of AsF5-intercalated TC-MLGNRs with perfectly smooth edges when compared to Cu. Using optimized thickness at each interconnect length, BWD analysis for neutral and AsF5-intercalated TC-MLGNRs with regard to their length is shown in Figure 5.6e,f, respectively. Similar to delay and EDP, the BWD of AsF5-intercalated TC-MLGNRs with ($p = 0.2$) is lower than that of Cu for all interconnect lengths. However, with smooth edges, AsF5-intercalated TC-MLGNRs exhibit 2.4× higher bandwidth density when compared with Cu interconnects. It is seen that AsF5-intercalation leads to overall improvement in the interconnect performance in terms of

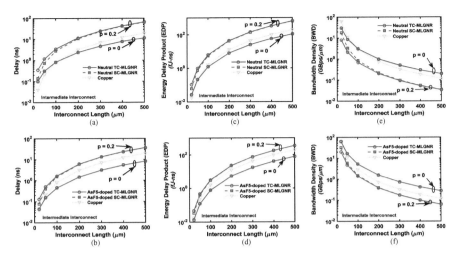

FIGURE 5.6
Variation in delay, EDP and BWD of (a–c) neutral, (d–f) AsF5-intercalated MLGNR for intermediate-level interconnects. For MLGNR interconnects, two cases are considered: (1) circles: $t = t_{opt}$ with top contacts; and (2) squares: $t = t_{opt}$ with ideal contacts. The analysis is done considering perfectly smooth edges ($p = 0$) and an edge-scattering probability of 20% ($p = 0.2$).

delay, EDP and BWD as compared to neutral TC-MLGNRs. However, only AsF5-intercalated TC-MLGNRs with perfectly smooth edges can outperform *Cu* at intermediate-level interconnects.

5.3.4 Global Interconnects

Lengths of global interconnects can be of the order of several millimeters. The drive capability and signal speed is increased by inserting repeaters. The size of repeaters is often much larger than the minimum-sized gates [33]. In our analysis, the size of repeater is chosen to be 100 times the minimum size for global interconnects. It must be noted that the use of different repeater sizes changes the absolute values of the performance metrics. However, the trends of the comparative results will remain the same. Typical interconnect lengths lie between 200 µm and 1000 µm at global level. At the ITRS technology year 2026, the *Cu* resistivity under the effect of various width-dependent scattering mechanisms is reported to be 12.22 µΩ.cm [26]. Also, *Cu* thickness (t_{ITRS}) is taken as 21.6 nm for global interconnects.

Figure 5.7a,b show the delay in neutral and AsF5-intercalated TC-MLGNRs interconnects with respect to length. It can be observed that neutral TC-MLGNRs exhibit higher delay as compared to *Cu*. From these plots it is clearly seen that AsF5-intercalated TC-MLGNRs with perfectly smooth edges exhibit lesser delay as compared to *Cu*. Its delay is higher than *Cu* in the presence of edge roughness. At 1000 µm interconnect length, AsF5-intercalated TC-MLGNRs ($p = 0$) show a 45.92% improvement in delay as compared to *Cu*.

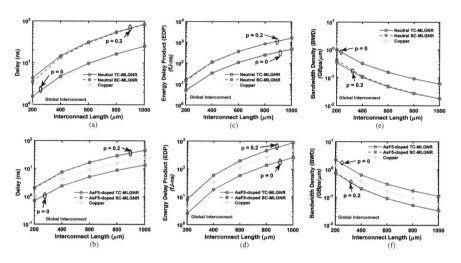

FIGURE 5.7

Variation in delay, EDP and BWD of (a–c) neutral, (d–f) AsF5-intercalated MLGNR for global interconnects. For MLGNR interconnects, two cases are considered: (1) circles: $t = t_{opt}$ with top contacts; and (2) squares: $t = t_{opt}$ with ideal contacts. The analysis is done considering perfectly smooth edges ($p=0$) and an edge-scattering probability of 20% ($p=0.2$).

Figure 5.7c,d represent the variation of EDP in neutral and AsF5-intercalated TC-MLGNRs with respect to their length. At 1000 μm interconnect length, the improvement in EDP is nearly 50% for AsF5-intercalated TC-MLGNRs with smooth edges. The performance in terms of BWD for neutral and AsF5-intercalated TC-MLGNRs at different lengths is shown in Figure 5.7e,f. AsF5-intercalated TC-MLGNRs with perfectly smooth edges exhibit 85.74% improvement in bandwidth density as compared to *Cu*. AsF5-intercalated TC-MLGNRs provide better performance than *Cu* only when edges are smooth. In the presence of edge roughness, their performance is still inferior to *Cu* interconnects. Higher in-plane and *c*-axis conductivity due to intercalation results in speed-up of the signal response and lower rise time, resulting in higher BWD.

The performance of AsF5-intercalated TC-MLGNRs has been summarized in Table 5.1 From Table 5.1, it can be seen that AsF5-intercalated TC-MLGNRs with perfectly smooth edges exhibit better performance in terms of delay,

TABLE 5.1

Comparative Summary of AsF5-intercalated TC-MLGNRs as Compared to *Cu* for Local, Intermediate and Global Interconnects

MLGNRs	Edges	Delay	EDP	BWD
AsF5/local	Perfectly smooth	**Better**	**Better**	**Better**
	Rough	Worse	Worse	**Better up to 3 μm**
AsF5/intermediate/global	Perfectly smooth	**Better**	**Better**	**Better**
	Rough	Worse	Worse	Worse

EDP and BWD as compared to *Cu* for local, intermediate and global-level interconnects. However, the presence of edge roughness leads to degradation in performance of AsF5-intercalated SC- and TC-MLGNRs. In the case of intermediate/global interconnects, its performance is worse than *Cu* for all performance metrics; while in case of local interconnects, EDP is lower and BWD is higher for AsF5-intercalated SC-MLGNRs as compared to *Cu* in the presence of edge roughness. However, there are promising chemical techniques are available that can fabricate GNRs with smoother edges [34].

5.4 Conclusions

The present work analyzes the performance of practically possible AsF5-intercalated TC-MLGNRs as a potential candidate to replace *Cu* for local, intermediate and global interconnects at deeply scaled technology nodes. The interconnect thickness of AsF5-intercalated TC-MLGNRs is optimized to obtain lowest delay, EDP and highest BWD when compared with neutral TC-MLGNRs and *Cu* interconnects for practical interconnect lengths. It is found that optimized *AsF5*-intercalated TC-MLGNRs with perfectly smooth edges demonstrate significantly higher speeds, lower energy-delay product and higher bandwidth density than *Cu* wires for local, intermediate and global levels. However, edge roughness in MLGNRs can degrade their performance significantly. Edge roughness increases the in-plane resistance of GNRs, thereby increasing the effective resistance of TC-MLGNRs. Edge roughness is primarily due to immature fabrication processes arising from patterning of narrow GNR interconnects from wider graphene sheets. However, there are promising chemical techniques available that can fabricate GNRs with smoother edges. Therefore, there is a need to find promising fabrication techniques that can be used to pattern ultra-thin GNRs with near-perfect smooth edges.

References

1. C. Subramaniam, T. Yamada, K. Kobashi, A. Sekiguchi, D. N. Futaba, M. Yumura, et al., One hundred fold increase in current carrying capacity in a carbon nanotube–copper composite, *Nat Commun*, vol. 4, pp. 2202-1–2202-7, 2013.
2. "International Technology Roadmap for Semiconductors," 2012.
3. X. Chen, D. Akinwande, K. J. Lee, G. F. Close, S. Yasuda, B. C. Paul, et al., Fully integrated graphene and carbon nanotube interconnects for gigahertz high-speed CMOS electronics, *IEEE Transactions on Electron Devices*, vol. 57, pp. 3137–3143, 2010.

4. H. Li, C. Xu, and K. Banerjee, Carbon nanomaterials: The ideal interconnect technology for next-generation ICs, *IEEE Design & Test of Computers*, vol. 27, pp. 20–31, 2010.
5. A. K. Nishad, A. Dalakoti, A. Jindal, R. Kumar, S. Kumar, and R. Sharma, Analytical model for inverter design using floating gate graphene field effect transistors, in *2014 IEEE Computer Society Annual Symposium on VLSI (ISVLSI)*, pp. 148–153, 2014.
6. R. Zhang, W. S. Zhao, J. Hu, and W. Y. Yin, Electrothermal characterization of multilevel Cu-graphene heterogeneous interconnects in the presence of an electrostatic discharge (ESD), *IEEE Transactions on Nanotechnology*, vol. 14, pp. 205–209, 2015.
7. Y. F. Liu, W. S. Zhao, Z. Yong, Y. Fang, and W. Y. Yin, Electrical modeling of three-dimensional carbon-based heterogeneous interconnects, *IEEE Transactions on Nanotechnology*, vol. 13, pp. 488–495, 2014.
8. V. R. Kumar, M. K. Majumder, N. R. Kukkam, and B. K. Kaushik, Time and frequency domain analysis of MLGNR interconnects, *IEEE Transactions on Nanotechnology*, vol. 14, pp. 484–492, 2015.
9. X. Du, I. Skachko, A. Barker, and E. Y. Andrei, Approaching ballistic transport in suspended graphene, *Nature Nanotechnology*, vol. 3, pp. 491–495, 2008.
10. A. A. Balandin, S. Ghosh, W. Bao, I. Calizo, D. Teweldebrhan, F. Miao, et al., Superior thermal conductivity of single-layer graphene, *Nano Letters*, vol. 8, pp. 902–907, 2008.
11. C. Xiangyu, L. Kyeong-Jae, D. Akinwande, G. F. Close, S. Yasuda, B. Paul, et al., 2009 High-speed graphene interconnects monolithically integrated with CMOS ring oscillators operating at 1.3GHz, in *IEEE International Electron Devices Meeting (IEDM)*, pp. 1–4.
12. P. Kumar, A. Singh, A. Garg, R. Sharma, 2013 Compact models for transient analysis of single-layer graphene nanoribbon interconnects, in *Computer Modelling and Simulation (UKSim), 2013 UKSim 15th International Conference on*, pp. 809–814.
13. A. K. Nishad and R. Sharma, Analytical time-domain models for performance optimization of multilayer GNR interconnects, *IEEE Journal of Selected Topics in Quantum Electronics*, vol. 20, pp. 17–24, 2014.
14. V. Kumar, S. Rakheja, and A. Naeemi, Performance and energy-per-bit modeling of multilayer graphene nanoribbon conductors, *IEEE Transactions on Electron Devices*, vol. 59, pp. 2753–2761, 2012.
15. A. K. Nishad and R. Sharma, Analytical time-domain models for performance optimization of multilayer GNR interconnects, *IEEE Journal of Selected Topics in Quantum Electronics*, vol. 20, pp. 3700108-1–3700108-8, 2014.
16. S. Rakheja, V. Kumar, and A. Naeemi, Evaluation of the potential performance of graphene nanoribbons as on-chip interconnects, *Proceedings of the IEEE*, vol. 101, pp. 1740–1765, 2013.
17. M. S. Dresselhaus and G. Dresselhaus, Intercalation compounds of graphite, *Advances in Physics*, vol. 51, pp. 1–186, Jan. 2002.
18. A. K. Nishad and R. Sharma, 2015 Performance analysis of AsF5-intercalated top-contact multi layer graphene nanoRibbon interconnects, in *2015 IEEE International Symposium on Nanoelectronic and Information Systems*, pp. 170–174.

19. L. Hong, X. Chuan, N. Srivastava, and K. Banerjee, Carbon nanomaterials for next-generation interconnects and passives: Physics, status, and prospects, *IEEE Transactions on Electron Devices*, vol. 56, pp. 1799–1821, 2009.

20. C. Xu, H. Li, and K. Banerjee, 2008 Graphene nano-ribbon (GNR) interconnects: A genuine contender or a delusive dream?, in *Electron Devices Meeting, 2008. IEDM 2008. IEEE International*, pp. 1–4.

21. X. Chuan, L. Hong, and K. Banerjee, Modeling, analysis, and design of graphene nano-ribbon interconnects, *IEEE Transactions on Electron Devices*, vol. 56, pp. 1567–1578, 2009.

22. K. S. Novoselov, A. K. Geim, S. V. Morozov, D. Jiang, Y. Zhang, S. V. Dubonos, et al., Electric field effect in atomically thin carbon films, *Science*, vol. 306, pp. 666–669, 2004.

23. V. Kumar and A. Naeemi, 2012 Analytical models for the frequency response of multi-layer graphene nanoribbon interconnects, in *IEEE International Symposium on Electromagnetic Compatibility (EMC)*, pp. 440–445.

24. C. Berger, Z. Song, T. Li, X. Li, A. Y. Ogbazghi, R. Feng, et al., Ultrathin epitaxial graphite: 2D electron gas properties and a route toward graphene-based nanoelectronics, *The Journal of Physical Chemistry B*, vol. 108, pp. 19912–19916, 2004.

25. Synopsys Raphael 2015. www.synopsys.com/silicon/tcad.html, ed, 2015.

26. W.-S. Zhao and W.-Y. Yin, 2012 Signal integrity analysis of graphene nano-ribbon (GNR) interconnects, in *IEEE Electrical Design of Advanced Packaging and Systems Symposium (EDAPS)*, pp. 227–230.

27. A. Naeemi and J. D. Meindl, Compact physics-based circuit models for graphene nanoribbon interconnects, *IEEE Transactions on Electron Devices*, vol. 56, pp. 1822–1833, 2009.

28. S. H. Nasiri, R. Faez, and M. K. Moravvej-Farshi, Compact Formulae for number of conduction channels in various types of graphene nanoribbons at various temperatures, *Modern Physics Letters B*, vol. 26, p. 1150004, 2012.

29. X. Wang, Y. Ouyang, X. Li, H. Wang, J. Guo, and H. Dai, Room-temperature all-semiconducting sub-10-nm graphene nanoribbon field-effect transistors, *Physical Review Letters*, vol. 100, p. 206803, 2008.

30. C. Berger, Z. Song, X. Li, X. Wu, N. Brown, C. Naud, et al., Electronic confinement and coherence in patterned epitaxial graphene, *Science*, vol. 312, pp. 1191–1196, 2006.

31. C. Jiang-Peng, Z. Wen-Sheng, Y. Wen-Yan, and H. Jun, Signal transmission analysis of multilayer graphene nano-ribbon (MLGNR) interconnects, *IEEE Transactions on Electromagnetic Compatibility*, vol. 54, pp. 126–132, 2012.

32. V. Kumar, R. Sharma, E. Uzunlar, Z. Li, R. Bashirullah, P. Kohl, et al., Airgap interconnects: Modeling, optimization, and benchmarking for backplane, PCB, and interposer applications, *IEEE Transactions on Components, Packaging and Manufacturing Technology*, vol. 4, pp. 1335–1346, 2014.

33. K. Banerjee and A. Mehrotra, A power-optimal repeater insertion methodology for global interconnects in nanometer designs, *IEEE Transactions on Electron Devices*, vol. 49, pp. 2001–2007, 2002.

34. L. Ci, Z. Xu, L. Wang, W. Gao, F. Ding, K. F. Kelly, et al., Controlled nanocutting of graphene, *Nano Research*, vol. 1, pp. 116–122, 2008.

6

Silicon-on-Chip Lasers for Chip-level Optical Interconnects

Il-Sug Chung

CONTENTS

6.1 Introduction

In the past decade, optical interconnects have been considered as a strong candidate to tackle increasing challenges in chip-level interconnects especially for computers – challenges that include energy consumption, interconnect density, and latency. Based on the ITRS roadmap, it is anticipated that in the year 2022, the off-chip and on-chip link energies need to be about 100 fJ/bit and 30 fJ/bit or less, respectively, which includes energies for a light source, a transmitter IC, waveguide loss, a detector, and a receiver IC [1]. To achieve this, it is essential to realize an energy-efficient light source as well as a detector which can directly drive a transistor without an energy-consuming trans-impedance amplifier (TIA). In this chapter, we will cover the light source – more specifically, state-of-the-art hybrid silicon-on-chip laser technologies – and aim to understand the frontier limits of light source technologies, and to answer several fundamental questions:

- Can one use forward error correction (FEC) while satisfying the link energy requirements of 100 fJ/bit or 30 fJ/bit?
- Is the smallest on-chip laser a way to go?
- Can we achieve a light source energy consumption of <10 fJ/bit without FEC?

In Section 6.2, the potential advantages of optical interconnects over electrical interconnects, and the advantages of on-chip lasers over off-chip lasers, are discussed. In Section 6.3, key parameters and quantities for laser analysis are defined, including definitions of terminologies and concepts. Answers to the first two questions will be discussed with respect to the energy budget for the entire link and the impact of noise. In Section 6.4, five state-of-the-art laser structures are reviewed and analyzed. Based on these, the last question is discussed.

6.2 Rationale for On-Chip Optical Interconnects with On-Chip Lasers

In this section, we discuss the expected advantages of optical interconnects over electrical ones for global on-chip interconnects where the link distance is about 1 cm. Then, three configurations of on-chip light sources are discussed in terms of link loss and energy efficiency, which include a combination of an external comb laser and on-chip modulators, a combination of on-chip lasers and on-chip modulators, and directly modulated on-chip lasers.

6.2.1 Potentials of On-Chip Optical Interconnects

As the strategy for improving computing power moves from increasing the performance of a single computing unit to increasing the number of computing units, the efficiency and performance of communications between computing units becomes a key factor in determining the performance of an entire computer system. For communications with a distance over meters, optical interconnects have fundamental advantages over electrical interconnects. Thus, data communications in data centers and clustered high-performance computer systems are already based on optical interconnects. In recent several years, there have been increasing discussions and studies on using optical interconnects even for chip-level communications with a much shorter interconnect distance of, e.g., a few centimeters [1–5].

Key metrics for interconnect technology are bandwidth density, energy efficiency, and latency [1]. First, the bandwidth density is defined as the bandwidth of a channel divided by the physical spacing between channels. This measure is becoming more important as a finite number of communication channels can be formed in the finite size of the perimeter of a chip. Even for 1-cm links inside a chip, optical interconnects may provide a bandwidth density which is a few times larger than electrical interconnects do [2]. Second, energy efficiency measures the energy consumed for sending and receiving a bit. For links longer than 1.5 cm, the energy efficiency of optical interconnects (with on-chip laser) is much smaller than that of electrical interconnects [2]. At this length of links, the latency of optical interconnects is comparable to that of electrical interconnects [2].

Though the overall performance of optical interconnects is expected to be considerably better than that of electrical interconnects, the driving force to move the chip link design from electrical to optical interconnects does not, for now, appear enough, since electrical interconnects can still meet the requirements on the key metrics introduced above. However, the need for optical interconnects is expected to be increasingly high in the near future. The number of cores is increasing to, e.g., a few thousand, which means the chip will be considerably larger, requiring longer links. In addition, the circuit transistor capacitance is becoming small enough to be directly driven by a photodetector, for example with a 22-nm technology node [2]. This eliminates the need for energy-consuming trans-impedance amplifiers (TIAs) in the receiver part of optical interconnects [1, 3]. Thus, the energy efficiency of optical interconnects can be even higher than that of electrical interconnects.

6.2.2 Light Sources for On-Chip Interconnects

In designing chip-level optical links, an energy-efficient high-speed light source is a key element since it influences the design as well as the performance of the entire optical link. For example, the characteristics of a light source determine the link power budget, the specifications of a detector, the maximum speed, the applicability of advanced modulation format (regarding linearity of laser output versus current), and the need for a modulator.

Figure 6.1 shows three possible configurations of light sources for on-chip interconnects, which can be either a laser alone or a combination of a laser and a modulator. In all configurations, wavelength-division multiplexing (WDM) is assumed, which is necessary to achieve the bandwidth of several Tb/s required for chip-level interconnects. The approach shown in Figure 6.1a employs an off-chip comb laser. A comb laser outputs multiple wavelengths with an even spacing. This light output is coupled into a silicon waveguide and each wavelength component is split though a passive de-multiplexer (de-MUX). Then, each wavelength component is modulated by a modulator, now carrying data. In this off-chip laser approach, fabrication of lasers can be done separately, not influencing fabrication conditions of electronics. Since lasers are thermally less influenced by the operating conditions of an electronics chip, the output power and lasing wavelength can be feasibly maintained at constant values. However, this approach requires alignment between an optical fiber and the chip, which may increase packaging cost. In terms of power saving, the off-chip approach is less attractive than the on-chip approach, as discussed below.

The approaches shown in Figure 6.1b,c use a combination of an on-chip laser and a modulator, and a directly modulated on-chip laser, respectively.

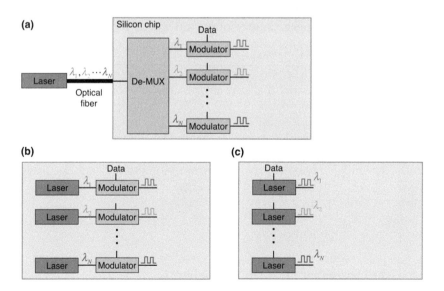

FIGURE 6.1
Three possible configurations of light sources for on-chip interconnects: (a) Combination of an external comb laser and on-chip modulators; (b) Combination of on-chip lasers and on-chip modulators; (c) Directly modulated on-chip lasers.

In these approaches, lasers can be monolithically integrated onto a silicon chip by using backend CMOS-compatible processing. Compared to the off-chip approach, a combination of an on-chip laser and an on-chip modulator has a power-saving merit of 10 to 20 dB, which are related to coupling loss, energy proportionality, and layout flexibility [2]. The directly modulated on-chip laser has a further power saving of 7 to 15 dB in relation to the absence of modulator insertion loss [6–8].

In [1], a difficulty in controlling wavelength and polarization, and a turn-on delay are pointed out as weaknesses of the on-chip laser approach. However, in recent on-chip lasers, for example in [24], wavelength control of individual lasers for WDM and control of polarization are feasible. Regarding the turn-on delay, directly modulated on-chip lasers are free from turn-on delay unless return-to-zero (RZ) format is used. Furthermore, in on-chip interconnects, the turn-on delay is less influential, since the required speed is relatively slow, i.e. several Gb/s, and the active region volume of on-chip lasers is typically small, leading to a short turn-on delay time for filling the active region.

However, the influence of optical feedback on laser performance can be significant for small cavity-volume lasers, and the thermal shift of lasing wavelength needs to be handled for WDM.

6.3 Basics for On-Chip Laser Analysis

In Section 6.3.1, quantities describing laser performance (threshold current, output power, 3-dB frequency) will be expressed in terms of key designing parameters (active region volume, injection efficiency, and confinement factor). In Section 6.3.2, key metrics for on-chip lasers (energy efficiency, noise, and bitrate) will be expressed in terms of quantities and parameters introduced in Section 6.3.1. Then, specifications on key metrics will be defined, based on the values from the literature. Finally, it will be discussed why FEC should be avoided and why a smaller laser is not always better.

6.3.1 Static and Dynamic Characteristics of Laser Diodes

Two representative graphs characterizing a laser diode are the output (P) and voltage (V) versus current (I) graph for static properties and the small-signal response spectrum for dynamic properties, as shown in Figure 6.2. In Figure 6.2a, threshold current (I_{th}) designates a point where the output starts to drastically increase. Differential quantum efficiency (η_d) is defined as [10]

$$\eta_d = \frac{q\lambda}{hc}\frac{dP}{dI} = \eta_o\eta_i \tag{6.1}$$

where:

- q is the electron charge
- λ is the lasing wavelength
- h is the Plank constant
- c is the speed of light.

The second equality explains that the differential quantum efficiency is determined by two efficiencies. The η_o measures the fraction of output power over all optical losses (mirror loss and internal loss) while the injection efficiency (η_i) quantifies the fraction of injected electrons arriving at the active region where the recombination of electrons and holes occurs, generating light. Thus, smaller η_i means more leakage current. The electrons in the active region are named as *carriers*. Since η_o rarely varies with increasing current, the change in η_d with increasing current will be used to measure a change in η_i in Section 6.4.

In Figure 6.2a, differential resistance, R_d, is defined as

$$R_d = \frac{dV}{dI}. \tag{6.2}$$

This will be used to calculate the RC constant of lasers in Section 6.4.

In Section 6.4, analysis and discussion will be focused on the impact of laser size (active region volume, V_a and optical mode volume, V_p) and current injection efficiency (η_i) on laser performance, rather than the impact of laser material properties such as material gain and non-radiative recombination. In preparation for Section 6.4, output P, current I, threshold current I_{th}, and 3-dB frequency f_{3dB} are given below in terms of V_a, Γ, and $\eta_i \left(= V_a / V_p \right)$, which are derived from rate equation models [10].

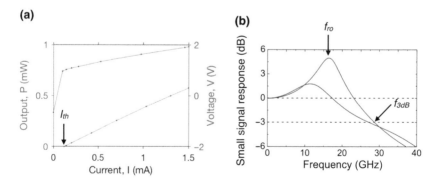

FIGURE 6.2

Static and dynamic characteristics of a laser diode: (a) Output power, P, and applied voltage, V, as a function of injection current, I; (b) Small signal response as a function of modulation frequency, f.

At steady state and well above threshold, the output power, P, is given by

$$P = \eta_0 \frac{hc}{\lambda} v_g g_{th} N_p (N_a) V_a$$

$$= \eta_0 \frac{hc}{\lambda} v_g \left(\alpha_i + \frac{1}{2L} \log \left(\frac{1}{R_1 R_2} \right) \right) N_p (N) V_a \qquad (6.3)$$

where:

g_{th} is the material gain at threshold
N_p is the photon density in the optical cavity
N_a is the carrier density in the active region
α_i is the internal optical loss including absorption and scattering
L is the cavity length
R_1 and R_2 are mirror reflectivity.

The current (I) and threshold current (I_{th}) are given by

$$I = I_{th} + \frac{qV_a}{\eta_i} v_g g_{th} N_p (N_a) = mI_{th} \qquad (6.4)$$

$$I_{th} = \frac{qV_a}{\eta_i} (R_{sp,th} + R_{nr,th}) \qquad (6.5)$$

where:

m is the ratio between I and I_{th}
v_g is group velocity in the cavity
$R_{sp,th}$ and $R_{nr,th}$ are spontaneous and non-radiative recombination rates at threshold, which are material properties of the gain material.

The 3-dB frequency (f_{3dB}) can be related to the relaxation oscillation frequency (f_{ro}):

$$f_{3dB} = \sqrt{2} f_{ro} = \left(\frac{v_g a \Gamma \eta_i}{qV_a} (I - I_{th}) \right)^{0.5}$$

$$= \left(\frac{v_g a \Gamma \eta_i}{qV_a} (m-1) I_{th} \right)^{0.5} \qquad (6.6)$$

where a is differential gain in the gain material.

6.3.2 Key Metrics for On-Chip Lasers

6.3.2.1 *Operating Energy: Energy per Bit*

The operating energy (also referred to as energy per bit) of an on-chip laser (E_{op}) is defined as static electrical energy supplied for generating a bit signal:

$$E_{op} = IV / B = IV / bf_{3dB}. \qquad (6.7)$$

Here, the bitrate B relates to the 3-dB frequency f_{3dB} by a factor b, which ranges from 1.3 to 2.0, depending on the type of modulation format and pre-emphasis. Since the voltage modulation amplitude is small, the dynamic energy consumption related to charging and discharging the junction capacitance is neglected in energy consumption calculation. Incorporating Equations 6.4 through 6.6 into Equation 6.7, the operating energy can be expressed in terms of V_a, Γ, and η_i:

$$E_{op} = \frac{V}{b}\frac{m}{\sqrt{m-1}}\frac{\sqrt{\frac{qV_a}{\eta_i}\left(R_{sp,th}+R_{nr,th}\right)}}{\sqrt{\frac{v_g a \Gamma \eta_i}{qV_a}}}$$

$$= \frac{Vq}{b\sqrt{a}}\frac{m}{\sqrt{m-1}}\sqrt{R_{sp,th}+R_{nr,th}}\left(\frac{V_a}{\eta_i\sqrt{\Gamma}}\right)$$

(6.8)

where:

   I is average injection current,
   V is average applied voltage
   B is the bitrate of a laser.

According to a study based on the ITRS roadmap [1], the system operating energy of 100 fJ/bit and 30 fJ/bit will be required for off-chip and on-chip interconnects, respectively, in the year 2022. Assumed that 20% of the system operating energy is used for a laser, the laser operating energy for off-chip and on-chip interconnects should be 20 fJ/bit and 6 fJ/bit, respectively. Since these target values are two or three orders of magnitude smaller than that of conventional DFB lasers for long-distance optical communications, which is several pJ/bit, disruptive innovations in laser design might be necessary.

Based on Equation 6.8, a laser with a smaller active region volume V_a is preferable for achieving a smaller operating energy. For example, the photonic crystal laser with an active region volume about a few hundred times smaller than that of DFB lasers, demonstrated a record small energy efficiency of 4.4 fJ/bit with about 2 µW output [9] (see Section 6.4.4 for details). Furthermore, it is anticipated that sub-fJ/bit operating energy might be achievable with even smaller active region volumes [3]. However, due to the laser intensity noise increasing with a smaller output power ($\propto V_a$), the active region volume of an on-chip laser (V_a) needs be larger than a certain value. Let us discuss this.

6.3.2.2 *Bit Error Rate: Applicability of FEC and Importance of Laser Output Power*

The bit error rate (BER) of an optical link measures the fraction of wrong interpretations of bits on the receiver side and is another important metric to consider. For chip-level interconnects, a BER of <10^{-12} is required.

Given a target BER value, the *receiver sensitivity* (P_{rec}) is defined as the minimum optical power to be received by a photodetector. The receiver sensitivity is determined by several factors: photodetector properties (responsivity, shot noise, and thermal noise), link properties (bitrate, extinction ratio, and link length), and light source property (intensity noise) [11]. The back-to-back receiver sensitivity can be estimated as a sum of the receiver sensitivity in the quantum limit ($P_{rec,0}$) which considers only the detector shot noise, and the power penalty, including degradations due to other factors than the propagation loss. Then, based on this receiver sensitivity and various optical losses (e.g. modulator insertion loss of 7 dB and silicon waveguide propagation loss of 3 dB/cm), the required laser output power can be estimated. Now let us think about the influence of the laser intensity noise.

The *relative intensity noise* (RIN) of a laser is defined as [10]

$$\text{RIN} = \frac{\delta P(t)^2}{P_0^2} \tag{6.9}$$

where:

$\delta P(t)^2$ is the time average of noise in laser output power
P_0 is the average laser output power.

This expression shows that a laser with a smaller output suffers from a larger RIN. At the quantum limit of photo-detection (no thermal noise, no dark current, 100% absorption efficiency, and no power penalty), the receiver sensitivity $P_{rec,0}$ is given by [11]

$$P_{rec,0} = -\frac{log(2\text{BER})}{2} \cdot \frac{hc}{\lambda} B \tag{6.10}$$

where it is assumed for simplicity that the zero bit carries no light. The real receiver sensitivity, P_{rec}, is much larger than $P_{rec,0}$. Their difference ($P_{rec} - P_{rec,0}$) is referred to as *power penalty*. The influence of the laser intensity noise is included in this power penalty. In case of conventional long-distance optical links based on a several-mW laser and a pin photodetector, the power penalty is known to be about 25 dB [11].

This power penalty increases with a smaller laser output, which is attributed to a larger relative intensity noise (c.f. Equation 6.9). For example, the

$P_{rec,0}$ value for achieving a BER of 10^{-12} in a 0.85-μm wavelength link at a bitrate of 17 Gb/s is given by

$$P_{rec,0} = -\frac{\log\left(2 \cdot 10^{-12}\right)}{2} \frac{6.63 \times 10^{-34} \cdot 3 \times 10^8}{0.85 \times 10^{-6}} 17 \times 10^9$$

$$= 54 \,(nW) = -42.71 \,(dBm) \tag{6.11}$$

In reality (c.f. Figure 6.4 of [12]), the required received power for a 0.1-mW output VCSEL achieving 10^{-12} BER is −11.7 dBm (=67 μW). The power penalty is about 31 dB (= −11.7 + 42.71), which is about 6 dB lager than the case of a laser of several mW output. This shows that nano-lasers with output of <10 μW cannot lead to a BER of <10^{-12} with currently available detector and waveguide technologies at room temperature. The BER can be reduced by performing a signal processing called *forward error correction* (FEC). However, the FEC may consume several hundred fJ/bit to a few pJ/bit [13]. Thus, considering the target energy/bit of 2–20 fJ/bit for lasers for on- and off-chip interconnects, the FEC needs to be avoided.

In all, the output power of a laser needs to be not less than several tens of μW to achieve a BER of 10^{-12}. Considering the coupling loss and waveguide propagation loss, an output power of >100 μW could be ideal.

6.3.2.3 Bitrate [1]

The bitrate of a laser or a modulator is expected to be similar to the clock speed, e.g. twice the clock speed at most, to avoid excessive energy consumption by serialization and deserialization. A study expects the optimal bitrates for on-chip interconnects to be 4–10 Gbit/s in the near term and 15 Gbit/s in the next decade (as for 2014), based on the ITRS 2011 Executive Summary. According to the ITRS 2007 roadmap, the clock speed of off-chip interconnects is expected to be 29.1 GHz in 2015 and 67.5 GHz in 2022. The bitrate of off-chip interconnect lasers could be about five times higher than that of on-chip interconnect lasers. Thus, lasers for on-chip interconnects do not need to be optimized for speed, while those for off-chip interconnects need to be.

6.3.3 Conclusion

The key metrics for lasers for chip-level optical interconnects include the energy/bit, BER of the entire optical link, and the bitrate. The requirement on the BER can be translated into that on the output power of lasers. The FEC needs to be avoided for meeting the requirement on the energy/bit, since its energy/bit can overwhelm the energy/bit requirement on the entire link. In addition, the size of a laser cannot be made very small to meet the requirement of the output power. The optimal size of a laser will be discussed in the next section. Based on discussions in this section, the criteria specification of

TABLE 6.1

Specification for Lasers for On-Chip Interconnects

Energy/bit	Output Power	Bitrate
<10 fJ/bit	>−11.7 dBm (67 µW)	A few to 15 Gbit/s

lasers for on-chip interconnects can be summarized as in Table 6.1, where the condition for laser output power assumes no propagation or coupling losses in the optical link from a laser to a detector.

6.4 State-of-the-Art On-Chip Lasers

Until recently, realizing a laser monolithically integrated on a silicon wafer has been a challenge, since the group IV semiconductors lattice-matched to a silicon wafer have typically an indirect bandgap, which prohibits efficient light generation. Recently, several methods have been reported for tackling this challenge, including the growth of a highly strained and highly n-doped Ge (group IV with a direct bandgap) on a silicon wafer [14], the growth of a group III-V gain material on a silicon wafer free from dislocations [15], and the low-temperature wafer bonding of a group III-V gain material onto silicon [16]. Figure 6.3 shows examples of laser structures based on these approaches. Among them, the hybrid approaches, i.e. the growth or the direct wafer-bonding of a III-V gain material onto a silicon wafer, provide almost the same electric-to-photonic conversion efficiencies as conventional III-V gain materials. In particular, the fabrication technologies of the wafer-bonding approach have matured; various laser structures with novel optical cavities have been reported in recent years, based on the wafer-bonding approach.

Here, we will review five on-chip laser structures with different optical cavities, aiming at understanding strengths as well as challenges of each laser structure. Four of them are hybrid lasers, based on wafer bonding.

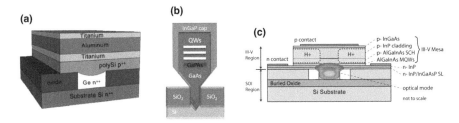

FIGURE 6.3

(a) Ge grown on Si [14]. (b) GaAs grown on Si [15]. (c) InP wafer-bonded to a SOI wafer (c.f. Section 6.3.2 for details) [16]. (Reprinted with permissions from Ref. [14–16], Optical Society of America.)

TABLE 6.2

Active Region Volumes of Lasers Reviewed in this Section

	Active Region Volume	**Cavity Length/Direction**
SEL	$10.0 \times (1.5\ \mu m)^3$	In-plane
VCSEL	$0.18 \times (0.85\ \mu m)^3$	Vertical
Hybrid VCL	$0.19 \times (1.5\ \mu m)^3$	Vertical
1D PhC laser	$0.04 \times (1.5\ \mu m)^3$	In-plane
LEAP	$0.005 \times (1.5\ \mu m)^3$	In-plane

This understanding will be the basis of discussions in Section 6.4.6. In this review, we pay attention to the size of a laser as a key parameter determining the performance of on-chip lasers. The active region volumes of lasers to be reviewed are summarized in Table 6.2.

6.4.1 Hybrid Silicon Evanescent Laser (SEL) [16, 17]

The silicon evanescent laser (SEL) is the first hybrid laser which emits light into a silicon waveguide. It features no need for accurate alignment during the wafer bonding, which removes one of obstacles to realizing III-V on-silicon lasers.

A schematic cross-section of an SEL is illustrated with a transverse optical mode profile in Figure 6.4a [16]. In this laser structure, the waveguide for transverse mode confinement is formed in the silicon layer of a silicon-on-insulator (SOI) wafer while the active material is in the III-V layer. The transverse optical mode is confined around the silicon waveguide since the refractive index of silicon, 3.47 at 1.55 μm wavelength, is higher than those of III-V layer materials, e.g. 3.17 for InP at 1.55 μm wavelength.

SELs are in-plane lasers. The terminations of III-V layers work as a mirror. Basically, the characteristics of SELs are very similar to those of conventional all III-V in-plane lasers. Since the reflectivity is not high (~30%), the cavity length is typically several hundred μm. One can make a DFB laser or a DBR laser by integrating surface grating typically onto the silicon layer. Characteristics of a recent version of hybrid DFB SELs with short cavity lengths of 100 μm and 200 μm are shown in Figures 6.4b–d [17], which is chosen for analysis in Section 6.4.6.

Like conventional in-plane lasers, the threshold current of SELs is tens of mA and the output power can be as high as >10 mW. If one reduces the cavity length, the threshold current could be lowered below 10 mA, as shown in Figure 6.4b. However, it appears challenging to further reduce it below 1 mA, due to lack of very high reflectivity mirrors.

6.4.1.1 Fabrication

Firstly, a silicon waveguide is formed on an SOI wafer. Then a III-V epitaxy wafer is wafer-bonded onto the SOI wafer and the substrate is removed from the III-V epitaxy wafer. The wafer bonding needs to be performed at

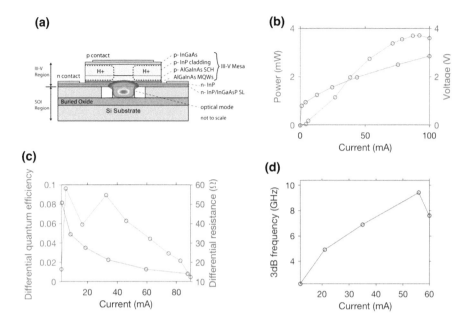

FIGURE 6.4

Silicon evanescent laser: (a) Schematic with an inset showing a simulated transverse mode profile (Reprinted with permission from Ref. [16], Optical Society of America.); (b) Output power and voltage versus injection current for two versions with cavity lengths of 100 μm and 200 μm; (c) Output spectrum of a 100-μm-long DFB laser; (d) Small signal modulation of a 200-μm-long DFB laser.

relatively low temperature, e.g. 300°C, for two reasons: (1) The wafer bonding is done at an elevated temperature and the bonded sample is cooled down to room temperature. This causes a stress due to the different thermal expansion coefficients of silicon and III-V layers. (2) Considering integration of lasers onto silicon electronic ICs, the bonding process should not influence the properties of silicon electronic ICs. This is one of the requirements for CMOS compatibility.

There are two types of popular wafer-bonding techniques: *direct wafer bonding* and *BCB wafer bonding*. The direct wafer bonding relies on Van der Waals force between flat surfaces of SOI and III-V wafers, which requires smooth surfaces with RMS roughness of less than 1 nm. BCB wafer bonding uses BCB as an intermediate layer between two wafers, which can be as thin as a few tens of nm. Both approaches provide strong bonding strength, wafer-scale bonding, and high yield.

These fabrication procedures apply to other hybrid lasers as well.

6.4.1.2 Strength and Challenge

The greatest strength of SELs is a relatively simple fabrication process. Since the position of optical modes is defined by the silicon waveguide, no accurate alignment is necessary during wafer bonding. All the structures in the III-V layers can be aligned to the silicon waveguide with lithography alignment accuracy. This significantly facilitates processing, reducing fabrication cost. Since cost is one of important considerations for switching from electrical interconnects to optical interconnects, this is an important advantage. Another strength is that all proven features for conventional in-plane lasers can in principle be adapted. For example, an optical feedback introduced to a DBF SEL boosts the 3-dB frequency up to 56 Gb/s, incorporating a photon-photon resonance with a higher frequency than the relaxation oscillation frequency [18].

A weakness of SELs is high threshold currents. With a threshold current of over 1 mA, it is difficult to meet the energy/bit requirement even for off-chip interconnects (<100 fJ/bit), as discussed further in Section 6.4. With a typical threshold for SELs of >10 mA, the energy/bit can be several pJ/bit, even combined with a power-efficient high-speed modulator.

6.4.2 Vertical-Cavity Surface-Emitting Laser (VCSEL)

6.4.2.1 General Features

As shown in Figure 6.5a, the laser structure consists of two distributed Bragg reflectors (DBRs) and a nominal cavity including a gain material between them. A VCSEL is a *vertical cavity laser*; the light generated from the gain material is vertically amplified between two mirrors. Since DBRs have very high reflectivity, e.g. >99.5%, the optical cavity can be as short as one or two times the lasing wavelength in material. This makes the active region volume small, leading to small threshold current, typically less than 1 mA. It also makes the optical volume small, allowing for very high speed. A bitrate of 56 Gb/s with NRZ modulation format was reported without FEC [19]. With DMT modulation format, a bitrate of 115 Gb/s and a BER of 10^{-3} were reported [20]. Popular wavelengths are 850 nm and 980 nm, which are widely used for optical interconnects in data centers and high-performance computers. VCSEL technologies at 1310 nm and 1550 nm have also been matured [19].

The VCSEL structure chosen for analysis here is one designed for very small energy/bit, not for very high speed [12]. The wavelength is 850 nm and the oxide aperture is 2 μm. Its static and dynamic characteristics are provided in Figure 6.5b–d. A record small energy/bit among VCSELs of 83 fJ/bit was demonstrated.

6.4.2.2 Strength and Challenge

VCSELs have the potential for surface-normal free-space interconnects between chips. The free-space chip-chip interconnect has been noted as an attractive option for off-chip interconnects, since it can dramatically increase

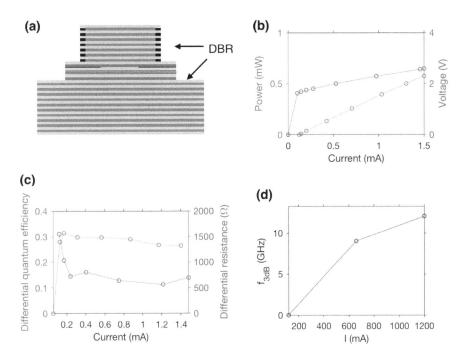

FIGURE 6.5
850-nm wavelength VCSEL [12]: (a) Schematic cross-sectional view; (b) Power and voltage versus current; (c) Differential quantum efficiency and differential resistance versus current; (d) 3-dB frequency versus current.

the off-chip interconnect bandwidth, efficiently solving the integration density issue, it does not require wavelength-division multiplexing (WDM), and it is free from timing issues [1]. VCSELs can meet the off-chip requirements of laser, i.e. energy/bit of <100 fJ/bit and bitrate of >50 Gb/s. The vertical emission direction of VCSEL and feasibility of integrating monolithic lens onto VCSEL chips are also ideal.

However, VCSELs are not fit for on-chip interconnects. Firstly, VCSELs cannot emit light laterally into an in-plane waveguide. Furthermore, it appears challenging to reduce the energy/bit of VCSELs to the on-chip requirement of <10 fJ/bit. The record small energy/bit of 83 fJ/bit was gained from a VCSEL with a 2 μm oxide aperture, which is already very small.

6.4.3 Silicon Vertical Cavity Laser (Si-VCL) [21–23]

The *silicon vertical cavity lasers* (Si-VCL) feature in-plane light emission from a vertical cavity as well as high speed potential, which are both enabled by using a *high contrast grating* (HCG) as a mirror. HCGs are a grating layer with near sub-wavelength periodicity, surrounded by low-index materials. They can be designed as a broadband reflector with a >99.9% reflectivity over

>100 nm wavelength range, or as a high-quality (Q)-factor resonator with a Q-factor approaching infinity. The thickness of HCG reflectors designed for 1.55 μm wavelength can be less than 0.5 μm, which is much thinner than that of DBRs for VCSELs.

As shown in Figure 6.6a, the Si-VCL consists of an HCG region formed in the Si layer of an SOI wafer, III-V layers including a gain material, and a dielectric DBR. The HCG region is connected to an in-plane silicon waveguide. In this design, the HCG plays two key roles. The first one is that the Si-HCG not only works as a mirror but also routes light from the vertical cavity laterally into the in-plane silicon waveguide, as shown in Figure 6.6b. This eliminates the obstacle to using vertical cavity lasers with on-chip interconnects. This vertical-to-lateral routing in the Si-HCG is quite different from the vertical-to-lateral coupling in conventional grating couplers. The Si-HCG routes only a small fraction of incident power, e.g. 0.3%, while reflecting most of it, e.g. >99.5%, which is necessary for reaching threshold condition in vertical cavity lasers.

The second role is to reduce the optical mode volume. The optical mode volume of a vertical cavity laser has contributions from the nominal cavity between two mirrors as well as the power penetration into mirrors. As shown in Figure 6.6b, there is no power penetration beyond the HCG layer while there is into the top DBR. This allows for a considerably smaller mode

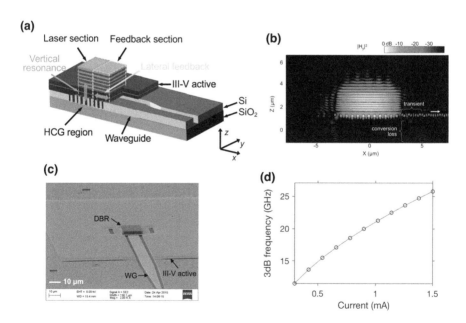

FIGURE 6.6
Hybrid vertical cavity laser based on high contrast gratings [21, 22]: (a) Schematic cross-section; (b) Simulated mode profile; (c) SEM image of a fabricated laser sample; (d) Measured 3dB frequency as a function of estimated injection level. (Reproduced from [21] with the permission of AIP Publishing and from [22] under CC license, https://creativecommons.org/licenses/by/4.0/.)

volume for HCG-based cavities than DBR-based cavities, leading to a higher confinement factor. For example, the confinement factor, Γ, of a fabricated Si-VCL shown in Figure 6.6c is 12%. For reference, typical DBR-based VCSELs have a confinement factor of about 3%. This has important implications for the interconnect light source. Since the speed of a laser is proportional to $\sqrt{\Gamma}$ (c.f. Section 6.3.1), the fabricated Si-VCL could have an Modulation current efficiency factor (MCEF) of 42 GHz/mA$^{1/2}$, which is a few times higher than that of DBR-based high-speed VCSELs.

The high MCEF of the Si-VCL has important potentials. At low injection levels, it may enable the achievement of a low energy/bit, which is desirable for on-chip lasers. At high injection levels, it may greatly reduce the heat generation, which is dominated by Joule heating around the active region. For example, a laser with a three-times larger MCEF than a reference laser generates 81 times less heat ($\propto I^4$) to achieve the same speed as the reference laser. This much less heat generation may significantly improve the lifetime reliability of microcavity lasers, which is of critical importance in optical modules for data center applications, especially at high bitrates.

However, all the features of Si-VCLs were demonstrated in optically pumped versions of lasers. Thus, it needs to be proven that they can be maintained for electrically pumped versions.

6.4.4 Lambda-Scale Embedded Active Region Photonic Crystal (LEAP) Laser [24]

The *lambda-scale embedded active region photonic crystal* (LEAP) *laser* features very small size (c.f. Table 6.2), and consequently record low energy/bit. As shown in Figure 6.7a, the optical cavity of a LEAP laser is realized by a 2D photonic crystal waveguide while its active region is implemented as a buried hetero (BH) structure including quantum wells, the red-colored region. The BH active region eliminates the surface recombination loss and the free-carrier absorption loss, which together increases threshold current. Thanks to the small active region volume V, the threshold current ($\propto V$) is as small as 4.8 µA (c.f. Figure 6.7b) and the MCEF ($\propto V^{-1/2}$) is as high as 60 GHz/mA$^{1/2}$ (=2.0 GHz/µA$^{1/2}$), which is to our knowledge the largest among all electrically injected laser diodes. This combination of low working current and high MCEF enables the LEAP laser to achieve an energy/bit of <10 fJ/bit, meeting the energy/bit requirement on lasers for on-chip interconnects.

In the LEAP laser, current is injected laterally through the current channels formed by ion diffusion and implantation (c.f. Figure 6.7a). This in-plane configuration of conductive regions results in a very small junction capacitance, C. A simple estimation anticipates,

$$C = \frac{3\,\mu m \times 0.23\,\mu m}{0.3\,\mu m} \times \varepsilon_0 3.2^2 = 0.2\,\text{fF}\,. \tag{6.12}$$

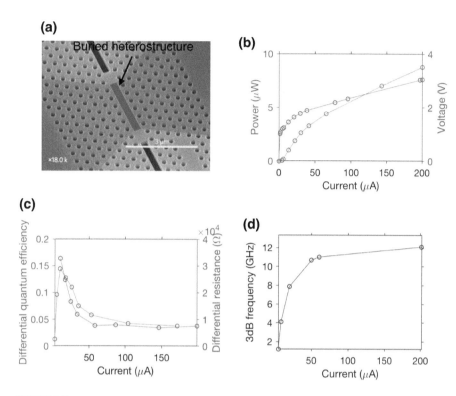

FIGURE 6.7
LEAP laser [24]: (a) SEM image of a fabricated laser sample (reprinted by permission from Springer Nature: [24], copyright 2013). The added green and purple color designate doped regions for carrier injection, while the red represents the buried hetero-structure; (b) Output power and voltage versus injection current; (c) Differential quantum efficiency and differential resistance versus current; (d) 3-dB frequency versus current.

Thus, though the differential resistance, R, is as high as 2×10^4, the inverse of RC constant $(=1/2\pi RC)$ is about 38 GHz. This value is much higher than the 3-dB frequencies of the entire frequency response shown in Figure 6.7d, which consists of the intrinsic (diode) response and extrinsic (parasitic elements) response. So the laser speed is not limited by parasitic elements.

A limitation of the current version of the LEAP laser is small output power. As shown in Figure 6.7b, the output power even at a high injection current of 200 µA is far below the required value of 67 µW for FEC-free detection (c.f. Table 6.1). This is attributed to the significant decrease of differential quantum efficiency with increasing current, as shown in Figure 6.7c, which relates to the current bypassing the buried hetero-structure at higher injection levels. This high-leakage current, i.e. low-injection efficiency, η_i, limits the 3-dB frequency at high injection levels, as shown in Figure 6.7d.

6.4.5 1D Photonic Crystal (1D PhC) Laser [25]

The *one-dimensional photonic crystal* (1D PhC) *laser* features compact size and sub-mW output. As shown in Figure 6.8a, the optical cavity of the 1D PhC laser uses a one-dimensional photonic crystal, sometimes also referred to as a nanobeam. This makes the footprint of this laser comparably small to that of the LEAP laser. For the electrical part, this laser relies on a surface passivation method to reduce the surface recombination loss, and the current is injected vertically through a grown p-i-n diode structure, as shown in Figure 6.8b. This vertical injection scheme prevents the leakage current issue from occurring, as it does in the LEAP laser. As shown in Figure 6.8d, the differential quantum efficiency ranges from 25% to 35% over the entire injection level, which is comparable to VCSELs. As a result, the output power may reach about 100 µW at 500 µA, which satisfies the power requirement for FEC-free detection.

However, the vertical configuration of conductive layers and the large N-contact area in the current version of the 1D PhC laser does not result in such a very small capacitance as in the LEAP laser. Thus, directly modulated speed could be limited by parasitics. Assuming the N-contact area outside the nanobeam is $2 \times 5\mu m \times 10\mu m$, a simple estimation anticipates,

$$C = \frac{0.6\mu m \times 15\mu m \times 0.98 + 2 \times 5\mu m \times 10\mu m}{0.234\mu m} \times \varepsilon_0 3.2^2 = 42 \ fF. \qquad (6.13)$$

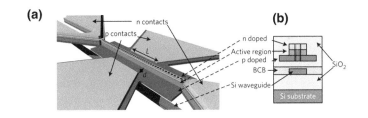

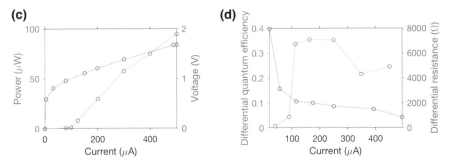

FIGURE 6.8
1D photonic crystal (1D PhC) laser [25]: (a) Schematic of the laser structure (reprinted by permission from Springer Nature: [25], copyright 2017); (b) Schematic cross-section; (c) Output and voltage versus current; (d) Differential resistance and differential quantum efficiency.

Then, with a differential resistance R of 2000 Ω read from Figure 6.8d, the inverse of RC constant, i.e. the extrinsic 3-dB frequency, is given by

$$\frac{1}{2\pi RC} = 1.9\,\text{GHz}\,. \tag{6.14}$$

Considering that the active region volume of the 1D PhC laser is considerably smaller than that of the VCSEL, this laser may achieve an intrinsic 3-dB frequency of >10 GHz. Thus, the speed of the 1D PhC laser is very likely to be limited by the parasitics.

6.4.6 Comparison Analysis of Five Lasers

In Figure 6.9a–g, the reviewed characteristics of the five lasers are compared to analyze the role of the laser size, i.e. active region volume. In Figure 6.9h, the characteristics of the fictitious LEAP, 1D PhC, and Si-VCL lasers are considered, assuming the weaknesses of the current versions of those lasers are overcome.

In Figure 6.9a, the threshold current I_{th} and active region volume V_a of each laser are plotted. The threshold current is smaller for a laser with a smaller active region volume V_a, as expected from Equation 6.5:

$$I_{th} = \frac{qV_a}{\eta_i}\left(R_{sp,th} + R_{nr,th}\right).$$

This validates the idea that the active region size needs to be smaller to make the electrical energy consumption (IV) and hence the energy/bit smaller.

Figure 6.9b plots the output power P of each laser as a function of injection current I. The output power of a laser with a smaller active region volume V_a is distributed in smaller values, as expected from Equation 6.3:

$$P = \eta_0 \frac{hc}{\lambda} v_g g_{th} N_p\left(N_a\right)V_a$$

where N_p is linearly proportional to I. This implies that the active region volume should be larger than a specific value to meet the required condition for output power in relation to the BER (c.f. Table 6.1). For example, the LEAP laser does not meet the output requirement for the FEC-free detection (dotted line in Figure 6.9b), even at high injection currents. Discussions on Figure 6.9a,b show that there is a trade-off relationship between the energy/bit and the output power, with the active region volume as a parameter. There should be an optimal active region volume to simultaneously satisfy the requirements in energy/bit and output power.

Figure 6.9c plots the differential quantum efficiency η_d of each laser over applied currents. The LEAP laser with a lateral current injection scheme has

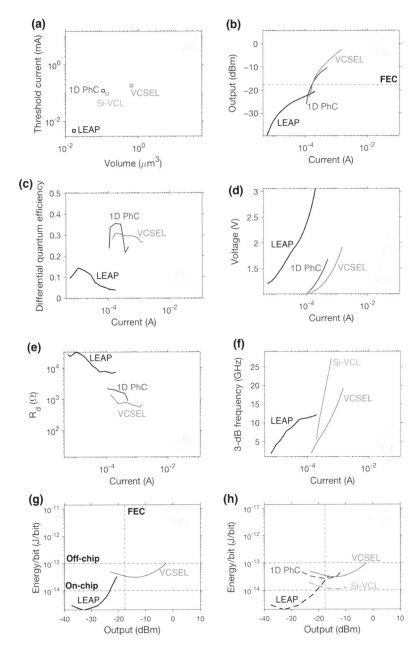

FIGURE 6.9

(a-g) Comparison of characteristics of the five lasers. Values are sampled from the experimentally measured values. (h) Fictitious comparison where it is assumed that the differential quantum efficiency of the LEAP laser is kept at 15%, the 3-dB frequency of the 1D PhC laser is not limited by the RC constant, and the electrically pumped version of the Si-VCL has a half confinement factor of the optically pumped version.

much smaller differential quantum efficiency values, especially at higher currents, compared to the 1D PhC and VCSEL, both with a vertical current injection scheme. This small η_d ($\propto dP/dI$) value of the LEAP laser limits its output power P, which becomes its weakness, as will be seen in Figure 6.9g.

Figures 6.9d,e plot the voltage V and differential resistance $R_d (= dV/dI)$ of each laser. This shows that smaller lasers have larger resistance. Especially, the LEAP and 1D PhC lasers have resistances larger than 10^3 Ω. Thus, unless the capacitance of these lasers is very small, their speed will be limited by the RC constant, which is the case for the 1D PhC laser.

Figure 6.9f plots the 3-dB frequency of each laser. As Equation 6.6 anticipates the 3-dB frequency is determined by several factors:

$$f_{3dB} = \left(\frac{v_g a \Gamma \eta_i}{q V_a} \left(I - I_{th} \right) \right)^{0.5}.$$

The LEAP laser does not have the highest 3-dB frequency though it has the smallest active region volume. This is attributed to the fact that its injection efficiency, η_i ($\propto \eta_d$), and the current injections $\left(I - I_{th} \right)$ are smaller than those of other lasers. However, about 10 GHz 3-dB frequency is enough for on-chip interconnects. Here, it is noted that the Si-VCL has much higher 3-dB frequency than the VCSEL though their active region volumes are similar. This shows that introducing a nano-structure into the optical cavity may significantly improve the speed of a laser beyond its limit in conventional designs.

Figure 6.9g plots two key metrics for on-chip lasers, i.e. energy/bit and output, with current as an intrinsic parameter. The vertical and horizontal dotted lines designate the FEC-free detection requirement in the laser output power and the energy/bit requirements for on/off-chip interconnects, respectively. The VCSEL satisfies both requirements for off-chip interconnects. However, none of the lasers simultaneously meets the requirements on two metrics for on-chip interconnects.

Figure 6.9h plots fictitious cases, assuming the LEAP laser, the 1D PhC laser, and the Si-VCL overcome their weaknesses. That is, the LEAP laser maintains a differential quantum efficiency of 15% over the entire injection currents, the 3-dB frequency of the 1D PhC laser is determined by its intrinsic diode response, not being limited by the RC constant, and the electrically pumped version of the Si-VCL has half of the confinement factor of the optically pumped version. It is noted that if the LEAP laser has a bit larger active region volume, its output power could reach the FEC free output while the 1D PhC laser may reach the on-chip energy/bit condition if its active region volume were a bit smaller, leading to slightly higher 3-dB frequency. From these observations, it could be concluded that the optimal active region volume for on-chip interconnects could be somewhere between the volumes of the LEAP and the 1D PhC lasers. The Si-VCL case of which the active region volume is similar to that of the VCSEL shows that the use of a functional nano-structure may help to further reduce the energy/bit value.

6.5 Conclusion

In designing lasers for on-chip interconnects, the FEC should be avoided due to its large energy/bit, which overwhelms the energy/bit budget of the entire link for a laser to a receiver. This condition puts a requirement that the laser output power should be larger than the responsivity of a receiver, which is close to 100 µW in cases of no propagation or coupling loss. Thus, the key metrics for lasers for on-chip interconnects are energy/bit, output power, and bitrate.

The energy/bit and output power are in a trade-off relation as the laser size increases. The energy/bit decreases (ideal) while the output power also decreases (not ideal) as the active region volume increases. Thus, there should be an optimal range of the laser size to simultaneously meet the requirements on energy/bit as well as output power. A laser cannot be made very small since it may reduce its output power below the receiver responsivity.

None of frontier on-chip laser technologies simultaneously meet the requirements on lasers for on-chip interconnects (<10 fJ/bit energy/bit and >67 µW output) as they are now, without employing the FEC. Analyses of those laser technologies anticipate that the ideal active region volume is larger and smaller than those of the LEAP and 1D PhC lasers, respectively. However, it is unclear whether the current limitations in the LEAP and 1D PhC lasers can be tackled in the near future even if the ideal active region volume is chosen.

References

1. D. A. B. Miller, Device requirements for optical interconnects to silicon chips, *IEEE Proc.*, vol. 97, pp. 1166–1185, July 2009.
2. Martijn J. R. Heck and John E. Bowers, Energy efficient and energy proportional optical interconnects for multi-core processors: Driving the need for on-chip sources, *IEEE J. Sel. Top. Quantum Electron*, vol. 20, p. 8201012, July/August 2014.
3. David A. B. Miller, Attojoule optoelectronics for low-energy information processing and communications, *J. Lightwave Technol.*, vol. 35, p. 346, February 2017.
4. Z. Zhou, B. Yin, and J. Michel, On-chip light sources for silicon photonics, *Light Sci. Appl.*, vol. 4, p. e358, August 2015.

5. M. Stucchi, S. Cosemans, J. Van Campenhout, Z. Tokei, and G. Beyer, On-chip optical interconnects versus electrical interconnects for high-performance applications, *Microelectron. Eng.*, vol. 112, p. 84, 2013.

6. F. Y. Gardes, D. J. Thomson, N. G. Emerson, and G. T. Reed, 40 Gb/s silicon photonics modulator for TE and TM polarisations, *Opt. Express*, vol. 19, p. 11804, June 2011.

7. A. Liu and M. Paniccia, et al., High-speed optical modulation based on carrier depletion in a silicon waveguide, *Opt. Express*, vol. 15, p. 660, January 2007.

8. A. Liu and M. Paniccia, et al., A high-speed silicon optical modulator based on a metal-oxide-semiconductor capacitor, *Nature*, vol. 427, p. 615, February 2004.

9. Koji Takeda and Shinji Matsuo, et al., Few-fJ/bit data transmissions using directly modulated lambda-scale embedded active region photonic-crystal lasers, *Nat. Photonics*, vol. 7, p. 569, 2013.

10. L. A. Coldren, S. W. Corzine, and M. L. Masanovic, *Diode Lasers and Photonic Integrated Circuits*, 2nd ed., Hoboken, New Jersey, Wiley, 2012, chap. 5.

11. G. P. Agrawal, *Fiber-Optic Communication Systems*, 4th ed., Hoboken, New Jersey, Wiley, 2010, pp. 161–171.

12. P. Moser, W. Hofmann, P. Wolf, J. A. Lott, G. Larisch, A. Payusov, N. N. Ledentsov, and D. Bimberg, 81 fJ/bit energy-to-data ratio of 850 nm vertical-cavity surface-emitting lasers for optical interconnects, *App. Phys. Lett.*, vol. 98, p. 231106, 2011.

13. R. S. Tucker and K. Hinton, Energy consumption and energy density in optical and electrical signal processing, *IEEE Photon. J.*, vol. 3, pp. 821–833, 2011.

14. R. E. Camacho-Aguilera, et al., An electrically pumped germanium laser, *Opt. Express*, vol. 20, p. 11316, 2012.

15. Y. Shi, et al., Optical pumped InGaAs/GaAs nano-ridge laser epitaxially grown on a standard 300-nm Si wafer, *Optica*, vol. 4, p. 1468, 2017.

16. A. W. Fang, et al., Electrically pumped hybrid AlGaInAs-silicon evanescent laser, *Opt. Express*, vol. 14, p. 9203, 2006.

17. C. Zhang and J. E. Bowers, et al., Low threshold and high speed short cavity distributed feedback hybrid silicon lasers, *Opt. Express* vol. 22, p. 10202, 2014.

18. A. Abbasi, et al., Direct and electroabsorption modulation of a III-V-on-silicon DFB laser at 56 Gb/s, *IEEE Sel. Top. Quant. Electron.*, vol. 23, p. 1501307, 2017.

19. D. M. Kutcha, et al., Error-free 56 Gb/s NRZ modulation of a 1530-nm VCSEL link, *J. Light. Technol.*, vol. 34, p. 3275, 2016.

20. M. Muller, et al., 1.3 μm short-cavity VCSEL enabling error-free transmission at 25 gbit/s over 25 km fibre link, *Electron. Lett.*, vol. 48, p. 1487, 2012.

21. I.-S. Chung and J. Mørk, Silicon-photonics light source realized by III-V/Si-grating-mirror laser, *Appl. Phys. Lett.*, vol. 97, p. 151113, 2010.

22. G. C. Park, et al., Ultrahigh-speed Si-integrated on-chip laser with tailored dynamic characteristics, *Sci. Rep.*, vol. 6, p. 38801, 2016.

23. G. C. Park, et al., Hybrid vertical-cavity laser with lateral emission into a silicon waveguide. *Laser Photon. Rev.*, vol. 9, L11–L15, 2015.

24. K. Takeda, et al., Few-fJ/bit data transmissions using directly modulated lambda-scale embedded active region photonic-crystal lasers. *Nat. Photon.*, vol. 7, 569–575, 2013.

25. G. Crosnier, et al., Hybrid indium phosphide-on-silicon nanolaser diode, *Nat. Photon.*, vol. 11, p. 297, 2017.

7

Integrated Photonic Interconnects for Computing Platforms

Nicola Andriolli, Isabella Cerutti, and Odile Liboiron-Ladouceur

CONTENTS

7.1 Recent Technological Trends in Point-to-Point Intra-Board Photonic Interconnects

This chapter provides a state-of-the-art overview of photonic integrated technologies for interconnects, by starting from point-to-point photonic interconnects used for intra-board communication and moving to more complex interconnection networks. Point-to-point interconnects enable optical communication from a single source to a single receiver. The architecture is therefore simple and avoids the need for a control beyond the transceiver. For these reasons, point-to-point photonic interconnects reached a high level of technological maturity and are being introduced in computing platform boards to overcome the limitations of electronics.

Compared to electrical links, photonic point-to-point links enable greater aggregated bandwidth-distance product, achieving higher communication capacity and longer reach. Telecommunications systems are efficiently exploiting these features [1, 2]. However, in computing platforms the number of shared elements and their physical separation forced photonic technology development to take a different scaling factor, making bandwidth density requirements as important as the aggregated bandwidth. For intra-board interconnects, bandwidth density can be defined with respect to the two dimensions of the surface board area (Gb/s per mm^2), or to the one dimension across the front edge of the board (Gb/s per mm). Energy efficiency is another key requirement for optical interconnects, where the power consumption is normalized to the transmitted/switched bit, using as a metric W/bit/s or J/bit. One picojoule of energy per bit corresponds to one milliwatt per gigabit per second.

A photonic point-to-point link consists of the optical source that is modulated by the associated electrical circuitry (e.g., driver), the optical channel, and the photodetector with its associated electrical circuitry (e.g., transimpedance amplifier). Photonic point-to-point links not requiring power-hungry clock recovery, serialiser-deserialiser (SerDes), or digital-to-analog/analog-digital converter (DAC/ADC) can achieve better energy efficiency. Hence, modulated data rates are often limited to the electrical line rate and are in some instances not resynchronized. Optical sources based on uncooled vertical-cavity surface-emitting lasers (VCSEL) are used, exploiting space-division multiplexing (SDM) by means of ribbons or multi-core fibers with one optical carrier per core. Directly modulated VCSEL sources have been shown to meet the electrical line rate of 56 Gb/s [3, 4] while being compact, low-cost, and energy-efficient.

In recent years, active optical cables (AOC) have been widely utilized at the edge of the servers in data center networks [5]. The introduction of optics within the computing platforms is also being further pushed inside the servers: the use of mid-board optical engines eliminates the bandwidth-limited electrical waveguides between the processor and the edge of the board [6, 7]. Solutions are now mature and standardization efforts are ongoing: the Consortium for On-Board Optics (COBO) is developing specifications for interoperable optical modules that can be mounted onto printed circuit boards (e.g., motherboards, daughterboards) [8]. Electro-optical backplanes in polymer and glass have been demonstrated to achieve a seamless optical connectivity from external fiber-optic networks to system-embedded optical interconnect architectures [9, 10].

As computing platforms scale, photonic point-to-point links must offer greater aggregated bandwidth density as well. As such, wavelength-division multiplexing (WDM) technology can be adopted in the transmission links. Alternatively, multi-core fiber as well as multi-mode modulation have recently attracted great interest in the research community [11, 12]. Complex modulation formats, e.g., pulse amplitude modulation with four levels (PAM-4) and multi-band carrier-less amplitude and phase modulation (m-CAP) [13, 14], are also a viable alternative for increasing the bandwidth between servers, at the cost of higher complexity and power dissipation for data processing.

Despite the current technological progress, the latency and power consumption scalability of the overall communication infrastructure with photonic point-to-point links is constrained by the signal conversions required between the optical and electrical domains. To overcome this issue, photonic technologies can be used to switch data directly in the optical domain within the interconnection networks for more power-efficient and time-of-flight constrained computing platforms.

7.2 Photonic Interconnection Networks

Photonic interconnection networks extend the capabilities of point-to-point interconnects by enabling the transport and exchange of data between multiple ports. The data organized in packets can thus be switched from any input port to any output port of the interconnection network. To realize such functionalities, fast photonic switching devices are required and need to be interconnected to realize high-throughput architectures. In addition, the switching devices need to be controlled for properly scheduling the packet switching, achieving low latency communications.

This section tackles these three aspects: (1) the integrated switching devices; (2) the interconnection network topologies; (3) the network control and scheduling. The first subsection discusses the most relevant switching devices that can be used as building blocks for realizing photonic integrated interconnection networks. An optimized design of such switching devices is presented as a use case for silicon photonics. The second subsection surveys the relevant topologies for interconnection networks. The last subsection is devoted to the scheduling and control of photonic interconnection networks and provides an overview of the challenging problems and proposed approaches.

7.2.1 Integrated Switching Devices

Interconnection networks can be designed using simple optical switches (e.g., 1×2 or 2×2) as building blocks for topologies with higher radix. Two of the most notable building blocks are the microring resonator (MR) and the Mach-Zehnder interferometer (MZI), which can be integrated using silicon platform and can be readily dynamically controlled with electrical signals.

The MR can be used to switch data between two parallel waveguides, realizing 1×2 or 2×2 elementary switches with a wavelength-dependent behavior. When the transmitted data is carried on a resonant wavelength of the MR, the data is guided (or dropped) to the other waveguide. Otherwise, the data passes through and remains on the same incoming waveguide. The use of MR as a switching element requires an optimized design to guarantee the required separation between resonant wavelengths (called free spectral range, dependent on the ring radius), an adequate pass bandwidth (dependent on the ring-waveguide coupling coefficient), a low insertion loss and a low crosstalk (improved with cascaded ring configurations [15]).

An example of a silicon MR is shown in Figure 7.1. The MR (microscopy image in Figure 7.1(a)) was fabricated with a 220-nm-thick silicon platform and has a radius of 10 μm and a cross-section of 480 nm × 220 nm, as sketched in Figure 7.1(b) [16, 17]. The schematic shows the cross-section of the ring itself and its metal contacts, together with the upper and lower coupling waveguides. A 2-μm-thick buried silicon oxide layer is placed below the silicon layer and a coating of silicon oxide cladding covers the integrated device. The MR is controlled by exploiting the thermo-optic effect. To efficiently tune the MR wavelength resonance, a coplanar heater is used. As shown in Figure 7.1(b), a 90-nm-high slab is added on the inner side of the microring. A section of 1 μm of the slab closer to the microring is kept undoped, while the innermost section of the slab is doped with phosphorus (n-doped) at a peak doping concentration of $5 \cdot 10^{20}$ cm^{-3}. The MR can be then thermally tuned by injecting a current, through metal pads and vias, in the conductive path created by the doped slab. The presence of a slab only on the inner side of the ring is beneficial for multiple reasons. First of all, the inner-side slab improves the bending loss since the optical mode is more confined compared to a classical slab. Moreover, the fully etched region between the MR and the coupling waveguide improves the thermal isolation of the ring, keeping the heat away from the coupling regions. At the same time, the undoped silicon slab section enhances the heat transfer from the doped slab to the ring waveguide, since silicon is characterized by a thermal conductivity higher than

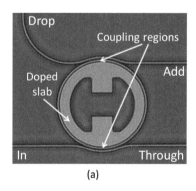
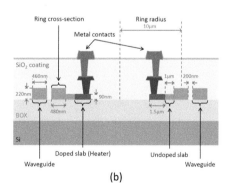

(a)   (b)

FIGURE 7.1

(a) Microscopy image of a ring resonator with the coupling regions and the inner side slab. (From F. Gambini et al., Experimental demonstration of a 24-port packaged multi-microring network-on-chip in silicon photonic platform, *Opt. Express* vol. 25, pp. 22004–22016, 2017, with permission); (b) Schematic of the microring cross-section. (From P. Pintus et al., Ring versus bus: A theoretical and experimental comparison of photonic integrated NoC, *J. Lightw. Technol.*, vol. 33, no. 23, pp. 4870–4877, Dec. 1 2015, with permission).

that of silica cladding. This ensures a lateral heat transfer with minimum heat diffusion through the cladding.

For an efficient switching, it is important to optimize the MR design to achieve a target waveguide-ring coupling coefficient. The target coupling coefficient was set to 10%, but such a high value is difficult to achieve for two main reasons: (1) the presence of the slab on the inner side of the microring moves the electromagnetic mode closer to the internal wall of the microring and away from the coupling waveguide; (2) technology limitations prevent the realization of very narrow inter-waveguide gaps. A typical guaranteed minimum gap is currently 200 nm.

To achieve the target coupling coefficient, the design of the MR input/output waveguides is optimized in two ways. First, the coupling waveguide width is narrowed from the standard width of 480 nm to 460 nm. This increases the mode field outside of the waveguide core, enabling a higher coupling with the ring waveguide. Second, the coupling region is lengthened by bending the coupling waveguide with an angle $\theta = 32°$, as shown in Figure 7.1(a).

The spectral response measured at the different output-ports of the MR is plotted in Figure 7.2(a). The through port has a typical response with notches occurring at the resonant wavelengths, showing a free spectral range (FSR) of approximately 9.62 nm. The drop signal, with Lorentzian shape, has a 3-dB bandwidth of approximately 37 GHz, while the backscattered signal worsens at resonance and is 20 dB below the dropped response. The resonance wavelength shift is reported in Figure 7.2(b) as a function of the dissipated electrical power. Measurements are obtained by injecting a current into the doped path of the ring and measuring the wavelength shift of the resonance. The measured data (red squares) are in good agreement with the results of the

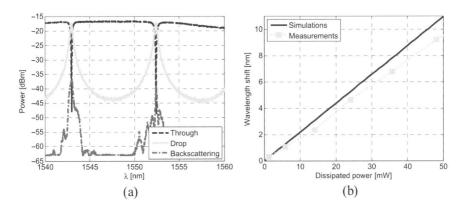

(a) (b)

FIGURE 7.2
(a) Spectral response of a 10-μm radius microring; (b) wavelength shift as a function of the electrical dissipated power for simulated (solid blue line) and measured data (red squares). (From F. Gambini et al., BER evaluation of a low-crosstalk silicon integrated multi-microring network-on-chip, *Opt. Express* vol. 23, no. 13, pp. 17169–17178, Jun. 2015, with permission.)

simulation (solid blue line), with a small discrepancy mainly caused by the doping level inaccuracies. The tuning efficiency of this MR is approximately 189 pm/mW. As such, to achieve a complete free spectral range shift, a power of 50.8 mW is required.

The other switching element that can be effectively realized in integrated photonics is based on the Mach-Zehnder interferometer (MZI) structure. The MZI structure offers a greater optical bandwidth at the cost of relatively larger chip area when compared to MR. The MZI consists of two input-ports (IN1 and IN2) and two output-ports (OUT1 and OUT2), as shown in Figure 7.3(a), with a 50/50 coupler after the input-ports to divide the optical power equally and another 50/50 coupler at the output to recombine the two optical signals. The top and bottom paths are of the same length (balanced) with a means to induce a phase difference between the two paths in a controlled manner. The phase difference effectively directs the combined optical signal at the second coupler to one of the two output-ports. As such, the MZI can be configured into two distinct states, which are illustrated in Figure 7.3(b). The bar state connects the top ports (IN1-OUT1) and the bottom ports (IN2-OUT2). Alternatively, the cross state connects the top input port to the bottom output-port (IN1-OUT2), while the bottom input port connects to the top output-port (IN2-OUT1). The two states are achieved by inducing the appropriate phase shift difference between the two paths, either thermally or through induced carrier changes. In the first approach, an ohmic metallic connection (Figure 7.4(a)) or even a doped silicon resistive element (Figure 7.4(b)) enables heat to be generated from an applied voltage that sufficiently changes the index of refraction, achieving the required phase shift.

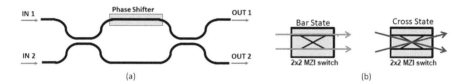

FIGURE 7.3
(a) Schematic of a 2×2 Mach-Zehnder interferometer (MZI) switch element with a phase shifter in one arm and 3-dB directional couplers; (b) Conceptual illustration of the bar and cross states.

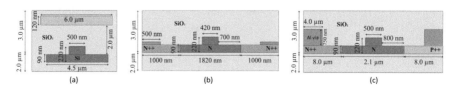

FIGURE 7.4
Cross-sectional view of a silicon photonic MZI-based switching element based on (a) thermal switching using a metallic heater; (b) thermal switching using a resistive junction; and (c) electro-optical switching using a p-i-n junction.

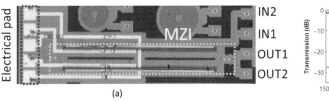
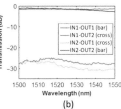

(a)

(b)

FIGURE 7.5

(a) Microscopic view of the 2×2 MZI switch element; (b) The normalized transmission spectra of the 2×2 MZI switch element with an applied voltage of 0.88 V (corresponding to a bar configuration).

Alternatively, p-i-n junctions are integrated structures (Figure 7.4(c)) that allow for fast carrier injection, leading to faster index change at the cost of higher insertion loss relative to thermal-based switches.

The microscopic image of an MZI element design example is shown in Figure 7.5(a). This switch was fabricated at the A*STAR Institute of Microelectronics (IME) in Singapore through a multi-project wafer (MPW) managed by CMC Microsystems [18]. The 2×2 MZI element consists of two 3-dB directional couplers connected by two waveguide arms with equal length for broadband operation. In this example, the length of the directional coupler is 3 μm, while the gap between the waveguides of the directional couplers is 200 nm. The width and height of the silicon waveguides are 500 nm and 220 nm, respectively, with a slab thickness of 90 nm. A 1-mm-long arm of the MZI is equipped with a lateral p-i-n diode acting as a phase shifter through carrier injection. A relatively long shifter is required to achieve a π phase shift with a low bias voltage. The cross-section of the MZI arm with the p-i-n diode is shown in Figure 7.4(c). Free-carriers are injected into the diode arm for high-speed switching. The heavily doped regions are 800 nm away from the waveguide edges to reduce the introduced insertion loss. When free-carriers are injected into the MZI arm with current flowing through the p-i-n diode, the phase difference between the two MZI arms is modified. When the phases of the two MZI arms are equal, the routing path IN1-OUT1 (IN2-OUT2) is established, leading to the bar state. When the phase difference of the two MZI arms is π, the light routing path is switched to IN1-OUT2 (IN2-OUT1) in the cross state. The MZI element exhibits a large optical bandwidth, as shown in Figure 7.5(b), reporting the power (normalized to remove the fiber-to-grating coupler loss) as a function of the input wavelength in the case of a bar configuration.

The measured transmissions of the designed 2×2 switch element for IN1 and IN2 light injections as a function of the applied bias voltage at the wavelength of 1530 nm is shown in Figure 7.6, normalized again to the grating coupler. A π phase shift voltage of $V_\pi = 0.2$ V at a bias of 0.88 V is required to switch between the bar and cross states. The bar and cross state extinction ratios (ERs) are approximately 16 dB and 28 dB, respectively. The currents

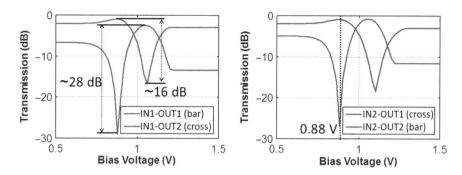

FIGURE 7.6

Normalized transmission of (left) IN1 input and (right) IN2 input as a function of the applied forward bias voltage.

for the applied voltages 0.88 V and 1.08 V are 1 mA (bar) and 14 mA (cross), respectively, which correspond to power consumptions of 0.88 mW and 15.12 mW, respectively. Thus, the switch requires 14.24 mW for a complete π-phase shift switching power. For IN1 light input (Figure 7.6 (a)), the insertion loss of the bar route IN1-OUT1 is approximately 1 dB over the wavelength range from 1500 nm to 1550 nm, while the crosstalk is less than -25 dB for the same wavelength range. For IN2 light input (Figure 7.6(b)), the insertion loss of the bar route IN2-OUT2 increases as the wavelength becomes longer. A possible reason is that the 3-dB directional coupler is wavelength dependent.

7.2.2 High-throughput Photonic Interconnection Topologies

Different switch topologies can be realized by means of 1×2 or 2×2 optical switching elements, as summarized in Figure 7.7 [19]. In bus and ring topologies, each switching element performs an optical add&drop. To achieve all-to-all connectivity, the bus (Figure 7.7(a)) must be operated with bidirectional transmission or two unidirectional buses are required. Ring topology exploiting similar add&drop switching elements (Figure 7.7(b)) offers all-to-all communication even when unidirectional, and avoids waveguide crossings.

The Spanke topology (Figure 7.7(c)) uses $1 \times n$ switches (resorting to a cascaded configuration of 1×2 switching elements) and passive n:1 couplers (realized with multi-mode interference devices or cascaded directional couplers) or active $n \times 1$ switches, providing lower optical loss at the expense of higher control complexity and power dissipation. Multistage topologies can also elegantly exploit 2×2 switching elements; in the literature the most common are Beneš, crossbar, and Spanke-Beneš. Beneš architecture (Figure 7.7(d)) is derived from a Clos switch in which the switching elements are 2×2. Such architecture has $(2\log n - 1)$ stages, each of them composed of $n/2$ switching elements. The number of crossed switching elements is the

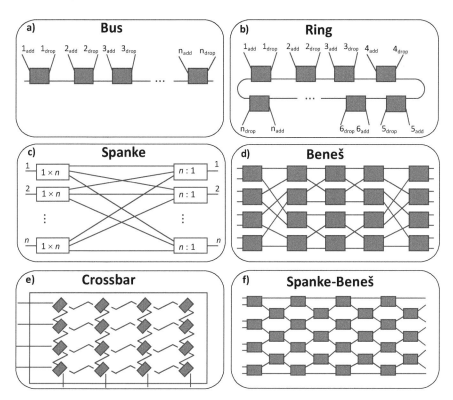

FIGURE 7.7
Optical interconnection network architectures.

same for every path and equal to the number of stages in the architecture, which has an insertion loss almost independent of the chosen input-output pair. In the crossbar architecture (Figure 7.7(e)), inputs and outputs are connected by means of a matrix: to connect input to output, the switching block in position must be set to bar state, whereas the other elements on the row or column are set to cross state. The routing algorithm is trivial, and the absence of interconnection crossovers eases fabrication and integrability. However, the number of switching blocks is n^2, growing quickly with the switch scale, and the insertion loss is highly dependent on the input-output pair (unless specific variations of the architectures are chosen aimed at path-independent loss [20]). Spanke-Beneš architecture (Figure 7.7(f)), also called n-stage planar, is a hybrid between the two previous architectures. Thanks to planarity, it overcomes the issues of wiring because intersections are avoided and, compared to crossbar, requires fewer switching blocks. Indeed, it consists of n stages and a total $n(n-1)/2$ switching elements. It is built by alternating a stage of $(n/2)$ switching blocks with a stage $(n/2-1)$ of switching blocks. The number of traversed elements is, however, path dependent.

7.2.3 Low-latency Dynamic Reconfigurability of Photonic Interconnection Networks

To enable low-latency communication between the ports of the interconnection network, a controller is needed to schedule the data transmissions based on the traffic pattern. The controller decisions can then be dynamically executed by electronically reconfiguring the switching elements. More specifically, the controller's aim is to schedule the transmissions of data packets and correspondingly to control the switching elements, to achieve low-latency, high-network throughput, and good scalability. For such reasons, synchronous data switching of fixed-size packets is typically considered: the electronic controller computes the best schedule and reconfigures the switching elements at fixed intervals of time or time slots.

Electronic control of a photonic interconnection network is challenging due to the different integrated technologies (i.e., electronic and photonic substrate) that need to be co-designed and co-packaged. Furthermore, the switching elements of the silicon photonic interconnection network are normally voltage-controlled with values outside the range of integrated electronics for customized electrical circuits (e.g., requiring tunable voltages instead of the typical voltage settings used in CMOS chips, as well as currents which might require specific drivers). Furthermore, the process variations within the chip require a one-time calibration of all building blocks. As such, the interface between the centralized controller and the switch matrix has to be carefully designed. Concerning packaging issues, the number of electrical pads (and corresponding package pins) required for the control of the switching elements affects scalability as well.

In addition, the optical architectures introduce novel issues in the scheduling problem with respect to the conventional schedulers used for electronic interconnection networks. Indeed, the scheduling problem must account for the behavior of each optical domain. This may limit the available switching configurations, leading to blocking behavior [21–23]. Also, the presence of multiple switching domains requires novel scheduling approaches that can achieve high scalability and delay performance. An example is the two-step scheduling framework proposed for multiplane architectures [22], requiring the solution of a conventional matching problem [24, 25] (i.e., for re-arrangeable non-blocking architectures) in each step.

Several scheduling algorithms have been proposed and studied for the matching problem encountered in the electronic interconnection networks. Optimal schedulers based on maximum weighted matching algorithms [24, 25] are too complex for a fast computation and a resource-efficient implementation in hardware. Sub-optimal algorithms based on maximal matching can trade the computational complexity with scheduling performance. In the class of maximal matching algorithms, longest-queue-first (LQF) is a heuristic approach that achieves good performance. Iterative implementations are advantageous for their ability to trade the optimality for the computational speed. iSLIP [26] is a well-known example of an iterative approach that can find a maximal matching with limited computational complexity and hardware size. Other scheduling algorithms have been proposed, such as the maximal matching with round-robin selection, which outperforms iSLIP at high loads (greater than 70%) [27], and the parallel hierarchical matching scheduler, which outperforms iSLIP for a low number of iterations [28]. Yet, among the various schedulers, iSLIP stands out for its low complexity and good performance [29], whereas the delay improvement of longest-queue-first approaches is still valuable, for instance, in two-step schedulers [22].

To achieve high network performance (in terms of packet latency and throughput) and high hardware performance (in terms of execution time and hardware resource utilization), an iterative and parallel implementation of such schedulers can be considered [23]. The parallel implementation is useful for lowering the execution time, which can be further reduced by stopping the iterations before achieving a complete convergence.

7.3 Case Study: Interconnection Networks in NANO-RODIN Project

This section presents two designed and fabricated photonic interconnection networks, one with the MR as the switching element and the other with the MZI as the switching element. These interconnection networks prototyped within the NANO-RODIN project [30] represent two significant case studies: their performance characterization is discussed, outlining the lesson learned toward best practices.

7.3.1 Microring-Based Topologies

The interconnection network based on MR is arranged in a ring topology (see Figure 7.7(b)), called multi-microring (MMR) [16, 17, 31]. The MMR interconnection network consists of a central MR and a number of smaller add&drop MR (local rings). The local rings are paired: each pair (Ti, Ri) consists of a ring Ti coupling the optical signal from the i-th transmitter to the central ring and of a ring Ri enabling coupling from the central ring to the i-th receiver. The central ring acts as a shared waveguide, allowing communication from any transmitter to any receiver. Two inherent advantages of this interconnection network are the absence of waveguide crossings and the filtering effect of the crossed rings, which improves the performance at the receiver [32].

A 12-ring MMR photonic interconnection network was designed (see the layout in Figure 7.8(a)), fabricated and packaged (see the photo of the fabricated photonic integrated circuit (PIC) in Figure 7.8(b)). Local rings are similar to the one described in Section 7.2.1: the ratio between the free spectral

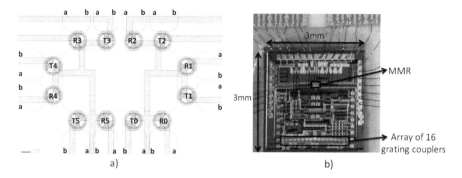

a)   b)

FIGURE 7.8

(a) Layout of MMR interconnection network; (b) Photograph of the fabricated silicon-based MMR interconnection network. (From F. Gambini et al., Experimental demonstration of a 24-port packaged multi-microring network-on-chip in silicon photonic platform, *Opt. Express* vol. 25, pp. 22004–22016, 2017, with permission).

range of the local ring and the free spectral range of the central ring has been chosen as 6, in order to match the local and central ring resonances. Parallel transmissions on multiple wavelengths (i.e., wavelength multiplexing) are possible with limited inter-channel crosstalk. Simultaneous transmissions between different port pairs (i.e., spatial reuse) are achievable on the same wavelengths with limited intra-channel crosstalk, provided that the different propagation paths are disjoint, as discussed in the following.

In this experiment, two data transmissions (with non-overlapping paths) are co-propagated in the central ring at the same wavelength, as shown in Figures 7.9(a)–9(e). The upstream transmission is one hop long. The downstream transmission is adjacent, with a hop length from one to five hops. All the transmissions take place at 1548.4 nm and only the local rings involved in signal add&drop have been tuned at the same wavelength. With this configuration, it is possible to assess the undesired interference caused by the upstream transmission. Figure 7.9(f) shows the bit error rate (BER) as a function of the received power when the hop length of the downstream transmission is increased. The continuous and dashed curves refer to the BER, respectively in the absence and presence of the upstream transmission T5→R0. The back-to-back curves (black) are reported as reference. In absence of the upstream transmission T5→R0 the measured BER even slightly outperforms the back-to-back BER: this behavior is due to the filtering effect of the rings, which act as adapted receivers. The power penalty related to the upstream transmission is limited to 0.5 dB at a BER = 10^{-9}. Figures 7.9(g) and (h) show the eye diagrams respectively of the one-hop and five-hop transmissions in the presence of the upstream transmission T5→R0. Clear and open eyes are obtained.

A limited penalty is also achieved in the case of bidirectional transmissions even when either the same source ring or the same destination ring is shared by simultaneous counter-propagating transmissions [31].

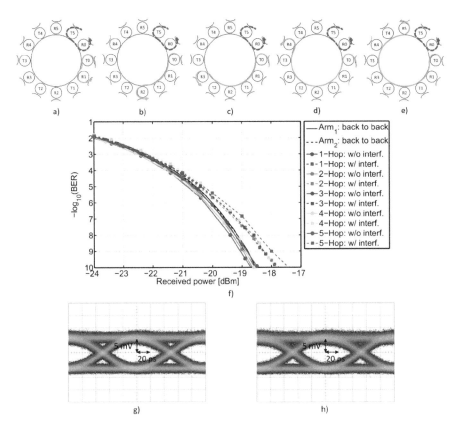

FIGURE 7.9

Two co-propagating transmissions. Transmission paths (blue) of one-hop (a), two-hop (b), three-hop (c), four-hop (d) and five-hop (e) transmission, with the one-hop upstream transmission (red) at the same wavelength. Bit error rate (BER) as a function of the received power in absence (continuous lines) and presence (dashed lines) of the upstream transmission, in the five different transmission patterns: one-hop (red), two-hop (green), three-hop (blue), four-hop (light blue) and five-hop (orange) (f). Eye diagrams for one-hop (g) and five-hop (h) transmissions in the presence of the one-hop upstream transmission. (From F. Gambini et al., Experimental demonstration of a 24-port packaged multi-microring network-on-chip in silicon photonic platform, *Opt. Express* vol. 25, pp. 22004–22016, 2017, with permission).

With an FPGA implementing the scheduler and controller, the switching behavior of the MMR PIC was experimentally tested using the setup depicted in Figure 7.10(a) [33]. An external cavity tunable laser (TL) generates a continuous-wave optical signal with an optical power of 4 dBm at the wavelength of 1548.6 nm. After the polarization controller (PC), the optical signal enters a Mach-Zehnder modulator (MZM) fed by a bit-pattern generator (BPG) at 10 Gb/s with a pseudo-random binary sequence (PRBS) of 2^{31}-1 bits. The modulated signal, amplified by an erbium-doped fiber amplifier (EDFA), filtered by an optical band-pass filter (OBPF), and controlled in polarization, is then injected into the packaged MMR, i.e. the device under

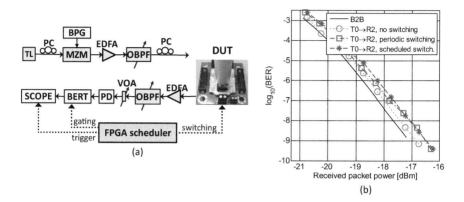

FIGURE 7.10

(a) Measurement setup with packaged MMR NoC and FPGA; (b) BER vs. received power at receiver R2. (From Y. Xiong et al., Demonstration of a packaged photonic integrated network on chip controller by an FPGA-based scheduler, *Optical Fiber Communication Conference and Exhibition (OFC)*, W1A.3, 2017.)

test (DUT). The output signal is amplified and filtered again, attenuated by a variable optical attenuator (VOA), and then detected by a photodiode (PD) for BER measurements and visualization in a sampling oscilloscope (SCOPE). A hardware implementation of the iSLIP algorithm for fixed-size data packets [25, 34] was run in an Altera DE2 FPGA board. At each time slot, the FPGA board schedules the packet transmissions, then controls the switching of the packaged PIC by tuning the individual rings, and triggers the BER measurements.

Figure 7.10(b) reports BER curves of the back-to-back (B2B) and of the transmission to receiver R2 without switching (i.e., no switching) and with switching controlled by the FPGA in either a periodic way (i.e., periodic switching) or a scheduled iSLIP-based way (i.e., scheduled switching). The scheduled switching has slightly higher BER, possibly due to the higher variance of the packet transmissions. However, the different BER curves are within 0.5 dB at BER of 10^{-9}, indicating that the variability falls within the measurement inaccuracy and thus validating the FPGA-controlled switching.

7.3.2 Mach-Zehnder-Based Topologies

Using the MZI-based switch as a building block, different topology configurations can be adopted to interconnect a relatively large number of ports [35, 36]. For a given topology, on-chip and coupling losses determine the number of ports. Port scalability is achieved by either optically recovering the optical loss through optical amplifications and/or regenerating the transmitted signal through an O/E/O repeater. The layout of the MZI structures and the required electrical pad will also have an important impact on the port scalability of a given topology.

In this case study, the realized interconnection network based on MZIs consists in a 4×4 optical switch with five integrated 2×2 MZI elements connected as a reduced Spanke-Beneš topology (Figure 7.11(a)). The eight input-/output-ports are denoted as Ix and Ox where x = 1, 2, 3 and 4. Though this topology leads to a blocking switch fabric, the Spanke-Beneš topology has no waveguide crossings and exhibits lower overall insertion loss. However, it suffers from non-uniform path loss. For example, the path I1 to O1 will have lower insertion loss compared to the path I1 to O4 with the optical signal propagating through three MZI building blocks compared to two. Further, the bar and cross states are likely to exhibit different loss behaviors such that even in a topology where each path may see the same number of MZI building blocks, the insertion loss of a given path will vary.

In the fabricated Spanke-Beneš 4×4 optical switch shown in Figure 7.11(b), light is coupled in and out of the switch chip through grating coupler arrays with a pitch of 127 µm. The electrodes of the MZI switch elements are connected out to electrical pads positioned along the chip bottom and upper edges for more efficient packaging. The size of each electrical pad is 80×80 µm² with a pitch of 150 µm. The footprint of the switch chip is 5.1×2.6 mm². To control the 4×4 optical switch, the electrical pads of the switch chip are wire-bonded to the chip carrier (Figure 7.11(c)). The controlling bias voltages are applied to the MZI switches via the marked wire-bonded pads.

The switching performance of the 4×4 optical switch is assessed using the experimental setup illustrated in Figure 7.12, including an FPGA-based controller which allows for an 8 Gb/s payload generator and error detector. Two continuous wave (CW) laser lights at the wavelength of 1530 nm are modulated by external modulators to generate a continuous optical on-off key (OOK) data stream. A 2^{31}-1 PRBS pattern generated by the FPGA is routed through the 4×4 MZI switch. The optical signals are then first amplified by an EDFA before coupling into input port I1 and I2 of the 4×4 switch. The amplification is required to compensate for the fiber-to-fiber insertion loss of the integrated chip and for the required optical power for proper bit error rate measurements. The light output from the switch chip is amplified by another EDFA injected into a photodetector (PD).

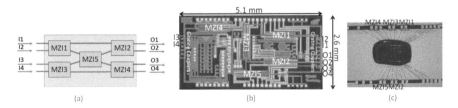

FIGURE 7.11
(a) Schematic of the Spanke-Benes MZI-based topology; (b) micrograph of the SiP fabricated chip; (c) SiP switch with MZI control pads wire-bonded to the carrier package.

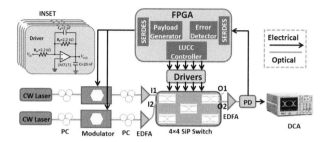

FIGURE 7.12
Schematic of the FPGA-based LUCC controller co-designed with the MZI-based multistage switch. CW: Continuous Wave; PC: Polarization Controller; EDFA: Erbium-Doped Fiber Amplifier; PD: Photodetector; DCA: Digital Communication Analyzer; INSET: interfacing driving circuit between the FPGA and each MZI of the optical switch. (From Y. Xiong et al. Co-design of a low-latency centralized controller for silicon photonic multistage MZI-based switches, *Optical Fiber Communication Conference and Exhibition (OFC)*, Th2A.37, 2017.)

In the scenario reported in Figure 7.13, the FPGA transmits a 2^{31}-1 PRBS signal at 8 Gb/s through the switch, but the FPGA switch controller is disabled, providing fixed switch configurations (depicted in Figure 7.13(a)) by means of static voltage signals, without any dynamic gating. A clear eye diagram is shown in the inset of Figure 7.13(b), demonstrating successful static routing through the optical switch. In Figure 7.13(b), the BER performance is reported for the two static routing paths (I2-O1 and I2-O2) depicted in Figure 7.13(a) as a function of the average received power in the photodetector. For the back-to-back (B2B) BER measurement, the switch is replaced by an attenuator with the same loss corresponding to the insertion loss associated with the respective channels. A negligible power penalty is observed for the tested routes (I2-O1 and I2-O2) compared to the B2B route.

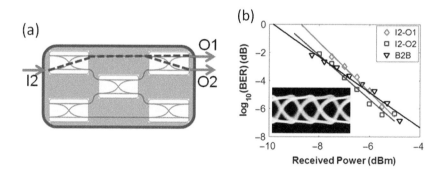

FIGURE 7.13
(a) Tested routing paths of the 4×4 MZI switch for BER measurement; (b) BER as a function of the average received optical power measured for routing path I2-O1, I2-O2, and the back-to-back configuration with respect to routing path I2-O1. (From Y. Xiong et al., Co-design of a low-latency centralized controller for silicon photonic multistage MZI-based switches, *Optical Fiber Communication Conference and Exhibition (OFC)*, Th2A.37, 2017.)

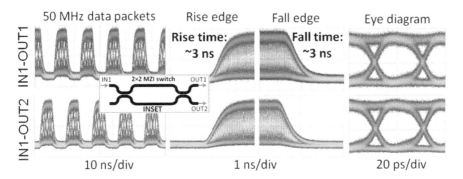

FIGURE 7.14

Switching response of a 10 Gb/s modulated optical signal to a 50 MHz square wave onto 2×2 MZI switch building block. The top row shows the optical signal at output 1 (OUT1) for the bar state and the bottom row for the cross state at output 2 (OUT2). The inset picture shows the schematic of the 2×2 balanced MZI switch. The bar and cross states are IN1-OUT1 and IN1-OUT2, respectively. (From Y. Xiong et al., Co-design of a low-latency centralized controller for silicon photonic multistage MZI-based switches, *Optical Fiber Communication Conference and Exhibition (OFC)*, Th2A.37, 2017.)

To evaluate the high switching speed at which the switch matrix can be configured, a 50 MHz electrical square-wave gating signal is applied to the MZI switch building block. The 10%/90% switching response is approximately 3 ns for both falling and rising edges (Figure 7.14). Difference in the rise and fall times may be attributed to the mobility difference of the electron and holes in the p-i-n diode.

We conclude the demonstration of the optical switch by considering two scenarios: (1) only I1 sends a link request for the destination O2; (2) both I1 and I2 send link requests for the destination O2, simultaneously. For the first scenario, Figure 7.15(a) illustrates that the 8 Gb/s payload is routed from I1 to O2. The received data packet is monitored by the DCA oscilloscope. As there is no conflict and I1 is granted access to its destination (O2), the controller completes its decision in one clock cycle (1 MHz for Altera Stratix IV). For the second scenario, a conflict occurs where the destination node O2 is the same for both requests by the two transmitters (I1 and I2 in Figure 7.15(b)). Figure 7.15(b) illustrates the successful contention resolution of the controller where the payload injected into I1 is given higher priority. The observable difference in optical power is due to the non-uniform optical insertion loss of the Spanke-Beneš topology for different routing paths. For this contention resolution, the controller successfully takes only one clock cycle to resolve the contention and to grant the access to I1 [37]. The switch configuration is further established within that same clock cycle. The data packet from I2 is delayed in the round-robin queue and finally transmitted after I1 completes its transmission to output-port O2.

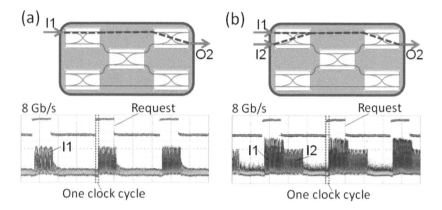

FIGURE 7.15
Validation of the optical 4×4 MZI switch: (a) only input I1 sends a link request for destination O2; (b) both input I1 and I2 simultaneously send link requests to destination O2, leading to a contention resolved by the controller within one clock cycle. (From Y. Xiong et al., Co-design of a low-latency centralized controller for silicon photonic multistage MZI-based switches, *Optical Fiber Communication Conference and Exhibition (OFC)*, Th2A.37, 2017.)

7.4 Conclusion

This chapter presented two case studies showing that low-latency, energy-efficient, automatically controlled interconnection networks can be realized using silicon photonic platforms. The case studies demonstrated that both MR and MZI are viable switching elements that can be used to realize more complex architecture with the desired performance. Both architectures achieved excellent experimental performance in static operation with data transmissions up to 10 Gb/s and in dynamic operations with FPGA-controlled switching.

These results laid down the basis for the exploitation of photonic integrated interconnection networks, going beyond the currently exploited point-to-point interconnects. The remaining challenging tasks concern the photonic-electronic integration and packaging, and the interconnection network control: significant progress is expected in this direction to push the transition from the demonstration stage to the exploitation stage.

Acknowledgments

The authors would like to thank M. Chiesa, S. Faralli, F. Gambini, P. Pintus, G. B. Preve, and Y. Xiong for their contributions to the presented activities.

References

1. S. Chandrasekhar, B. Li, J. Cho, X. Chen, E. C. Burrows, G. Raybon, and P. J. Winzer, High-spectral-efficiency transmission of PDM 256-QAM with parallel probabilistic shaping at record rate-reach trade-offs, *ECOC* 2016.
2. A. Turukhin, O.V. Sinkin, H. G. Batshon, H. Zhang, Y. Sun, M. Mazurczyk, C. R. Davidson, et al., 105.1 Tb/s power-efficient transmission over 14,350 km using a 12-core fiber, *OFC* 2016, paper Th4C.1.
3. F. Karinou, C. Prodaniuc, N. Stojanovic, M. Ortsiefer, A. Daly, R. Hohenleitner, B. Kögel, and C. Neumeyr, Directly PAM-4 modulated 1530-nm VCSEL enabling 56 Gb/s/λ data-center interconnects, *IEEE Photonics Technol. Lett.*, vol. 27, no. 17, pp. 1872–1875, Sept. 1, 2015.
4. D. M. Kuchta, T. N. Huynh, F. E. Doany, L. Schares, C. W. Baks, C. Neumeyr, A. Daly, B. Kögel, J. Rosskopf, and M. Ortsiefer, Error-free 56 Gb/s NRZ modulation of a 1530-nm VCSEL link, *J. Lightwave Technol.* vol. 34, no. 14, pp. 3275–3282, Jul. 15, 2016.
5. C. DeCusatis, Optical interconnect networks for data communications, *J. Lightwave Technol.*, vol. 32, no. 4, pp. 544–552, Feb. 15, 2014.
6. H. Nasu, Short-reach optical interconnects employing high-density parallel-optical modules, *IEEE J. Sel. Top. Quantum Electron.*, vol. 16, no. 5, pp. 1337–1346, Sept.–Oct. 2010.
7. F. Coppinger, D. Langsam, A. Page, and M. Verdiell, The benefit of mid-board optic and other flyover technology, *2017 IEEE Optical Interconnects Conference*, Santa Fe, NM, 2017, pp. 17–18.
8. http://cobo.azurewebsites.net
9. R. C. A. Pitwon, et al., FirstLight: Pluggable optical interconnect technologies for polymeric electro-optical printed circuit boards in data centers, *J. Lightwave Technol.*, vol. 30, no. 21, pp. 3316–3329, Nov. 1, 2012.
10. L. Brusberg, S. Whalley, R. C. A. Pitwon, F. R. Faridi, and H. Schröder, Large optical backplane with embedded graded-index glass waveguides and fiber-flex termination, *J. Lightwave Technol.*, vol. 34, no. 10, pp. 2540–2551, May 15, 2016.
11. B. J. Puttnam, T. A. Eriksson, J.-M. Delgado Mendinueta, R. S. Luís, Y. Awaji, N. Wada, M. Karlsson, and E. Agrell, Modulation formats for multi-core fiber transmission, *Opt. Express*, vol. 22, pp. 32457–32469, 2014.
12. C. Simonneau, A. D'Amato, P. Jian, G. Labroille, J.-F. Morizur, and G. Charlet, 30Gbit/s 3 × 3 optical mode group division multiplexing system with mode-selective spatial filtering, in *Optical Fiber Communication Conference 2016*, paper Tu2J.3.
13. K. Zhong, X. Zhou, T. Gui, L. Tao, Y. Gao, W. Chen, J. Man, et al., Experimental study of PAM-4, CAP-16, and DMT for 100 Gb/s short reach optical transmission systems, *Opt. Express*, vol. 23, pp. 1176–1189, 2015.
14. M. Iglesias Olmedo, T. Zuo, J. B. Jensen, Q. Zhong, X. Xu, S. Popov, and I. Tafur Monroy, Multiband carrierless amplitude phase modulation for high capacity optical data links, *J. Lightwave Technol.*, vol. 32, no. 4, Feb. 15, 2014.
15. P. Pintus, P. Contu, N. Andriolli, A. D'Errico, F. Di Pasquale, and F. Testa, Analysis and design of microring-based switching elements in a silicon photonic integrated transponder aggregator, *J. Lightwave Technol.*, vol. 31, no. 24, Dec. 15, 2013, pp. 3943–3955.

16. P. Pintus, F. Gambini, S. Faralli, F. Di Pasquale, I. Cerutti, and N. Andriolli, Ring versus bus: A theoretical and experimental comparison of photonic integrated NoC, *J. Lightwave Technol.*, vol. 33, no. 23, Dec. 1, 2015, pp. 4870–4877.

17. F. Gambini, S. Faralli, P. Pintus, N. Andriolli, and I. Cerutti, BER evaluation of a low-crosstalk silicon integrated multi-microring network-on-chip, *Opt. Express*, vol. 23, no. 13, Jun. 2015.

18. Y. Xiong, F. Göhring de Magalhaes, B. Radi, G. Nicolescu, F. Hessel, and O. Liboiron-Ladouceur, Towards a fast centralized controller for integrated silicon photonic multistage MZI-based switches, *Optical Fiber Communication Conference and Exhibition (OFC)*, W1J.2, Anaheim (USA), March 19–23, 2016, pp. 17169–17178.

19. P. G. Raponi, N. Andriolli, I. Cerutti, D. Torres, O. Liboiron-Ladouceur, and P. Castoldi, Heterogeneous optical space switches for scalable and energy-efficient data centers, *J. Lightwave Technol.*, vol. 31, no. 11, Jun. 1, 2013, pp. 1713–1719.

20. T. Goh, A. Himeno, M. Okuno, H. Takahashi, and K. Hattori, High-extinction ratio and low-loss silica-based 8×8 strictly nonblocking thermooptic matrix switch, *J. Lightwave Technol.*, vol. 17, no. 7, Jul. 1999, pp. 1192–1199.

21. C. Develder, J. Cheyns, E. Van Breusegem, E. Baert, D. Colle, M. Pickavet, and P. Demeester, Architectures for optical packet and burst switches, in *European Conf. on Optical Communication*, vol. 4, 2003, pp. 100–103.

22. P. G. Raponi, N. Andriolli, I. Cerutti, and P. Castoldi, Two-step scheduling framework for space-wavelength modular optical interconnection networks, *IET Commun.*, vol. 4, no. 18, pp. 2155–2165, 2010.

23. I. Cerutti, N. Andriolli, P. Pintus, S. Faralli, F. Gambini, P. Castoldi, and O. Liboiron-Ladouceur, Fast scheduling based on iterative parallel wavelength matching for a multi-wavelength ring network-on-chip, in *Int. Conf. on Optical Networks Design and Modeling (ONDM)*, 2015, pp. 180–185.

24. R. K. Ahuja, T. L. Magnanti, and J. B. Orlin, *Network Flows: Theory, Algorithms, and Applications*. Prentice Hall, 1993.

25. A. L. Stolyar, Max weight scheduling in a generalized switch: State space collapse and workload minimization in heavy traffic, *Annals of Applied Probability*, vol. 14, no. 1, pp. 1–53, 2004.

26. N. W. McKeown, The iSLIP scheduling algorithm for input-queued switches, *IEEE/ACM Trans. on Netw.*, vol. 7, no. 2, pp. 188–201, Apr. 1999.

27. W. Kabaciński, A. Baranowska, and L. Rubik, FPGA implementation of the MMRRS scheduling algorithm for VOQ switches, in *Int. Telecommunications Network Strategy and Planning Symp.* (NETWORKS), 2010, pp. 1–6.

28. F. J. González-Castaño, C. López-Bravo, M. Rodelgo-Lacruz, and R. Asorey-Cacheda, Decoupled parallel hierarchical matching schedulers, *Int. J. Commun. Syst.*, vol. 20, no. 3, pp. 365–384, 2007.

29. H. Bilbeisi and L. Mason, Time-slotted scheduling for agile all-photonics networks: Performance and complexity, in *Int. Symp. on Performance Evaluation of Computer and Telecommunication Systems (SPECTS)*, 2008, pp. 249–255.

30. http://tecip.sssup.it/Projects/RODIN/

31. F. Gambini, P. Pintus, S. Faralli, M. Chiesa, G. B. Preve, I. Cerutti, and N. Andriolli, Experimental demonstration of a 24-port packaged multi-microring network-on-chip in silicon photonic platform, *Opt. Express*, vol. 25, pp. 22004–22016, 2017.

32. A. Parini, G. Bellanca, A. Annoni, F. Morichetti, A. Melloni, M. J. Strain, M. Sorel, et al., BER evaluation of a passive SOI WDM router, *IEEE Photon. Technol. Lett.*, vol. 25, no. 23, pp. 2285–2288, 2013.

33. Y. Xiong, N. Andriolli, S. Faralli, F. Gambini, P. Pintus, M. Chiesa, R. Ortuno, and O. Liboiron-Ladouceur, Demonstration of a packaged photonic integrated network on chip controller by an FPGA-based scheduler, *Optical Fiber Communication Conference and Exhibition (OFC)*, W1A.3, Los Angeles (USA), March 19–23, 2017.

34. J. A. Corvera, S. M. G. Dumlao, R. S. Reyes, P. Castoldi, N. Andriolli, and I. Cerutti, Hardware implementation of an iterative parallel scheduler for optical interconnection networks, Photonic Networks and Devices, in *Proc. Advanced Photonics Conf.*, 2016, pp. NeM3B-4.

35. B. G. Lee, et al., Silicon photonic switch fabrics in computer communications systems, *J. Lightwave Technol.*, vol. 33, no. 4, pp. 768–777, Feb. 15, 2015.

36. K. Suzuki et al., Ultra-compact 8×8 strictly-non-blocking Si-wire PILOSS switch, *OSA Optics Express*, vol. 22, no. 4, pp. 3887–3894, Feb. 2014.

37. Y. Xiong, F. G. de Magalhaes, G. Nicolescu, and O. Liboiron-Ladouceur, Co-design of a low-latency centralized controller for silicon photonic multi-stage MZI-based switches, *Optical Fiber Communication Conference and Exhibition (OFC)*, Th2A.37, Los Angeles (USA), March 19–23, 2017.

8

Ultra-Low Power SiGe Driver-IC for High-Speed InP Mach-Zehnder Modulator

Jung Han Choi

CONTENTS

8.1 Introduction

Recently there has been an increasing demand for high-performance optical transceivers using InP-based Mach-Zehnder modulators (MZMs) for reduced cost, lower power consumption, and greater flexibility to support more advanced modulation formats [1]. In addition, there are stringent industry trends toward modules with small form-factor, highly integrated opto-electrical submounts with control electronics, and higher-speed operation. Figure 8.1 displays the evolution of optical transceiver modules [2]. As the footprint of various modules gets smaller, they support higher speed than previous transceiver generations. Specifically, Figure 8.2 illustrates the power requirements of each optical module according to standardization organizations [3, 4]. It is obvious that the evolution of optical transceivers needs higher integration of optics and electronics with smaller form-factor and lower power consumption.

Currently, some standards are under discussion for future optical transceiver formats [5]. Among them, two optical module standards seem attractive ones for future data center applications [6]: OSFP (Octal Small Form-factor Pluggable module) [7] and QSFP-DD (Quad Small Form-factor Pluggable Double Density) [4]. OSFP's first proposal is to use 8-lanes with 50 Gb/s PAM-4 resulting in 400 Gb/s physical links. It possibly enables 4-lanes×100 Gb/s PAM-4=400 Gb/s and 8-lanes×100 Gb/s PAM-4=800 Gb/s in the future. One feature of this system is that it is agnostic with respect to different I/O protocols and it allows 12.8 Tb/s per switch slot (=32×400 Gb/s) [6]. In contrast to the OSFP, the QSFP-DD form-factor allows 14.4 Tb/s aggregate bandwidth per single switch slot since QSFP-DD is slightly narrower than OSFP [6]. Even though the market still requires mainly 100 Gb/s transceivers, it is certain that future optical transceivers will take form-factors like QSFP-DD and/or OSFP for future data center applications.

The important message from these discussions is that future optical and electrical components in the transceivers should fulfill lower-power and higher-speed operation with a smaller footprint. One approach, which the Fraunhofer Heinrich-Hertz Institute (hereafter Fraunhofer HHI) has used, is the co-design of optics and the electronic driver for the transmitter. Briefly speaking, the optics and the driver-IC are designed together in terms of impedance for the low-power transmitter. Two different output configurations at the driver-IC are considered to achieve lower power and good signal

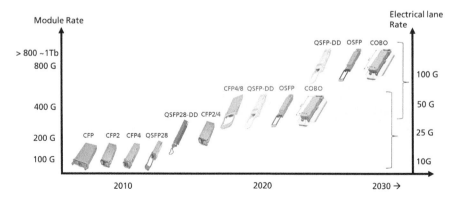

FIGURE 8.1
Evolution of optical transceiver modules [2].

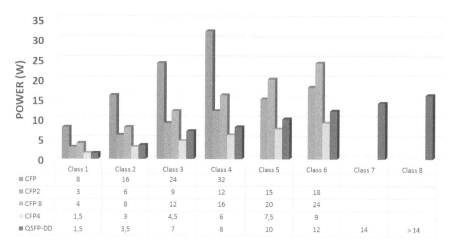

FIGURE 8.2
Power comparison of various optical modules.

integrity: open-collector and back-terminated ICs. Their quantitative comparisons will be addressed, and details about the co-design concept and examples will be presented in the following sections.

8.2 Driver-IC Design for the Mach-Zehnder Modulator (MZM)

8.2.1 Operation Principle of MZM

The Fraunhofer HHI InP Mach-Zehnder modulator (MZM) uses a periodically capacitively loaded traveling-wave electrode (TWE) structure [8, 9] (see Figure 8.3a). By having a periodic capacitive loading in the TWE structure, the velocity of the electrical wave can be matched to that of the optical wave in the semiconductor. The capacitive loading is done by reverse-biasing of the *pin* diode and its electrical equivalent circuit is illustrated in Figure 8.3b. The *pin* structure includes a multi-quantum well in its intrinsic region. Both MZM arms share the common n-InP layer and by serial push-pull operation one can achieve chirp-free output [8]. Several design parameters are actually affecting the electro-optical performance of the MZM. These parameters and their effects on the device performances and trade-offs are summarized in [8, 9] and [10]. Lower MZM impedance actually increases the required driver-IC power in order to get the necessary output voltage for driving the MZM. Lower characteristic impedance of the MZM allows the reduction of V_π without compromising other modulator specifications, e.g. bandwidth and insertion loss [9]. An optimized low- and non-50 Ω impedance for the modulator plays a significant role in achieving the lowest power and this fact is actually the core concept of the co-design approach.

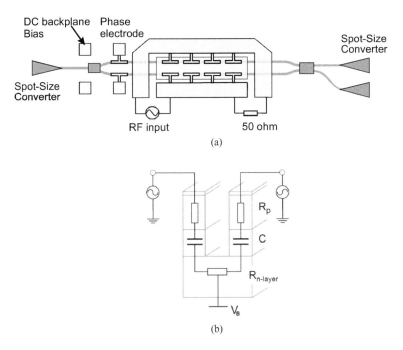

FIGURE 8.3
InP Mach-Zehnder modulator structure [8, 9].

Figure 8.4a displays the relation between the driving voltage and the MZM impedance for different bandwidths. It is obvious that a lower impedance in the MZM leads to a lower driving voltage for the same bandwidth performance. The corresponding power consumption is proportional to $\sim V_\pi^2$. However, Figure 8.4b proves that a higher impedance of the MZM results in lower power consumption for the same drive voltage. It can be noted that the power reduction is proportional to $\sim 1/Z_{MZM}$, where Z_{MZM} means the characteristic impedance of the MZM. It is necessary to make a trade-off between the power consumption and the bandwidth of the modulator. The optimized impedance of the MZM therefore results in the reduction of the power consumption of the driver-IC.

This impedance engineering methodology is called a *co-design* technique between the driver-IC and the MZM. It should be noted that the impedance matching will be accomplished following the result of impedance engineering. The performance of the HHI driver-ICs are compared with other ICs reported with regard to power and speed. They are illustrated in Figure 8.5. Obviously, a very low-power driver-IC for a high-speed InP MZM operating up to 56 GBd could be achieved. In the following sections, details about the low-power IC design in terms of architecture, packaging, and system performances will be discussed.

Table 8.1 displays examples of the MZM and driver-IC schemes including termination ICs. The possible modulation formats for each configuration are

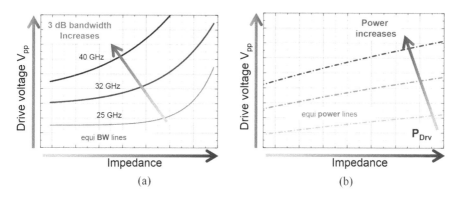

FIGURE 8.4
The relations between drive-voltage and the MZM impedance for (a) equal 3 dB bandwidth and (b) equal power consumption [9, 10].

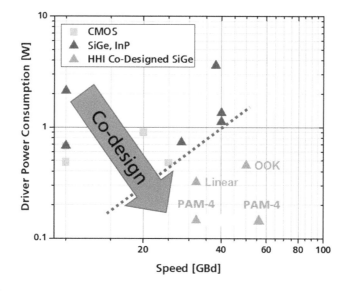

FIGURE 8.5
Power consumption of the MZM driver-IC developed at the Fraunhofer HHI compared to others [11, 12].

also provided. Depending on the MZM input and output electrode configurations, the position of the driver-IC and the termination IC shall be determined correspondingly if there is no need for an RF interposer between the MZM and the driver-IC.

For single MZM using a driver-IC and a termination IC (the first column in Table 8.1), either the amplitude or the phase modulation is feasible. For example, on-off keying (OOK), pulse amplitude modulation (PAM), and phase shift keying (PSK). For both IQ MZM and distributed-feedback (DFB)

TABLE 8.1

Examples of modulator and driver-IC schemes and corresponding modulation formats

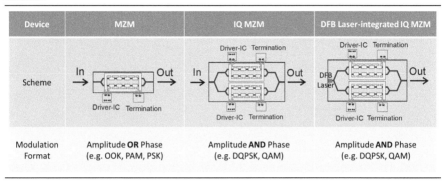

Device	MZM	IQ MZM	DFB Laser-integrated IQ MZM
Scheme			
Modulation Format	Amplitude **OR** Phase (e.g. OOK, PAM, PSK)	Amplitude **AND** Phase (e.g. DQPSK, QAM)	Amplitude **AND** Phase (e.g. DQPSK, QAM)

laser-integrated IQ MZM chip (the 2nd and 3rd columns in Table 8.1), the amplitude and the simultaneous phase modulation are possible, such as differential quadrature phase shift keying (DQPSK) and quadrature amplitude modulation (QAM). The MZM and driver-IC configurations are dependent on the modulator electrodes.

8.2.2 Driver-IC Architecture

8.2.2.1 Terminations of the Driver-IC

In Figure 8.6, two possible termination configurations for the driver-IC are illustrated in combination with the MZM. In the back-termination architecture (Figure 8.6a) reflected signals due to impedance mismatch are absorbed by back-termination. On the other hand, if the output collector is directly connected to the modulator pads and the termination components are located at the end of the modulator output, then it is called open-collector architecture (Figure 8.6b). The RF-broadband coils shown in Figure 8.6 decrease the supply voltage to the driver-IC resulting in the lower power consumption. However, these can be omitted if the DC bias is supplied at the center-tap of the termination resistance at the MZM end. For example, the termination IC can be divided into two 25 Ω broadband resistances and the center-tap can be located at the center of these two resistances, so that the supply voltage is common to the two resistances without RF-coil. It is noted that the Fraunhofer HHI InP MZM has the characteristic impedance of differential 50 Ω. In the case of the center-tap biasing, the power consumption increases due to the higher supply voltage compared to the case of RF-broadband coils.

Concerning signal integrity, the back-termination driver-IC exhibits better performance since this architecture is robust to impedance mismatch between the driver-IC and the MZM due to the absorption of back-reflected signals. In fact, if Au wire-bondings are used for the interconnection, the

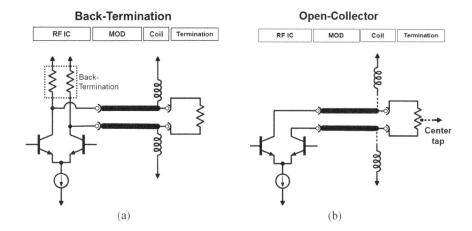

FIGURE 8.6
Architectures of driver-IC output stage for the MZM: (a) back-termination and (b) open-collector.

impedance mismatch cannot be neglected when the driver-IC operates at high speed, i.e. > 32 GBd. Compared to the wavelength of the fundamental frequency for 32 GBd NRZ signal, the electrical length of these wire-bondings can be made short enough to be considered as a simple inductor. They are in the range 100~250 pH if Au double-wires with a diameter of 20~25 μm are used. The inductance is dependent on the distance between the pad at the IC and the modulator. About 10 Ω in magnitude can be calculated when 100 pH wire inductance is calculated at 16 GHz (fundamental frequency of 32 GBd NRZ signal). Even when these wire-bonding inductances are present, the back-termination driver-IC can absorb the reflected signal, resulting in better signal integrity compared to the open-collector driver-IC architecture.

In contrast to the back-terminated IC, the power supply for the output stage in the open-collector (open-drain in CMOS transistor) IC is supplied through the termination resistance at the modulator end. Therefore, one disadvantage for this open-collector driver-IC is that without DC biasing at the collector (for example, using a bias-Tee), one cannot test the open-collector IC as an off-the-shelf-component. To avoid this problem and to measure the exact output voltage from the driver-IC, an impedance conversion-IC, 25 → 50 Ω (single-ended), has been developed at the Fraunhofer HHI. Details can be found in [14].

In Table 8.2, the power efficiencies for two different driver-IC architectures having different interconnection methods with the MZM are compared [15]. If the open-collector driver-IC is biased through the RF-broadband coil and AC-coupled to the MZM, the maximum efficiency will be regarded as 100% as a reference. On the other hand, the open-collector architecture will have only up to 50% efficiency if it is DC-coupled to the MZM and a DC voltage to the output stage is provided through the MZM termination. (see the second row with green shading in Table 8.2)

TABLE 8.2

Comparison of driver-IC power efficiency for different architectures and electrical connections to the MZM

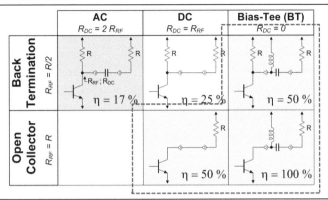

The same efficiency can be expected for the back-termination architecture when it uses the RF-broadband coil and capacitor for bias-Tee. It should be noted that the cost for the module package, footprint, and complexity of the module will be increased considerably for this configuration. The electrical interconnection solutions within the dashed line in Table 8.2 enable particularly low-power operation, and open-collector topology is believed to be the most attractive one among them. The efficiencies given are apparently the theoretical or maximum achievable efficiencies. In reality they are probably somewhat lower.

8.2.2.2 Electro-optic (EO) Performance Comparison of Driver-IC Architectures

For a comparison of the electro-optic (EO) performances of the back-terminated and open-collector driver-ICs, chip-on-carriers (CoCs) must be fabricated. The results of the comparison are illustrated and summarized in Figure 8.7. Obviously, the power consumption is about 30% smaller for the open-collector driver-IC configuration compared to the back-terminated driver-IC. The RMS error vector magnitude (EVM) for QPSK modulation using IQ MZM at the same output voltage is better for the back-termination IC than for the open-collector IC. This result is expected, since the back-termination IC includes the output termination resistance at the output stage of the driver-IC, so that it can absorb the reflected signals due to impedance mismatch.

8.2.2.3 Comparison between Single-Ended and Differential Driving

In Table 8.3, two electrical driving schemes, namely, single-ended and differential, are compared for an IQ MZM [13, 15]. When the IQ MZM is driven by single-ended electrical signals, there can be crosstalk between the I- and

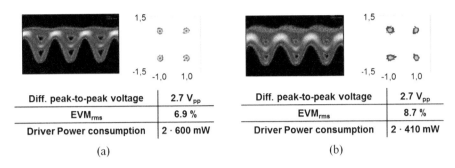

Diff. peak-to-peak voltage	2.7 V$_{pp}$
EVM$_{rms}$	6.9 %
Driver Power consumption	2 · 600 mW

(a)

Diff. peak-to-peak voltage	2.7 V$_{pp}$
EVM$_{rms}$	8.7 %
Driver Power consumption	2 · 410 mW

(b)

FIGURE 8.7
Comparison of EO performance: (a) back-Termination and (b) open-collector architectures [15].
Source: © [2015] IEEE. Reprinted with permission from[15].

Q-modulator and frequency chirp can happen as well. However, differential signals do not induce crosstalk and frequency chirp at the IQ MZM. As a consequence, EO eye diagrams for 32 GBd and QPSK using the differential signals will be better than those for the single-ended case. Concerning driver electronics, the single-ended output circuit requires a higher collector-emitter breakdown voltage (BV$_{CE}$, BV$_{DS}$ for CMOS) than the differential output solution, therefore III-V devices are preferable to Si devices, since III-V electronics do in general show higher breakdown voltages than their Si counterparts. For the differential schemes, Si devices like SiGe HBTs are suitable for the generation of the driving voltage required for the MZM. Finally, it should also be mentioned that differential signals have an advantage over single-ended signals with regard to signal integrity.

TABLE 8.3

Comparison of different electrical driving schemes for IQ MZMs [13]

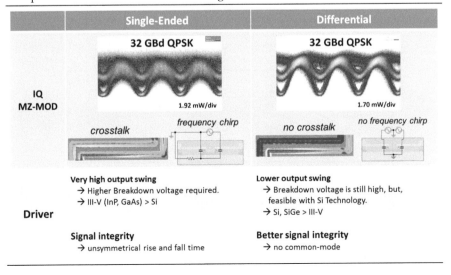

	Single-Ended	Differential
IQ MZ-MOD	32 GBd QPSK 1.92 mW/div crosstalk · frequency chirp	32 GBd QPSK 1.70 mW/div no crosstalk · no frequency chirp
Driver	Very high output swing → Higher Breakdown voltage required. → III-V (InP, GaAs) > Si Signal integrity → unsymmetrical rise and fall time	Lower output swing → Breakdown voltage is still high, but, feasible with Si Technology. → Si, SiGe > III-V Better signal integrity → no common-mode

8.3 Design Examples

8.3.1 Back-Terminated Driver-IC

8.3.1.1 Design

Figure 8.8a displays the output stage of a back-terminated driver-IC. It includes two 25 Ω termination resistances to absorb the reflected signal from the MZM. In order to reduce the Miller capacitance at the amplification transistor, cross-capacitances are added, which decreases the input capacitance at the base of the transistor, and thus enhances the bandwidth of the post amplifier. Functional block diagrams are illustrated in Figure 8.8b. The driver-IC is composed of two amplification stages including two buffer stages. The power supplies for each stage are provided independently and the current source for each stage is controlled in an analog way. In addition to single-channel driver-ICs we also developed a twin-channel driver-IC. Photos for single- and twin-channel are shown in Figure 8.9. The sizes of the single- and twin-channels are 1.05×0.9 mm², and 1.05×1.35 mm², respectively.

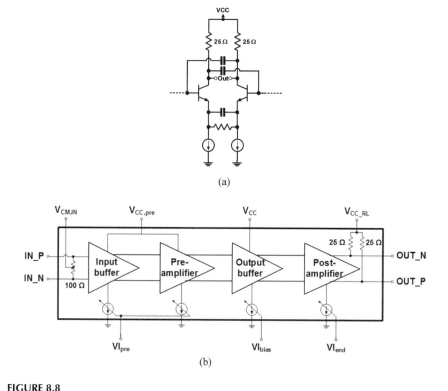

FIGURE 8.8
(a) Output stage and (b) functional block diagram for back-terminated MZM driver-IC.

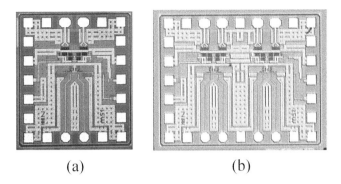

(a)   (b)

FIGURE 8.9
Photographs of (a) single- and (b) twin-channel driver-ICs.

8.3.1.2 Electrical Characterization

Since the output stage of the driver-IC is terminated with 25 Ω, measurements using standard 50 Ω instruments lead to incorrect results (Table 8.4). As discussed in Section 8.2.2.1, an impedance conversion-IC is exploited to make an electrical submount. Since the impedance conversion provides output impedance matching of the back-terminated IC, i.e. 25 Ω, the outputs from the driver-IC do not suffer from reflections. The output impedance of the impedance conversion-IC is also matched to the instrument impedance, so that back-reflections from the instrument are suppressed. However, it should be noted that the impedance conversion-IC is a passive network and exhibits −6.5 dB differential insertion loss. The summary of back-terminated driver-IC is given in Table 8.4.

S-Parameter results are shown in Figure 8.10. Differential S-Parameters were measured when $P_{in,diff}$= −1 dBm. S-Parameter results do not include the impedance conversion-IC, and for the S-Parameter measurements of the driver-IC it is directly in contact with the RF GSSG (ground-signal-signal-ground) probe. Since the output differential impedance is 50 Ω (single-ended 25 Ω), the output impedance of the network analyzer is configured by using the control software to have output differential 50 Ω and common-mode 12.5 Ω.

Figure 8.11 illustrates the de-embedded electrical eye at 28 Gb/s. Since the impedance conversion-IC needs to be connected with the driver-IC for the time-domain eye, a de-embedding process has been carried out in order to remove the frequency response of the impedance conversion-IC. The rise and fall time are measured to be 12.2 ps and 12.4 ps, respectively, and RMS and peak-peak jitter are 828 fs and 4.6 ps, respectively.

8.3.1.3 Electro-Optical Measurements of CoC and Transmission Experiments

Figure 8.12 provides a CoC illustration with two different biasing solutions. One option is supplying the bias directly to the center-tap of the termination

TABLE 8.4

Summary of back-terminated driver-IC measurement

Parameter	Min	Typ	Max	Unit	Conditions
Bandwidth		28		GHz	$P_{in,diff}=1$ dBm
Power		620		mW	without coil, with coil: 490 mW
Data rate			32	Gb/s	
Rise/fall time		12.5		ps	20%–80%
Gain*		16.0		dB	Differential S_{21} $Z_{in,diff}=100\ \Omega$, $Z_{Load,diff}=50\ \Omega$
Group delay distortion*			±8	ps	
Jitter (rms)		828		fs	
Jitter (p-p)		4.6		ps	
Differential input signal		700		mV$_{PP}$	AC-coupled
Differential output signal		3100		mV$_{PP}$	$2\times25\ \Omega$ load
P_{1dB}		13.4		dBm	output-referred, $Z_{Load,diff}=50\ \Omega$
THD		6		%	1GHz, 3 V$_{PP}$ output conditions
CMRR*	18.6			dB	up to 20 GHz
Input reflection*	DC<f <8 GHz 8 GHz <f <24GHz 24GHz <f <BW		−24 −13.6 −13	dB	Differential input
Output reflection*	DC<f <8 GHz 8 GHz <f <24GHz 24 GHz <f <BW		−9 −3 −3	dB	Differential output
Operation temperature		40		°C	

* denotes that measurements were carried out at room temperature condition, 23°C. Unless noted, measurement temperature was 40°C.

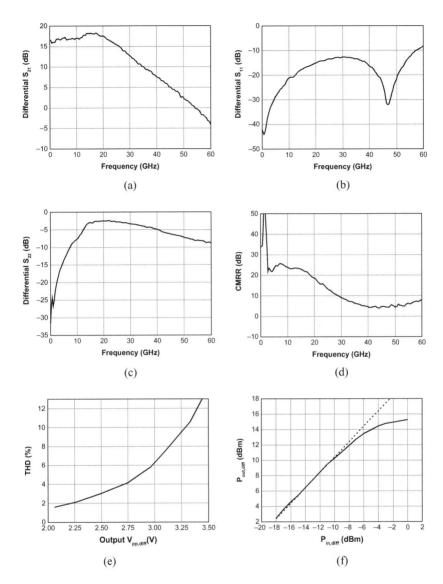

FIGURE 8.10
S-Parameter measurement results: (a) S_{dd21}, (b) S_{dd11}, and (c) S_{dd22} for $P_{in,diff} = -1$ dBm (23 °C), $Z_{in,diff} = 100\ \Omega$, $Z_{Load,diff} = 50\ \Omega$; (d) Common-mode rejection ratio (CMRR) at 23 °C; (e) total-harmonic distortion (THD) at 1 GHz, 40 °C; (f) 1-dB compression at 1 GHz (40 °C).

resistance, which should be made by two 25 Ω resistances. The second variant uses an RF-broadband coil. Both (center-tap and RF-broadband coil) enable the saving of DC power of the driver-IC. Figure 8.13 presents the photograph of a CoC using a back-terminated, twin-driver-IC. The active device shown in Figure 8.13 is a DFB laser-integrated IQ MZM. Details are discussed in [16].

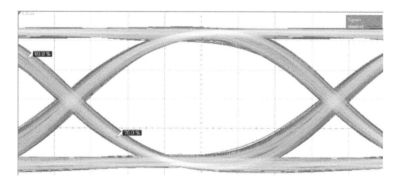

FIGURE 8.11
De-embedded electrical eye waveform at 28 Gb/s (5 ps/div, 700 mV/div, 40 °C). Eye includes wire-bond inductance at differential outputs.

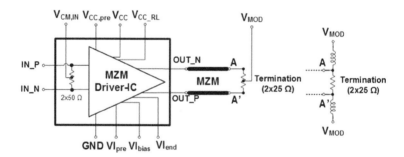

FIGURE 8.12
Chip-on-carrier diagram. Two different biasing concepts using either the center-tapped (left) or RF-broadband coil (right) are presented.

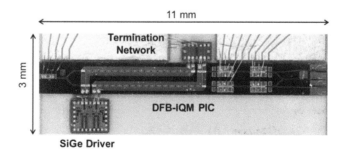

FIGURE 8.13
Photograph of optical transmitter with a distributed-feedback laser-integrated IQ MZM and SiGe-based twin differential driver-IC [16].
Source: © [2015] IEEE. Reprinted with permission from [15].

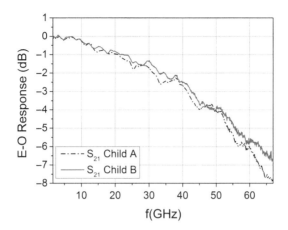

FIGURE 8.14
Small-signal EO responses of the two nested child modulators [16].
Source: © [2015] IEEE. Reprinted with permission from [15].

The small-signal EO 3-dB bandwidths of the DFB-IQM PIC are 41 GHz and 43 GHz, respectively, as shown in Figure 8.14.

Transmission experiments have been carried out up to 2.5 km using standard single-mode fiber (SSFM) at 28 GBd. Results are given in Figure 8.15 for back-to-back (B2B) and 2.5 km. EVM values of 10.4% and 12% were obtained for B2B and 2.5 km, respectively. Corresponding QPSK optical eyes were also measured and are given at the right-hand side of each EVM result. The measurements were done with 100 mA DFB DC current, 2^{31}-1 pseudo-random binary sequence (PRBS) lengths at 20 °C temperature. The power consumption of the twin driver-IC was 0.96 W for a differential output voltage swing of 3 V_{pp}. Bit error rate (BER) smaller than the hard-decision forward error correction (HD-FEC) limit, 3.8×10^{-8}, was measured at an OSNR larger than 14 at 32 GBd. An SSMF transmission up to 9 km was achieved before passing the HD-FEC threshold. At longer transmission distances, the performance was limited by pulse broadening caused by fiber dispersion [16, 17].

8.3.2 Open-Collector Driver-IC

8.3.2.1 Design

Figure 8.16a displays the output stage of the open-collector driver-IC. Similar to the design shown in Figure 8.8, cross-capacitances are used for reducing the Miller capacitance, so that higher bandwidth can be obtained. The collector of the output stage is not loaded with any lumped element, but open-state. It is why this topology is called the open-collector configuration (open-drain for CMOS). The output transmission line having the characteristic impedance of 25 Ω is designed and laid out. Since the Fraunhofer HHI InP MZM has the characteristic impedance of single-ended 25 Ω, the output transmission has the same impedance as the modulator. It should be noted

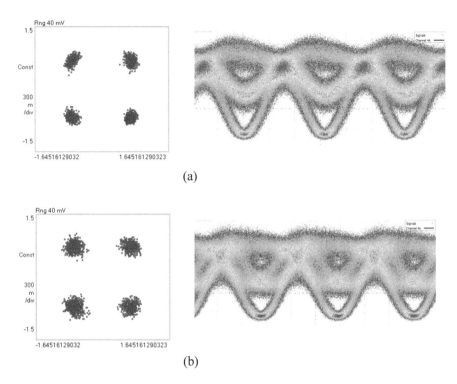

FIGURE 8.15
QPSK constellations at 28 GBd for (a) B2B, EVM = 10.4%, and optical eye (right), (b) after 2 km
SSMF, EVM = 12%, and optical eye (right) [16].

that in the CoC the double wire-bondings are used for the interconnection,
which leads to the inductance of about 150 pH. This packaging inductance
is electro-magnetically simulated and is included in the IC design simulation test-bench. The bonding inductance can induce the inductive peaking,
hence possibly increases the bandwidth. However, one should pay attention
to the stability factor of the driver-IC including the inductance, avoiding the
unwanted oscillation.

A functional block diagram of the open-collector output stage is presented
in Figure 8.16b. The lowest power consumption is achieved with one input
buffer for 50 Ω matching (differential 100 Ω) and a single amplification stage.
The output stage includes peak detectors for monitoring the output voltage
from the circuit. Corresponding results are provided in Section 8.3.2.2.

8.3.2.2 *Electrical Characterization*

Photographs of the fabricated driver-ICs are shown in Figure 8.17a,b. The
two ICs have different output directions, designated as north and south. The
concept for the CoC is also shown in Figure 8.17c. The advantage of this CoC

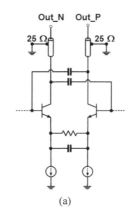

(a)

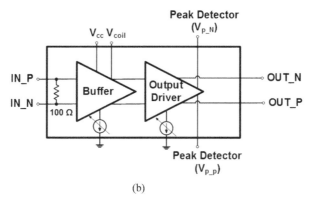

(b)

FIGURE 8.16
(a) Open-collector output stage and (b) its functional block diagram.

is that RF inputs to the driver-IC come from the same direction, which simplifies the RF interposer and its packaging. The driver-IC is measured using a differential GSSG RF-probe. It was measured without impedance conversion-IC for S-Parameters, since the network analyzer (Rohde&Schwarz ZVA) supports non-standard differential impedance, such as differential 50 Ω, and common- mode 12.5 Ω, in its control software. For the time-domain measurement, however, using the impedance conversion-IC is essential.

Figure 8.18 illustrates characteristics of open-collector driver-IC including differential S-Parameter results and nonlinearity measurements, e.g. total-harmonic distortion (THD), P_{1dB}, as well as peak-detector results. Differential 3 dB bandwidth was 30 GHz and S_{21} gain was 17.5 dB for $Z_{Load,diff}$=50 Ω. Whole measurement results are summarized in Table 8.5. For the P_{1dB} and THD measurements, an electrical sub-mount, which combines the driver-IC and the impedance conversion-IC, was used. For the peak detector, measured output power is converted into differential output voltage using $P_{out,diff} = 10 \cdot \log (V_{pp,diff} 2/(8 \cdot Z_{Load,diff} 1mW))$.

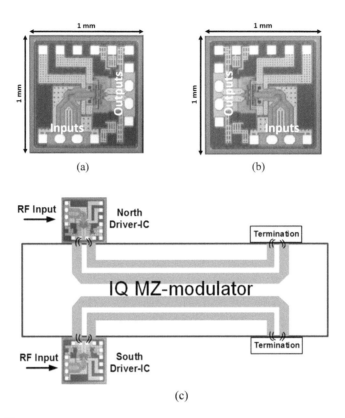

FIGURE 8.17
90°-bended open-collector driver-ICs with different output directions: (a) north and (b) south driver-IC; (c) CoC concept diagram.

Figure 8.19 illustrates time-domain measurement results. As discussed in Section 8.2.2.1, time-domain measurements can be performed using an impedance conversion-IC as shown in Figure 8.19a. DC supply to the open-collector IC has been provided using the center-tap V_{tap} and shall be adjusted considering the voltage drop through the 25 Ω resistance at the impedance conversion-IC. The electrical eye at 32 GBd, 40 °C is given in Figure 8.19, after carrying out the de-embedding of the impedance conversion-IC. It should be noted that the eye includes wire-bond inductance at differential outputs.

8.3.2.3 Electro-Optical Measurements of CoC

Figure 8.20a shows the CoC using an InP IQ MZM with two termination InP ICs at the MZM ends and two open-collector driver-ICs. It is measured using two on-wafer differential GSSG RF probes to drive the ICs. The phases of differential RF signals from the signal generators are tuned to have the same phase for differential inputs to the driver-IC. For this purpose, the variable RF delays are connected to the generator. The EO measurement setup is

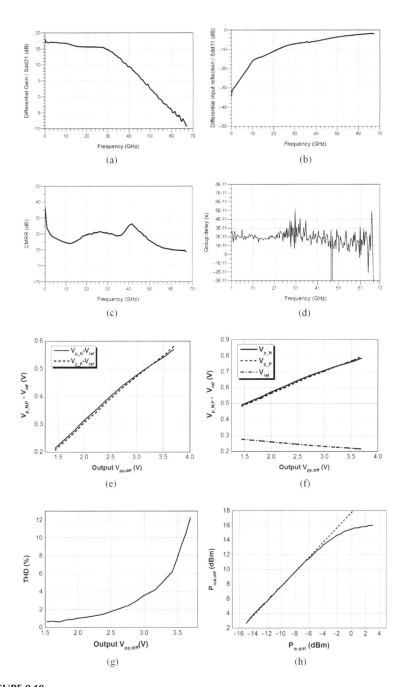

FIGURE 8.18

S-Parameter measurement results: (a) S_{dd21}, (b) S_{dd11} for $P_{in,diff}= -5$ dBm (23 °C), $Z_{in,diff}=100 \Omega$, $Z_{Load,diff}=50 \Omega$, (c) CMRR and (d) group delay at 23 °C; (e) and (f) output peak-level detectors (40 °C, sinusoidal input); (g) THD at 1 GHz, 40 °C; (f) 1-dB compression at 1 GHz (40 °C).

TABLE 8.5

Summary of measurement results for open-collector driver-IC

Parameter	Symbol	Min	Typ	Max	Unit	Conditions
Bandwidth	BW		30		GHz	Differential S_{21}
Power	P		270		mW	
Data rate	DR		32		Gb/s	
Rise/fall time	t_r/t_f		10		ps	20%–80%
Gain*			17.5		dB	Differential $Z_{in,diff}=100\ \Omega$, $Z_{Load,diff}=50\ \Omega$
Group delay distortion*	GD			±5	ps	
Jitter (rms)			523		fs	
Jitter (p-p)			3.47		ps	
Differential input signal	$V_{IN,P} - V_{IN,N}$		600		mV_{pp}	AC-coupled
Differential output signal	$V_{OUT,P} - V_{OUT,N}$		3000		mV_{pp}	$2 \times 25\ \Omega$ load
P_{1dB}	P_{1dB}	13.6		14.4	dBm	output-referred, $Z_{Load,diff}=50\ \Omega$
THD	THD		3.7		%	1GHz, 3 V_{pp} output conditions
CMRR*	CMRR		14		dB	up to 20 GHz
Input reflection*	Sdd11	DC<f<8 GHz		−19	dB	Differential input
		8 GHz < f < 24 GHz		−9		
		24 GHz < f < BW		−8		
Output peak-level detector			170m		$V/V_{pp,diff}$	$Z_{Load,diff}=50\ \Omega$, each output (V_{p_N}, V_{p_P}) referenced to V_{ref}.
Operation temperature			40		°C	

* denotes that measurements were carried out at room temperature condition, 23 °C. Unless noted, measurement temperature is 40 °C.

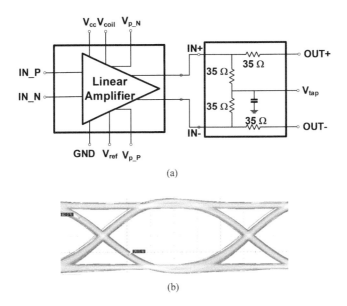

FIGURE 8.19
(a) Schematic illustration of electrical sub-mount for time-domain measurements.
(b) De-embedded electrical eye waveform at 32 Gb/s (5 ps/div, 700 mV/div, 40°C).

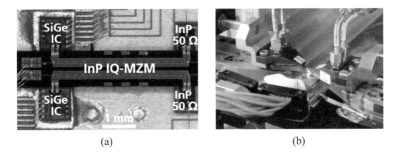

FIGURE 8.20
(a) Photograph of the optical transmitter with an InP IQ MZM and SiGe driver-ICs. (b) On-wafer EO measurement setup.

shown in Figure 8.20b. The CoC is located on the TEC and the temperature is set to 45 °C. The frequency response of the IQ MZM is given in Figure 8.21 for two nested child modulators. The small-signal EO 3-dB bandwidths are 28 GHz and 31 GHz, respectively.

EO measurement results at 32 GBd for QPSK and 16-QAM are depicted in Figure 8.22a, and b, respectively. The power consumption for each driver-IC was 267 mW. The input differential signal was about 500 mV$_{pp,diff}$. For these measurements, the equalizer at the optical modulation analyzer (OMA, Keysight) is activated. Very good EVMs at both modulator formats are obtained for the PRBS length of 2^{31}-1. For 16-QAM, the energy efficiency of

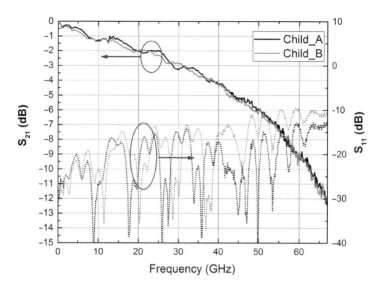

FIGURE 8.21
IQ MZ-modulator PIC small-signal EO responses.

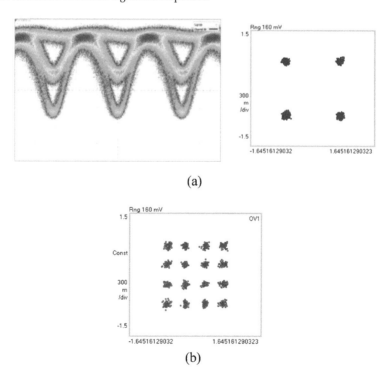

(a)

(b)

FIGURE 8.22
(a) QPSK EO eye (left) and constellation diagram (right) at 32 GBd. EVM: 6.3%. (b) 16-QAM constellation diagram at 32 GBd, EVM: 6.42%.

the CoC corresponds to 4.17 pJ/bit excluding the TEC power. We estimate about 12.5–14.7 pJ/bit assuming the required TEC power ranging from 1070 mW to 1340 mW when the CoC is mounted on the TEC.

8.4 Conclusion and Perspective

This chapter describes the co-design methodology for the InP MZM and its driver-IC to achieve the low power. The low-power TOSA in modern optical communication plays a significant role in satisfying power requirements of various optical transceiver standards and producing competitive products. In fact, the optics and driver-IC have been separately designed and packaged, leading to higher power consumption. The co-design in the TOSA means the impedance engineering between the MZM and the output impedance of the driver-IC. We explained how one determines the characteristic impedance of the MZM and discussed its relations with the MZM bandwidth and power.

With regard to the driver-IC architecture, the open-collector IC can be regarded as a lower-power solution than the back-terminated driver-IC. The Fraunhofer HHI has demonstrated very good EO performance using the open-collector IC. It must be mentioned here that for a faster baud rate (> 56 GBd) with high-order modulation format, the precise electrical model of the package between the modulator and the open-collector IC must be predicted and provided in order to avoid impedance mismatch and signal reflections. The back-termination driver-IC is more robust than the open-collector in terms of impedance mismatch since the back-termination absorbs the reflected signal; however, it consumes more power. It is believed that in future optical communications the co-design of optics and electronics will become common design methodology to accomplish the low-power TOSA and achieve a good signal integrity.

Acknowledgments

The author would like to thank Dr. Gerrit Fiol, Mr. Klemens Janiak, and Dr. Herbert Venghaus for their proof-readings and suggestions to improve the chapter contents. He is grateful to Prof. Dr. Martin Schell for his support of this work. He also gives his appreciation to Dr. Karl-Otto Velthaus, Dr. Heinz-Gunter Bach, Dr. Norman Wolf, and Dr. Lei Yan for their work on and investigation of co-design and driver-IC design.

References

1. David A. B. Miller, Attojoule optoelectronics for low-energy information processing and communications, *J. Lightwave Technol.*, vol. 35, no. 3, Feb. 2017.
2. Adam Carter and Dan Tauber, To a terabit and beyond, *Fiber Systems*, Issue 19, Spring 2018.
3. [Online]. Available: www.cfp-msa.org.
4. [Online]. Available: www.qsfp-dd.com.
5. [Online]. Available: www.lightwaveonline.com/articles/2017/12/surveying-the-new-optical-form-factors-for-400-gigabit-ethernet.html.
6. [Online]. Available: www.fiber-optic-transceiver-module.com/qsfp-dd-vs-osfp-the-wave-of-the-future-400g-transceiver.html.
7. [Online]. Available: http://osfpmsa.org/index.html.
8. Haitao Chen, Development of an 80 Gbit/s InP-based Mach-Zehnder modulator, Dissertation, Technical University of Berlin, 2007.
9. Karl-Otto Velthaus, et al., Impedance-engineered low power MZM/driver assembly for CFP4-size pluggable long haul and metro transceiver, in *Proc. ECOC 2014*, Tu11.1, Oct. 2014.
10. B. Gomez Saavedra, et al., First 105 Gb/s low power, small footprint PAM-8 impedance-engineered TOSA with InP MZM and customized driver IC using predistortion, in *Proc. OFC 2015*, Th4E.4, Mar. 2015.
11. Martin Schell, et al., DAC-free generation of M-QAM signals with InP segmented Mach-Zehnder modulators, in *Proc. OFC* 2017, W4G.4, Mar. 2017.
12. Jung Han Choi, et. al., Ultra-low power SiGe 2-bit DAC driver for InP-IQ Mach-Zehnder modulator, in *Proc. ECOC 2017*, W1F.2, Sept. 2017.
13. M. Rausch, et al., A performance comparison of single-ended- and differential driving scheme at 64 Gbit/s QPSK modulation for InP-based IQ-Mach-Zehnder modulators in serial-push-pull configuration, in *Proc. ECOC 2015*, doi:10.1109/ECOC.2015.7341728, Oct. 2015.
14. [Online]. Available: www.hhi.fraunhofer.de/fileadmin/PDF/PC/ICD/20150907_Impedance_Conversion_IC.pdf
15. Norman Wolf, et. al., Electro-optical co-design to minimize power consumption of a 32 GBd optical IQ-transmitter using InP MZ-modulators, in *Proc. CSICS 2015*, 10.1109/CSICS.2015.7314499, Oct. 2015.
16. Sophie Lange, et al., Low power InP-based monolithic DFB-laser IQ modulator with SiGe differential driver for 32 GBd QPSK modulation, *J. Lightwave Technol.*, vol. 34, no. 8, pp. 1678–1682, Apr. 15 2016.
17. Sophie Lange, et al., Low power InP-based monolithic DFB-laser IQ modulator with SiGe differential driver for 32 GBd QPSK modulation, in *Proc. ECOC 2015*, 10.1109/ECOC.2015.7341851, Sept. 2015.

9

Efficient and Low-Power NoC Router Architecture

Gul N. Khan

CONTENTS

9.1 Introduction

Network-on-Chip (NoC) architecture provides a communications infra-structure for the cores of a multi-core System-on-Chip (SoC). The NoC resources are connected to the SoC cores enabling them to communicate among each other concurrently by sending messages asynchronously. NoC systems improve the scalability and power efficiency of complex SoCs as compared to other conventional communication systems. Figure 9.1a illustrates an SoC including some IP cores which are connected through a 4×5 mesh NoC architecture. The NoC includes a network of switches (routers) which are interconnected by data links. Figure 9.1b illustrates a typical NoC router that consists of some input- and output-ports, an arbiter and a cross-bar switch [1]. The input- and output-ports can be simple data buses that connect a router to its channels, but at least one of them should include a circuit to perform buffering and traversal of incoming flits. In all the designs presented in this paper, the input- ports only utilize buffering organization, and the output-ports are simple data buses. A most viable communication

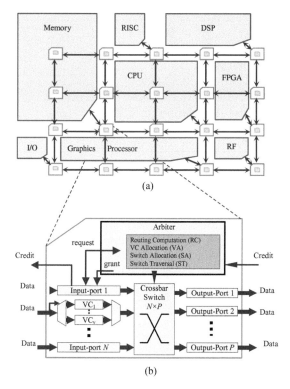

(a)

(b)

FIGURE 9.1

(a) SoC with 4×5 Mesh NoC architecture (b) NoC router architecture with N inputs, P outputs and v VCs.

mechanism employed in NoCs is packet-based wormhole routing [2]. The messages in wormhole routing are organized as multiple packets where each packet consists of some flits. A flit is a basic unit of data that is generally transferred at clock rate. The first flit of a packet is called the header flit and holds the route information of its associated packet. The remaining flits are called body flits except the last flit, that is called the tail flit. The body and tail flits contain data and can contain two pieces of information: tail state and VC identification.

When the header flit of a packet passes through a route consisting of routers, the route path is reserved for that packet. The route path remains reserved until all the packet flits pass through it. Such a kind of data flow does not always provide optimum performance. In fact, messages from other sources that share their routes or part of their routes with the reserved route must wait until the route becomes free. These waiting conditions continue even when there is no communication through the reserved route. This means that the data flow is not always populated. In other words, the reserved route is sometimes idle and blocks other packets from passing through it. This kind of data flow incurs higher latency and lower utilization of shared NoC resources. One way of alleviating this problem and improving NoC throughput is to utilize Virtual Channel (VC).

In VC organization, the flits of one packet can interleave with the flits of another packet over a physical channel by using a rotating flit-by-flit arbitration. The routing path of each flit can be guaranteed because the flits belonging to a packet are attached with an identical VC identification (*VC-ID*) tag at each router. Then these flits become differentiable at downstream routers. The *VC-ID* tags exist in all the flits of a packet, and these tags are identical when the flits enter a router. Figure 9.2a illustrates the conventional micro-architecture of a VC decoder that is a simple de-multiplexer. In the figure, the *VC-ID* is connected to the selection port of the multiplexer and causes the incoming flit to go to the associated VC buffer. In fact, the flits of a packet are always stored in the same VC buffer.

9.1.1 Static and Dynamic VC Organization

One of the typical forms of buffering is static, i.e. the numbers of VCs and their buffers stay constant during communication (Figure 9.2a). Various studies have shown that for a large number of static VCs, communication load is difficult to balance across them [2]. Some VCs remain idle while the others are overloaded. Therefore, it is better to allocate more buffer storage to busy VCs and less to idle VCs. Moreover, static VC buffers are the expensive components of NoC routers and they become more expensive for larger flit size or when the VC buffer depth becomes larger. The two above drawbacks of static VCs necessitate an adaptive VC organization to achieve VC flow control with maximum buffer utilization. A well-known adaptive VC organization is called Dynamically Allocation Multi Queue (DAMQ).

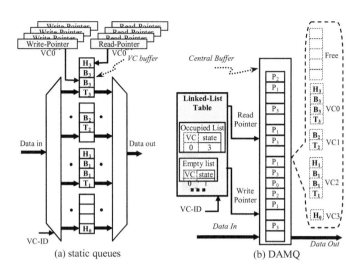

FIGURE 9.2
Input- ports with static and dynamic queues.

Most of the DAMQ organizations are table-based [3, 4]. In table-based organization, a central buffer includes multiple VC queues, and a table is utilized in order to keep the flits of each VC in a First Come, First Served (FCFS) order as illustrated in Figure 9.2b. In fact, the table keeps the address of incoming flits in a FCFS order. The dynamic VCs of DAMQ buffers improves the port buffer utilization through sharing its buffer slots among all the VCs of the port and allocating more buffer slots to the active VCs. Large VC buffer depth will keep more flits of a packet and leads to a free route of the packet in wormhole NoC communication. A larger number of free routes reduces contention and, eventually, improves overall NoC performance. Despite the performance merits of DAMQ organizations, they have a number of limitations, listed below:

- The first problem is the complex hardware due to the linked list and dynamic queue management [5, 6].
- Setup limitation is a problem with some DAMQ mechanisms. For example, limitations in the minimum buffer space of each VC, number of VCs and number of flits per packet are observed for some DAMQ schemes.
- Head of Line (HoL) blocking is a problem in the communication of some DAMQ schemes. Assuming that a VC (queue) can receive more than one packet, if the header packet faces a blockage, the other packets in the VC must wait until the blockage is removed.
- There are interventions among the VCs of a DAMQ-based input-port that can lead to higher traffic congestion as compared to static VCs [7].
- Each flit arrival/departure has a large delay due to the complex design of DAMQ-based VCs.

9.1.2 Timing Problem of Table-Based Adaptive Virtual Channels

The flit arrival/departure of a dynamic input-port can be easily compared with the static VC input-ports. Figure 9.2 shows the architectures of these input-ports. The control logic of the static input-port is simpler, where each VC can be configured by using a parallel FIFO buffer, as shown in Figure 9.2a [7]. The number of VCs is equal to the number of FIFOs, as each FIFO represents a VC. The *read-pointer* and the *write-pointer* point to the location of a FIFO, where a flit is read or written, respectively. A pointer works like a simple counter, which is incremented circularly and continuously for each read and write operation. The flit arrival/departure is also simpler in a static input-port. If arbitration takes one step, the arrival/departure of flits in a squeezed pipelined scheme consumes two clock edges, as illustrated in Figure 9.3a. At the entrance of an input-port, the arriving flit is decoded according to its *VC-ID* (VC identification) and by means of the first de-multiplexer. Then it waits to be latched in the FIFO buffer (VC) before the first clock edge. At the first clock edge, the flit is stored in the VC where a request corresponding to that flit is simultaneously issued to the arbiter. At the second clock edge, the arbiter allocates the address for the crossbar switch (output) and ID for the downstream router VC, then issues a *grant* signal. The *grant* signal causes the flit to travel out of the router. For proper operation of the decoder at the entrance of the input-port, the *VC-ID* should be issued before latching the flit in the buffer. Assuming that the flit and its *VC-ID* are transferred at the same clock transition, each flit arrival/departure requires a two-clock event delay in the static router. We have assumed that the FIFOs are dual-port, where the arrival of one flit can coincide with the departure of another flit.

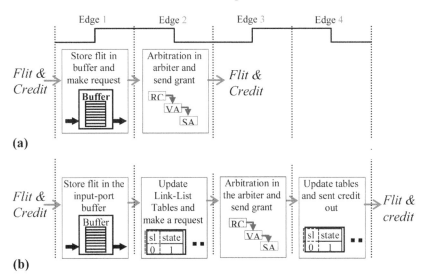

FIGURE 9.3
Static versus conventional dynamic router pipelines.

In the case of dynamic input-ports, the VC buffers are allocated dynamically, resulting in complex control logic. A linked-list-based DAMQ has been employed as a conventional DAMQ in many research projects [4, 7–9]. Using this mechanism, a central buffer (queue) maintains multiple VCs, and the data flow is directed by Linked-List tables as illustrated in Figure 9.2b [4]. The *read-pointer* and *write-pointer* are updated based on the contents of the Linked-List tables. In a squeezed pipeline design, if arbitration takes one step, the arrival/departure of flits will take four clock edges, as illustrated in Figure 9.3b. A header flit arrives at the input-port and waits to be latched in a VC (buffer). Then the flit is latched into the input-port buffer at the first clock edge. On the second clock edge, the linked-list tables are updated according to the *VC-ID*, which leads to a request signal being issued to the arbiter. The arbiter assigns a proper address for the cross-bar switch (output) and ID for the VC at the third clock edge and then issues a *grant* signal. The *grant* signal causes the flit to exit the router, and the Linked-List tables are updated at the fourth clock edge. In a linked-list DAMQ-based VC organization, the *read-* and *write-pointers* cannot be updated at read or write events; instead, they are updated one clock event after the read and write (i.e. after updating the tables). However, in the static queue mechanism, the read and write pointers can be incremented at the read or write events. This causes the pipeline stages in traditional DAMQ (or table-based) input-ports to take two more clock events than those of static input-ports. The same communication characteristics are expected for other table-based DAMQ input-ports such as ViChaR [3]. The VC organization approach presented in this chapter does not employ tables, and its flit arrival/departure delay is equal to that of static VCs but with all the advantages of dynamic VCs.

9.1.3 Data Flow Arbitration

After buffering a flit in a VC of the input-port, the VC issues a request to the arbiter for accessing the shared resources. The arbiter can perform arbitration and allocation through four pipelined stages, as illustrated in Figure 9.4. Each flit of a packet proceeds through Routing Computation (RC), Virtual channel Allocation (VA), Switch Allocation (SA), and Switch Traversal (ST) stages [1]. The RC and VA stages process the header flit (once per packet). The body and tail flits of a packet pass through RC and VA without any processing, whereas SA and ST stages operate on each flit of the packet. For wormhole routing, the destination information of a packet is embedded in the header flit that passes through all the pipeline stages. In our design, the first three stages, i.e. RC, VA and SA, proceed in parallel in one clock cycle, as illustrated in Figure 9.4. However, in past designs, each stage used to take one or more clock cycles. The ST is the last stage and it can be overlapped with the first stage of the following flit [10]. Among the four arbiter stages, VA and SA stage modules perform arbitration.

FIGURE 9.4
Pipeline stages of a VC-based arbiter for a four-flit packet that takes five cycles to be arbitrated with no communication stall.

9.1.4 NoC Switch Allocator

The SA module can be designed as a unit to maximize matching among the requesters and the resources. However, such designs are difficult to be parallelized or pipelined, and too slow for the applications where latency is critical [10]. Latency-sensitive applications typically employ fast and fair matching designs such as separable SA modules. A separable SA (switch allocator) is significantly less complex due to separate requesters for some groups. Moreover, the critical path delay of a separable SA can be improved by choosing a fair and fast arbitration among each group. The separable SA modules are usually utilized and investigated as part of NoC research. In a separable SA, two sets of arbiters can perform arbitration: one for the inputs and the other for outputs, as illustrated in Figure 9.5. In an input-first separable SA, arbitration is first performed to select one request at each input-port. The outputs of these *input-arbiters* become the inputs to a set of output arbiters to select a single request for each output-port [10]. In order to ensure fairness and to perform arbitration in a single iteration, a Round-Robin (RR) scheme can be utilized, as shown in Figure 9.5. An input-first separable SA micro-architecture employed in our NoC router design is shown in Figure 9.6. The *input-arbiter*s perform arbitration among the VCs of each input-port of the router where the decoder modules generate the requested outputs of the winner VCs of input-ports. Each bit of the decoder output corresponds to an output-port, where an active bit of the decoder output shows the requested output by the winner VC of the relevant input. The second set of arbiters, *output -arbiter*s, perform arbitration among the winner input-ports for the output-ports.

NoC structures are viewed as a suitable solution to meet the wiring challenges of multi-core and many-core systems. NoC designs that consume minimal power and IC area along with higher performance are required for SoC design, especially for low-power applications. Current NoC system and router designs are not optimal. The main problem with the current NoC

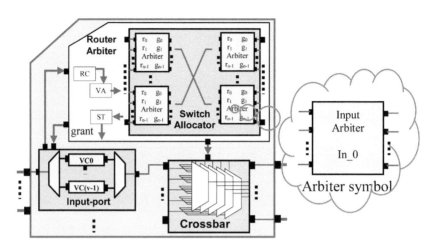

FIGURE 9.5
Wormhole *v*-VC router with Switch Allocator.

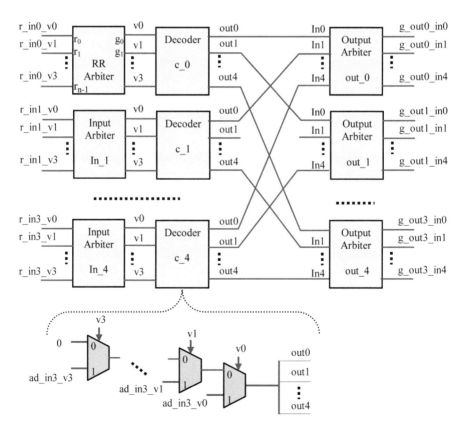

FIGURE 9.6
Micro-architecture of a 5×5 separable SA.

design is related to lower performance during high contention and traffic congestion. The router is the key component of an NoC, and its efficient design and implementation will improve the overall NoC performance. NoC performance and energy budget depends heavily on the usage of buffer resources by the routers. One must utilize the buffer intelligently for NoCs by employing adaptive or dynamic VCs as they have maximum buffer utilization. Current DAMQ buffer design suffers a number of problems, such as complexity, lower buffer utilization, setup limitation, and HoL blocking. Arbitration is another important activity in NoC routers. It also has a few problems, such as complexity, lower speed, weak fairness, traffic starvation, and pipelining difficulty. These arbitration drawbacks and points related to current NoCs have motivated us to investigate the high-performance components of NoC as well as routers including input-port organization and arbitration. The remainder of this chapter is organized as follows. The past research and NoC router design-related techniques are described in Section 9.2. The router and NoC architecture is described in Section 9.3, whereas the router and its various mechanisms are evaluated by experimental results in Section 9.4. The conclusions are drawn at the end of the chapter.

9.2 Overview and Past Works

DAMQ is a unified and dynamically allocated buffer structure that was originally presented by Frazier and Tamir [11]. It is a single storage array that maintains multiple FIFO queues. Nicopoulos *et al.* introduced a centralized shared buffer-based Virtual Channel Regulator (ViChaR) to implement a DAMQ mechanism [3]. In ViChaR, the information of incoming buffer is saved in a table and two trackers. The VC control table module holds the VC slot IDs of all the current flits and it can grow very large for small flit-size big packets. In other words, the ViChaR approach suffers from setup limitations and complexity. Xu *et al.* have presented an application that employs ViChaR architecture [6] where VCs are assigned based on the designated output-port of a packet in an effort to reduce the HoL blocking. Unlike ViChaR, this design uses a small number of VCs. Their buffer design is similar to ViChaR except that each VC can store multiple packets and fixed number of VCs employed. Their buffer design uses a smaller arbiter and hybrid (in between static and dynamic) VC allocation scheme.

A 'self-compacting buffers' approach is presented by Park *et al.* to implement DAMQ switching elements [12]. There is no reserved space dedicated for any VC, and a VC can receive any number of flits to occupy the whole channel buffer. Data in the self-compacting buffer is stored in a FIFO manner within the region for each VC. When a flit is to be inserted in the middle of the buffer, the required space is created by moving down all the flits residing below the insertion address. Similarly, when a read operation is conducted from the top of a region, the data removed from the buffer may result in empty holes in the middle and the data below the read address is shifted up to fill the hole. This approach has some drawbacks. Firstly, when the buffer capacity is increased, its fall-through time also increases, leading to higher latency [13]. Due to which, the latency of the self-compacting buffer depends on its depth rather than the number of stored items. Another drawback is its higher dynamic power consumption due to data shifts in the buffer. Frias and Diaz have proposed an interesting buffer architecture to alleviate the above drawbacks of the self-compacting buffer [14]. They proposed a new cell that has the capability of performing all the required data moves. The novelty of their approach is at the transistor level rather than the gate level.

Evripidou *et al.* have presented a new version of the linked-list mechanism, which mimics the DAMQ organization presented by Frazier and Tamir. The linked-list buffer is expensive in terms of hardware but it leads to higher performance [11]. Our implementation of the Linked-List-based DAMQ (LLD) mechanism is presented here to illustrate the complexity and cost of the LLD mechanism as compared to the proposed VC architectures presented in this chapter. Five lookup tables are used to implement the LLD input-port organization as shown in Figure 9.7. The *VC-State* and *Slot-State* tables keep a Boolean value for each VC and slot (empty/occupied). The *Header-List* table keeps the addresses of slots that contain the header flits of VCs. The *Tail-List* table keeps the addresses of slots that point to the tail flits of VCs. The *Linked-List* table keeps the addresses of the next slot of each slot, or it links

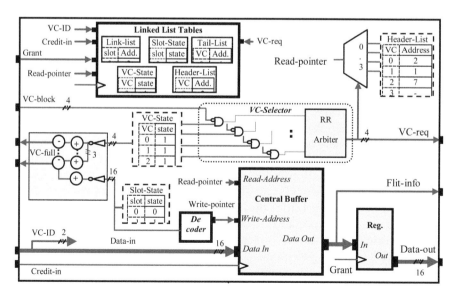

FIGURE 9.7
16-bit 16-slot 4-VC LLD input-port micro-architecture.

the addresses of slots that are associated with each VC in a FIFO manner. The *Slot-State* table has a record of the occupied slots in the central buffer. The pipelined communication illustrated in Figure 9.3b shows the case of LLD routers; flit-data is stored for two clock events and transferred at two clock events. The tables are updated at the negative clock edge, and the signals are detected and issued at the positive clock edge.

The flit arrival and departure is detected by *credit-in* and *grant* signals, respectively. Assume that a VC is empty and ready to accept data. Upon the arrival of a flit for the VC, three things happen simultaneously. First of all, the corresponding bit of the VC is set in the *VC-State* table indicating an occupied VC. Then the content of the *write-pointer* is stored in the *Tail-List* and *Header-List* tables. Finally, the corresponding bit is set in the *Slot-State* table. The *write-pointer* is then updated and it will point to the next free slot to accommodate the incoming flit. When another incoming flit tries to reside in the VC, three things occur at the negative clock edge. The content of the *write-pointer* is stored in a location of the LLD table where the *Tail-List* table points to it and the *write-pointer* is stored in the *Tail-List* table. Finally, the *Slot-State* table is updated, which leads to the updating of the *write-pointer*. When a flit exits from the VC, three types of event take place. If the header and tail addresses are the same (the last flit), the corresponding bit is reset in the *VC-State* indicating the empty VC. Otherwise, the location of the LLD table is identified by the *Header-List* table, and it is stored as the new header address of the VC in the *Header-List* table. Finally, the corresponding bit of the *Slot-State* table is reset, which causes the *write-pointer* to be updated.

The *VC-Selector* module in the router input-port selects a VC for arbitration. It issues the request signal, *VC-req*, and is used to generate the *read-pointer* address as illustrated in Figure 9.7. The *VC-Selector* module has a logic circuit that operates on the contents of *VC-State* table and creates the VC availability signal (*VC-ava*). The *VC-block* signal is reversed and ANDed with its corresponding bit in the *VC-State* table as illustrated in Figure 9.7. The Round-Robin (RR) arbiter module selects a VC request among the active VC requests and follows RR arbitration. A credit signal (*VC-full*) is required for each VC to close it in the case of a full condition. The *VC-full* module, shown in Figure 9.7 reserves a slot for each VC. Assume that S0 to S15 represent the cell states of the *Slot-State* table, and V0 to V3 represent the cell states of the *VC-State* table. The state of *VC-full* signal is determined by the following logic equation.

$$\text{VC-full}[i] = \left(\left(S_0 + S_1 + \ldots + S_{15} \right) - \left(V_0 + V_1 + \ldots + V_3 \right) - V_i \right)$$

(9.1)

$$\text{where } i \in \{0 \ldots 3\}$$

A circuit-based DAMQ input-port architecture, Effective Dynamic Virtual Channel (EDVC) is presented by Oveis-Gharan and Khan [15]. The EDVC mechanism utilizes the common features of the DAMQ input-port to create a dynamic flow control. The EDVC mechanism has some drawbacks. First of all, the *read-pointer* and *write-pointer* architectures grow large with the size of the input-port buffer. For example, if the size of the input-port buffer increases n times, the multiplexers employed in EDVC *read-pointer* and *write-pointer* will grow exponentially (n^2 times), which will in turn increase the hardware overhead and the critical path delay, which has a direct effect on the speed of the EDVC router. The second drawback of EDVC design is its higher latency at lower flit-injection rates. The authors later presented a Rapid Dynamic Queue (RDQ) approach to overcome these drawbacks [16]. RDQ VC structure and organization is similar to EDVC mechanism where RDQ demonstrates much higher performance under different injection rates and consumes economical hardware in all the configurations of an input-port buffer.

9.2.1 Heterogeneous NoC Router

A number of researchers have focused on the design and organization of routers that have direct impact on the NoC power, performance and area [17–19]. Virtual Circuit Switching (VCS) is proposed by Jiang *et al.* [17]. Their approach intermingles the Circuit Switching (CS) and Packet Switching (PS) to obtain low latency and power consumption in NoCs. They have also proposed a path allocation algorithm to determine VCS connections and CS connections in a mesh-connected NoC. VCs in the VCS approach are exploited to form a number of VC connections by storing the interconnect information in routers. Flits can directly traverse the routers by using only the Switch Traversal (ST) stage. The main advantage of the VCS approach is that it can have a similar router pipeline as a circuit switching router, and can have multiple VCS connections to share a common physical channel. Basically, VCS connections cooperate with PS connections and CS. When flits on CS or VCS connections arrive at routers, crossbar switches are immediately configured so that the CS or VCS flits can bypass directly to the ST stage. When there is no CS or VCS flit, the corresponding ports of crossbar switches are released to PS connections. The VCs of routers in VCS connections are preconfigured in such a way that they are only connected to particular downstream virtual channels. Crossbar switches are preconfigured during the switch allocation (SA) stage before VCS flits require to pass through. As VCS connections are established over virtual channels, a physical channel can be shared by n VCS connections where n is the number of virtual channels. Other communications competing for the same physical channel must be conducted through the packet switching. There are some drawbacks of the VCS approach. First of all, VCS router architecture requires more hardware as compared to conventional wormhole routers because they must have the circuitry for both packet and circuit switching. Moreover, extra hardware is needed to distinguish VCS packets from PS packets. Moreover, despite the extra hardware of VCS, the approach is only efficient in deterministic communications suitable to application-specific NoCs. However, our proposed NoC routers in this chapter are efficient for any type of NoC communications. Moreover, they improve the performance while their architectures are simpler and accommodate lower hardware overhead.

Another heterogeneous NoC router architecture has been introduced by Ben-Itzhak *et al.* [18]. They exploit a shared-buffer technique to handle the bandwidth heterogeneity of an NoC link. Their approach reduces the number of shared buffers required for a conflict-free router without affecting the performance. Reducing the number of shared buffers also reduces the crossbar size and overall it has lower chip area and power consumption. The heterogeneous router supports different capacities and different numbers of VCs for each link while keeping the router frequency fixed. Their router architecture utilizes serial-to-parallel converters in order to store the incoming flits in different input buffer slots at each link clock cycle, and parallel-to-serial converters can be used in order to transmit several flits in a Time-Division Multiplexing (TDM) way depending on the link frequency. The shared-buffer technique presented by Ben-Itzhak *et al.* optimizes the number of shared buffers related to arrival and departure conflicts discussed by Ramanujam *et al.* [19]. The drawbacks of the VCS approach discussed earlier can be considered for the heterogeneous approach. In other words, the heterogeneous router architecture is tailored for specific types of NoC communication. However, NoC has emerged as a scalable and global communication architecture to address the SoC design challenges. Its features enable it to be easily expandable and it can provide services for a variety of SoC communications. The heterogeneous feature of this approach violates the scalability and reusability of NoC and additional research and investigation for heterogeneous NoC routers is required.

9.2.2 Router Buffer Organization

A number of researchers have focused on the design and organization of router buffers due to their close relationship with NoC power, performance and area. Complex router architectures are efficient for certain NoC configurations or data flow circumstances. CUTBUF NoC router architecture has been presented by Zoni *et al.* [20], where VCs are dynamically assigned to an input-port depending on the actual input-port load and reusing of each queue by packets of different types. Their approach significantly reduces the number of physical buffers in routers, saving area and power without affecting the NoC performance. It is assumed that a conventional VC can be re-allocated to a new packet only if the tail of its last allocated packet has left. However, a reserved VC in CUTBUF protocol is released when either the tail flit traverses the same pipeline router stage, or when the related packet gets blocked. In this way, the CUTBUF scheme increases the buffer utilization as well as preventing HoL blocking, where a VC can be re-allocated to a new packet under certain specific conditions. The newly allocated packet to an occupied

VC will not be blocked because the previous packet is guaranteed to traverse the router. To implement this mechanism, the following condition is checked to allow a new packet to allocate a non-empty VC. The earlier packet will be leaving the router when its tail flit is stored in the input buffer VC, and the downstream router has enough credits to store all the flits of the VC.

A conventional NoC protocol has also been investigated in the Packet-Based Virtual Channel (PBVC) approach [21]. A VC in a PBVC scheme is reserved when a packet enters the router and released when the packet leaves. A VC will hold the flits of only one packet at a time that will remove the HoL blocking. The PBVC technique is more efficient for DAMQ-based schemes where an input- or output-port employs a centralized buffer whose slots are dynamically allocated to VCs. The experimental results have shown that for HoL-specific traffic, the average latency and throughput are improved in the PBVC approach as compared to conventional DAMQ-based NoCs. Yung-Chou and Yarsun have also presented a new DAMQ-based buffer organization called DAMQ-MP that can accommodate multiple packets greater than the available number of VCs [22]. Their approach can solve certain on-chip communication issues, such as heavy network congestion or short packets, to improve the performance. They have introduced the DAMQ-MP data flow as follows. Whenever the tail flit of a packet enters the input buffer via one of the VCs, this packet releases the VC by alerting a notification signal to the upstream router. Therefore, a new packet from the upstream router can be sent out and enters the free VC. In other words, the new packet does not need to wait until the tail flit of the previous packet gets out of the VC. This approach can save a lot of time, especially when the routing computation for the arriving packets is high. The above data flow organization means that the DAMQ-MP is efficient in various cases, including small packets, large buffer capacity, heavy-congested traffic (such as a saturated network), and a small number of VCs. As one may notice, the DAMQ-MP approach is the reverse of the PBVC approach. The DAMQ-MP approach lets a VC accept a new packet when the VC is not empty, whereas the PBVC approach prevents a VC from accepting a new packet when the VC is not empty. In fact, both approaches are efficient under different schemes and traffic patterns.

RoShaQ, another interesting router architecture that allows the sharing of multiple buffer queues, has been presented by Tran and Baas [23]. In this approach, each input-port allocates one buffer queue and shares all the remaining queues. The router architecture maximizes buffer utilization by sharing multiple buffer queues among input-ports. Buffer sharing reduces the packet stall time at input-ports. The RoShaQ is able to achieve higher throughput when the network load is heavy. For light traffic load, the RoShaQ router achieves low latency by allowing packets to effectively bypass the shared queues. In this way, their proposed router achieves higher performance when the traffic load becomes heavy and for low-load traffic. A RoShaQ NoC is also deadlock-free, as for a light load the packets normally bypass the shared queues and RoShaQ acts like a wormhole router to meet the deadlock-free condition. In the case of a heavy load, when a packet cannot win the output-port, it is allowed to occupy only a shared queue which is empty or contains packets having the same output-port. Clearly, in this case the RoShaQ works as an output-buffered router which is also shown to be deadlock-free. The approach combines two switching mechanisms, VC-based and simple wormhole, to improve the performance. When the first packet enters an input-port, it is serviced through wormhole switching without involving VC pipelines. However, when the second packet enters (assume the first packet is still there), it is serviced through VC-based switching involving VC pipelines. RoShaQ routers must accommodate both circuits related to wormhole switching and VC-based switching, leading to more hardware. Moreover, an extra circuit is needed to distinguish wormhole packets from VC-based packets. Therefore, the RoShaQ router architectures become more complex, with more hardware as compared to conventional VC-based routers. The performance improvement of RoShaQ is due to more hardware overhead rather than an efficient or novel architecture.

9.2.3 NoC Router Arbiters

Arbiters are commonly used to allocate and access shared resources. Whenever a resource (such as a buffer, channel or a switch-port) is shared, an arbiter is required to assign the access to the resource at a particular time. A wormhole *v*-VC router was presented in Section 9.1, where the *router arbiter* module has a *switch allocator* consisting of two sets of simple arbiters. A simple arbiter arbitrates among a group of requesters for a single resource, as illustrated in the form of a symbol for an *n*-input in the right side of Figure 9.5. The arbiter accepts *n* requests ($r_0, r_1,..., r_{n-1}$), arbitrates among the asserted request lines, selects an r_i for service, and then asserts the corresponding grant line g_i. Arbiters can be categorized on the basis of fairness (weak, strong or FIFO) arbiters [1, 24]. For a weak fairness arbiter, every request is eventually granted. The requests of a strong fairness arbiter are granted equally often. The requests of FIFO fairness are granted on a first come, first served basis. In terms of priority, the arbiters can be grouped into fixed and variable priority architectures. For a fixed-priority arbiter, the priority of requests is established in a linear order. Round Robin (RR) is a well-known variable priority arbiter. The functionality of an RR arbiter can be explained in this way: a request that is most recently granted will have the lowest priority in the next arbitration cycle. The RR arbiters are simple, easy to implement, and they are also starvation-free. When the input requests are large in number, the structure of an RR arbiter grows, requiring larger chip area, higher power consumption, and longer critical path delay [1].

The micro-architectures of popular RR arbiters such as Matrix and Round-Robin (RoR) have been extensively presented and evaluated [1, 25, 26]. Matrix and RoR arbiters have the same functional behavior and perform strong fairness arbitration. The Matrix arbiter is claimed to be useful for a small number of inputs as it is fast and economical. Fu and Ling evaluated and compared the RoR and Matrix arbiters in terms of resource, performance and power consumption for an FPGA platform [25]. They concluded that the Matrix arbiter consumes more resources for the same power consumption and it can process data quickly as compared to the RoR arbiter. The same evaluation is made by Oveis-Gharan and Khan [26]. They also presented a strong fairness Round-Robin arbiter design, IRR. Their approach is simpler, faster and consumes less hardware as compared to the RoR and Matrix arbiters. The IRR arbiter is employed later in our NoC router design presented in Section 9.3. A High-speed and Decentralized Round-Robin Arbiter (HDRA) has been presented by Lee *et al.*, and is illustrated in Figure 9.8 [27].

Each boxed circuit represents a filter circuit whose main components are a D-type flip-flop and a multiplexer. The circuit filters out the input without request or the one whose request has already been granted in the current arbitration cycle. The unfiltered inputs with their requests participate in the arbitration again in the next cycle by resetting their corresponding D-type flip-flops to low (or at zero), which is done by enabling the *ack* signals from high to low level. The HDRA arbiter will reset itself asynchronously by the *self-rst* input. The *sys-rst* indicates the system reset signal and is used initially before each arbitration cycle for all the requests. A four-input HDRA arbiter has a simpler architecture because the *act, rnext* and *self-rst* are connected together. The HDRA architecture has been used in various applications, e.g. in the router model of simulation framework implemented by Guderian and others [28].

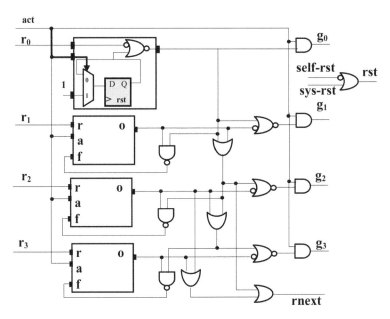

FIGURE 9.8
Four-input HDRA architecture.

9.3 Low-Power NoC Router RDQ-IRR

The novel NoC router micro-architecture presented in this chapter consists of input-ports, an arbiter and a crossbar switch, as illustrated in Figure 9.9b. The router employs VC organizations shown in Figure 9.9c, which is an amended version of Rapid Dynamic Queue (RDQ) employed in input-port modules [16]. An Index-based Round-Robin (IRR) arbiter is employed for arbitration, as shown in Figures 9.9a.

9.3.1 RDQ: Rapid Dynamic Queue Input-Port

The RDQ mechanism utilizes the common features of a DAMQ input-port to create a dynamic flow control [16]. For example, The VC identification (*VC-ID*) is saved with the flit in the input-port buffer, as shown in Figure 9.10, which is used to issue the VC *request* signals for router arbiter. A small *Slot-State* table (having Boolean values) is used to manage the input-port buffer mechanism. Each *Slot-State* bit corresponds to a buffer slot representing its occupancy state. The *read-pointer* points to the occupied locations of buffer per clock cycle. The *write-pointer* points to empty location of buffer until a write occurs. When a flit faces blockage, its VC will stop issuing requests and receiving new flits. Moreover, the address of the header flit of a blocked VC is stored in the *Blocking -Header* table. The highlighted locations in Figure 9.10 illustrate that the VC2 is blocked, and its header flit address is stored in the *Blocking -Header* table. To maintain the order of flits associated with the blocked VC, the following steps are performed. When the blockage of a VC (e.g. VC2) is removed, and the IRR *read-pointer* points to the header flit (location 2), the VC request become free, leading the *read-pointer* to read the whole buffer once and send out the flits of the freed VC (VC2). Then the VC starts accepting any new flits. *VC-full* and *VC-block* signals assist during the blockage conditions, as illustrated in the blocking circuit associated with a VC shown in Figure 9.11. Each VC has a *VC-full* signal, which is issued to the upstream router and represents the state of the VC to accept a new flit. The *VC-block* signal comes from the arbiter, indicating that the VC flit cannot exit the router. An asserted *VC-block* signal causes the VC to stop request-ing (*stop-req* signal is asserted) the arbiter and will not receive any new flit. The RDQ mechanism also prevents a data flow condition where a reserved empty VC can become closed. Various conditions associated with the VC closing and requesting operations are listed in Table 9.1.

The closing and requesting operations are actuated by the *VC-full* and *VC-req* signals respectively. There are a number of conditions determin-ing the state of the *VC-block*. As mentioned earlier, the *VC-block* becomes set when the VC cannot succeed in winning a free VC of the downstream router input-port. Consider the first condition, where the *VC-block* is reset, indicating no blockage in communication. For the second condition, the

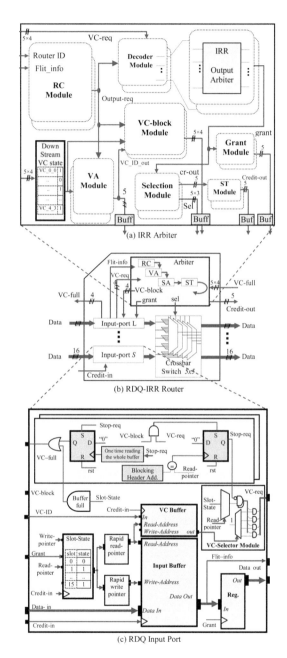

FIGURE 9.9
RDQ-IRR NoC router micro-architecture.

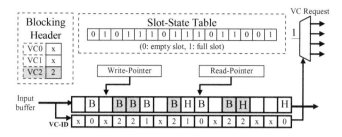

FIGURE 9.10
RDQ input-port operation.

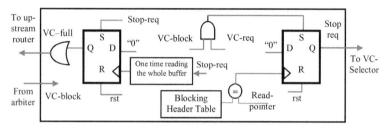

FIGURE 9.11
Blocking circuit associated with a VC.

TABLE 9.1

VC Closing and Requesting Operations of RDQ Mechanism

Cond.	VC-Block	Stop-Req	VC-Req	VC-Full
1	0	0	X: Normal 1: *read-pointer* points to a flit 0: *read-pointer* does not point to a flit of the VC.	X: Normal 1: buffer is full. 0: buffer is not full.
2	1	0	0: *read-pointer* points to no flit of the blocked VC.	X: Normal 1: buffer is full. 0: buffer is not full.
		1	0 to1 then returns 0: *read-pointer* points to a flit of blocked VC. The flit location is stored in the Blocking Header table.	1: stop incoming flit for the VC
3	1 to 0	1: *read-pointer ≠ Header flit Address*	0: stop request	1: stop incoming flit for the VC
		0: *read-pointer = Header flit Address*	Normal	1: stop incoming flit for the VC
		0: *read-pointer points to the occupied slots*	Normal	Normal

VC-block becomes set. In that case, the *read-pointer* does not point to a flit of the VC, and the VC is open for the incoming flits. As soon as the *read-pointer* points to a flit of the VC, the *VC-req* switches from 0 to 1 and leads to the *stop-req* signal being set and consequently the *VC-req* returns to 0. Meanwhile, the flit location is stored in the *Blocking -Header* table. In other words, the router stops receiving flits for that VC as well as requesting the arbiter. The order of VC flits inside the port buffer is kept until the blockage exists. In the third condition, the *VC-block* switches from 1 to 0, i.e. the VC can succeed in the arbiter. In this case, the *read-pointer* does not point to a flit of the blocked VC, the VC issues no request to the arbiter and receives no incoming flit. The *read-pointer* continues pointing to the occupied slots until it reaches the header flit of the blocked VC, as illustrated in the upper part of Figure 9.9c. At this point, the VC can only issue a request to the arbiter so that its flits can exit the router without allowing any new flit to enter the VC. The *read-pointer* continues to point to the occupied slots until it reads the whole buffer. When it points to the least significant occupied slot of the buffer two times, the closing and requesting operations of the VC return to the normal condition.

9.3.2 NoC Router Arbitration

The router arbiter can perform arbitration through four pipelined stages: Routing Computation (RC), Virtual- channel Allocation (VA), Switch Allocation (SA), and Switch Traversal (ST), as illustrated earlier in Figure 9.4. The structure of RC can be a simple multiplexer, as shown in Figure 9.12, due to XY routing considered for communication in a 2D mesh NoC. The destination ID, *Flit_info* (stored in the header flit of a packet), is compared with the ID of the current router. Then, the requested output-port is generated at the RC output according to XY routing algorithm. The structure of VA can be a fixed-priority arbiter, as shown in Figure 9.13. The first free VC of the downstream router input-port is selected in an ascending order. The SA module includes decoder, IRR *output-arbiter* modules, and the post-SA circuits (*Selection* and *Grant* modules) as shown in Figure 9.9a. The decoder module was introduced in Section 9.1.4 above, and the remaining modules are discussed in the following sections.

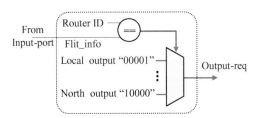

FIGURE 9.12
RC generates a free Output VC.

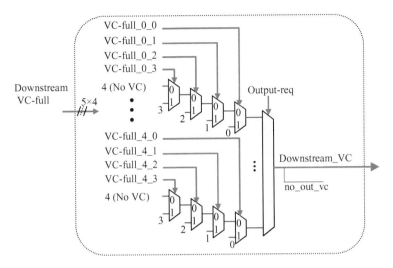

FIGURE 9.13
VC Allocator generates a free downstream VC.

9.3.2.1 VC Arbitration

In our router design, the VC arbitrations are implemented in the input-port by means of *VC-Selector* modules. We employ a central buffer to store all the VC flits of an input-port and arbitration is performed in one clock event. When a *grant* signal is issued to an input-port, the *read-pointer* is pointing to the winner VC flit, or the winner flit is loaded to the output-port of the buffer. The input-port micro-architectures of Figures 9.7 and 9.9c illustrate this scheme, where the *VC-Selector* chooses a VC for requesting to the arbiter, and simultaneously the flit of the VC is loaded to the buffer output. The *VC-req* signals in the LLD input-port of Figure 9.7 are used to generate the *read-pointer*, and these signals for the RDQ input-port of Figure 9.9c are employed to cater for blocking. One may think that the *VC-Selector* modules can be accommodated in the switch allocator, and the *VC-req* signals can be fed back to the input-port. The problem of this design is the dependency of input-port and arbiter on each other in terms of hardware and speed, as part of the critical paths of input-ports is shared with the arbiter. To prevent such sharing, we implement the VC arbitration as part of the input-port, and remove the *input-arbiter* modules from SA, as illustrated in Figure 9.14. In this way, the input-port and arbiter will operate independently.

9.3.2.2 Post-Switch Allocation

The *Selection* module generates the credit and selection address of the crossbar multiplexers related to an output-port of the router. Its micro-architecture is presented in Figure 9.15. When g_in3_out1 is asserted, the circuit generates the number 3 at the Sel output that leads input-port 3 to be connected

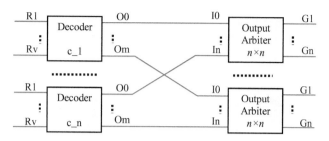

FIGURE 9.14
$n \times m$ SA micro-architecture for n inputs, m outputs and v VCs per input-port.

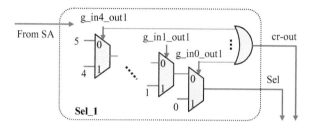

FIGURE 9.15
Output-port 1 of selection module.

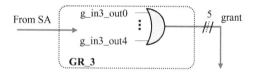

FIGURE 9.16
Input 3 of grant module.

to output-port 1 of the crossbar switch (see Figure 9.5). Moreover, a credit signal, *cr-out*, is issued to the ST module. Then the ST module issues a credit, *credit-out* signal to the first output-port to store the flit in the associated downstream VC at two clock events later. Figure 9.16 shows the Grant module that generates the *grant* signal associated with the third input-port of the router. The asserted g_in3_out1 allows the flit of third input-port to transmit via the crossbar switch and pass through the first output-port.

9.3.2.3 IRR Output Arbiter

An Index-based Round-Robin (IRR) arbiter employs a least recently served priority scheme and achieves strong fairness arbitration [26]. The IRR arbiter is employed by the *output -arbiter* of the SA module (Figure 9.9a). The IRR arbiter has smaller arbitration delay, lower chip area and it also consumes less power as compared to other RR arbiters. Figure 9.17 shows the

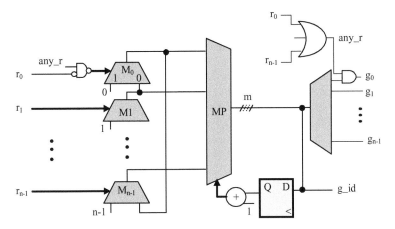

FIGURE 9.17
n-input IRR arbiter, where $m = log_2(n)$.

micro-architecture of an IRR arbiter that takes one clock cycle for arbitration. The index of the winner asserted request is switched to the output as the index of *grant* and *g_id*. Then the *g_id* is decoded to create the *grant* signals. If the next index of the granted request is employed for the next priority selection, the current granted request receives the least priority, and its next request receives the highest priority among all the requests. To accomplish it, the *g_id* array is stored in a register, whose output is incremented and connected to the selection port of the multiplexer (MP), as shown in Figure 9.17. In this way, the arbiter follows either the least recently served priority or a Round-Robin scheme. To keep the priority unchanged, the priority register output is fed back into a multiplexer to cater for no request. It guarantees strong fairness arbitration for our IRR design.

9.3.2.4 FIFO Arbitration and VC Selector

The selection of a VC from the input-port VCs in LLD [4] and ViChaR [3] is implemented in the *VC- Selector* module by employing a Round-Robin (RR) arbiter to ensure fairness and avoid traffic starvation, as shown in Figure 9.7. However, the selection of a VC from the input-port VCs in an RDQ mechanism happens in the data flow process. It is based on the location of *read-pointer* and the state of data in the input-port, as indicated in Figure 9.9c. The *VC-req* is updated according to the location of the flit data in the input-port buffer. This feature of RDQ *VC- Selector* requires the arbitration among the input-port VCs to follow a FIFO fairness priority [24].

9.3.2.5 Rapid Read and Write Pointers

The *rapid-read-pointer* module only points to the occupied locations and the *rapid-write-pointer* points only to the empty locations of the input-port buffer shown in Figure 9.9c. An RR priority scheme is followed, where a slot that is just read has the lowest priority for the next cycle [29]. In this way, each asserted bit of *Slot-State* table is pointed per clock cycle in an ascending and circular order. Figure 9.17 can also be used to illustrate the micro-architecture of an n-bit *rapid-read-pointer*, where the requests $(r_0, r_1 ..., r_{n-1})$ and *g_id* are substituted with the *Slot-State* content and *rapid-read-pointer* output respectively. A similar *rapid-read-pointer* architecture is proposed for *rapid-write-pointer*, except the places of input-ports of each M_0 to M_{n-1} multiplexers are exchanged. Moreover, the *rapid-write-pointer* output is directly connected to the selection port of *MP*. In this way, the *rapid-write-pointer* module points to a *Slot-State* empty slot and stop there until the slot becomes full, then it jumps to next empty slot on the following clock cycle and in an RR priority order.

9.3.3 Pipelined RDQ-IRR Router

The buffering pipeline of the RDQ-IRR router consumes two clock events in a squeezed scheme if the arbitration takes one clock event (step), as illustrated in Figure 9.3a. We mentioned earlier that the LLD and ViChaR port models are table-based and their router pipeline takes four clock events for one clock event arbitration, as shown in Figure 9.3b. The arbitration stages in our RDQ router implementation follow the timing diagram of Figure 9.4. The pipeline stages for LLD and ViChaR routers take two clock events longer than that of the RDQ router. Each head flit of a packet must proceed through the stages of Routing Computation (RC), Virtual- channel Allocation (VA), Switch Allocation (SA), and Switch Traversal (ST). The micro-architectures of the aforementioned stages are described here.

9.3.3.1 Pipeline Stages Micro-architectures

The arbiter sub-modules, such as RC, VA, Decoder, *output -arbiter*, Selection (see Figure 9.15) and Grant (see Figure 9.16) illustrate that the routing from the inputs of RA to the outputs of Selection, Grant or VA modules are not sequential. In other words, the outputs of Selection, Grant or VA modules are not sequential. Therefore, the RC, VA and SA stages can determine the winner VC and winner input-port of each output-port in a single iteration. In this way, the data flit related to winner VCs can exit the router at the following clock event, and the VCs can make new requests. There is no intermediate register among the paths from the inputs of RC to the outputs of SA, and the registers related to the *output -arbiter*s only keep the state of SA for the following clock event (Figure 9.17 and 9.14). In this way, all the arbitration stages are performed in one clock event. At the end of the SA stage, if the output of the *output -arbiter* associated with a VC is active, the associated downstream router VC, *grant*, and crossbar addresses are issued. The following additional actions are performed for the head and tail flit:

- If the request is from a header flit, the associated RC and VA outputs are also reserved.
- If the request is from a tail flit, the reservations are also removed.

The *credit_out* (credit) signals associated with *grant*s are issued at the ST stage to store the data in the downstream input-port at the following clock event. The *cr-out* signal, generated in the *Selection* module of Figure 9.15 is repeated in the ST module at the following clock event.

9.3.3.2 RDQ-IRR Router Arbiter

In conventional DAMQ routers (such as LLD), when the requested output of a VC is blocked (i.e. no output credit), the arbiter issues a block signal that causes the input-port to select any other available VC for service. No output credit means the corresponding *VC-full* signal of the downstream router is set. However, the RDQ-IRR arbiter issues the block signal under two conditions, i.e. either on losing switch arbitration to some other input-port or having no output credit. This approach requires some extra hardware in the arbiter of the RDQ router as compared to an LLD router. The LLD arbiter updates a table at each clock cycle, where the table consists of an array of registers where each bit represents the blocking state of a VC of the input-ports. Figure 9.18a shows a typical LLD blocking circuit of one-bit register associated with VC1 of input-port 3. If there is no output credit at each clock cycle, the *VC-block* signal is asserted, otherwise it is de-asserted. In our RDQ-IRR approach, the blocking circuit requires extra hardware (one OR gate per VC), as shown in Figure 9.18b. When the requested output is closed or the input VC loses arbitration to other input-ports, the *VC-block* is asserted; otherwise, it remains de-asserted.

A specific situation can arise in our RDQ-IRR NoC, when a packet blockage condition travels back to relevant up-stream routers related to the blocked packet. However, our VC organization maintains the order of flits associated with the blocked packet. The blockage situation can lead to chain-blocking and a back-pressure is created in the NoC [16]. The implementation of this situation has direct effect on the performance of RDQ-IRR NoC, specifically at low injection rates.

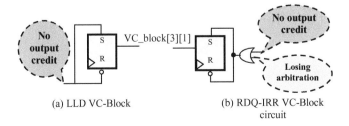

(a) LLD VC-Block   (b) RDQ-IRR VC-Block circuit

FIGURE 9.18
VC-block module associated to VC1 of input-port 3.

9.3.4 Simpler Communication in RDQ-IRR Router

The RDQ-IRR data flow mechanism can be easily explained if there is no contention in the NoC communication. In such a situation, the RDQ-IRR buffer works like static VCs. We employ asynchronous communication among the routers. The following functions describe the working of RDQ-IRR given in Figure 9.9 according to the timing diagram of Figure 9.3a.

Flit Arrival State: Clock-edge 1

1. A *Credit-in* signal causes the incoming flit (e.g. flit F1) and its *VC-ID* to be saved in a slot pointed by the *write-pointer*. The corresponding bit of the *Slot-State* table is also set.
2. When the *read-pointer* points to an occupied slot, a request signal is issued by the *VC-Selector* module according to the *VC-ID*. The *read-pointer* also causes the flit to appear at the output of the central buffer and the flit information (*flit-info* signals) is read by the arbiter.
3. The RC module of the arbiter reads the flit information and determines its requested output-port.
4. The VA module reads the RC output and determines if a free VC in the downstream router input-port is available.
5. If a VC is available then the SA module (decoder and *output -arbiter*) reads the RC output and *VC-req* signals and performs arbitration among the winner VCs for the output-ports.
6. The *Selection* module reads the SA output and determines the associated address for the crossbar switch. The *Grant* module also reads the SA output and sets the *grant* signals for the winner VCs.
7. The *VC-block* module generates the *VC-block* signals under two conditions:
   - Losing switch arbitration to some other input-ports (i.e. *VC-req* is set and its associated *grant* is reset).
   - No output credit. It is determined by RC, VA and *Downstream-VC-state* outputs as illustrated in the *VC-block* module in Figure 9.9.
8. The ST module reads the *cr-out* signals and keeps their record to issue *credit-out* signals after two clock events.

Flit Departure State: Clock-edge 2

1. A *grant* signal causes the associated flit to exit the input-port and the corresponding bit of the *Slot-State* table is reset.
2. The *Sel* signals cause the crossbar to switch the input-ports to their associated output-ports.

3. The *VC-ID-out* signals carry the *VC-ID* of the flit.

4. In case of no output credit or loss of arbitration, the *VC-block* signals cause the associated VC to become blocked (no *grant* signal).

Credit and Next Flit Arrival State: Clock-edge 3

1. The *credit-out* signals are issued by the ST module so that the transferred flits are stored in the downstream router input-ports' buffers.

2. Steps (1) to (8) of Flit Arrival State are repeated for the next incoming flits.

9.3.5 RDQ NoC Design

NoC architectures are commonly presented in Globally Asynchronous Locally Synchronous (GALS) design style [30], and we have also followed the GALS style in our router design for 2D-mesh NoC. For GALS-based NoCs, routers that are locally synchronous are easier to design where the NoC architectures are globally asynchronous. It means that routers are independent of the NoC in terms of their clock, and the faster clock rates of routers lead to a faster NoC. The NoC router circuitry can be divided into synchronous and asynchronous components. The speed of the asynchronous components is almost proportional to the semiconductor technology of silicon. The state of synchronous components changes with their clock signal, and the speed of these circuits depends on their maximum possible clock rates (f_{max}). The f_{max} is determined by the critical path, which is related to the slowest logic path in the circuit. If a router consists of some pipelined components, the synchronous components determine the speed of the router. The synchronous components of an NoC router are those that utilize synchronous data buffers (such as registers or RAM). The crossbar switch has an asynchronous architecture, as illustrated in Figure 9.5, and it does not affect the speed of the router. However, input-port and arbiter utilize synchronous buffers to temporarily store data flits or information and affect the speed of a router. These two components are investigated in terms of their pipelined stages and critical path delay in the following sections.

We propose an efficient and fast DAMQ router architecture called RDQ-IRR that employs RDQ-based input-ports and IRR arbiters in NoC routers. The features of our approach are listed below:

- The arbitration among the VCs of RDQ-IRR routers follows a FIFO style.
- The arbitration in switch allocation module of RDQ-IRR routers ensures strong fairness and avoids traffic starvation.
- The RDQ-IRR router arbiter architecture benefits from easy separation or pipelining. It is very fast, which makes it a proper choice for NoC applications where latency is critical.

- The flit arrival/departure in an RDQ-IRR router is faster than that of other table-based DAMQ routers. In addition to saving one clock cycle for each flit arrival/departure at the input-port, the critical path delays of RDQ-IRR router components are lower as compared to those of other DAMQ routers.

- In addition to higher performance metrics, RDQ-IRR routers are simpler and faster, as well as consuming less power and chip area.

9.4 RDQ-IRR-based NoC Experimental Results

We set up eight types of NoCs based on the conventional LLD, ViChaR [3, 4, 11] and novel EDVC [15] and RDQ approaches discussed in this paper. The hardware structure and performance of these NoCs are evaluated and compared to illustrate the efficiency of our RDQ-based approaches. The NoC architectures follows GALS scheme and the links between routers are assumed to have no effect on the performance and hardware requirements of the NoCs. The architectural details of eight NoCs are presented. The RDQ-IRR NoC employs EDVC-based RDQ input-port [15] and IRR arbiters [26] in its routers. The remaining NoCs utilize LLD [4, 11] and ViChaR [3] input-ports and one of the RoR, Matrix or HDRA arbiters in their router structures, and they are called LLD-RoR, LLD-Matrix and LLD-HDRA, as well as ViChaR-RoR, ViChaR-Matrix and ViChaR-HDRA, according to the input-ports and arbiters utilized in their routers. The crossbar switch module has identical structures for all the eight NoCs due to the same input-port buffer width (16 bits) and same 2D-mesh NoC topology.

9.4.1 RDQ-IRR-based NoC Hardware Requirements

We have evaluated all the above-mentioned eight NoCs in terms of main hardware characteristics such as power consumption, chip area, and speed, which are determined by Verilog implementation and employing Synopsys Design Compiler. ASIC technology libraries such as 15 nm NanGate are used to obtain the evaluation results [31]. The setup constraints and semiconductor technology parameters, such as global operating voltage of 0.8 V and time period of 1 *nsec*, are applied for all the NoC components evaluated. The width of the slot buffer is equal to the flit size of 16 bits. The hardware characteristics of various modules of all the NoC routers are evaluated, and the results are presented in Tables 9.2 and 9.3. Table 9.2 results are organized according to number of VCs and buffer slots utilized in each input-port. For example, the input-port (*port_LLD_HDRA_4 v_4 s*) related row in the table provides chip area, power and critical path delay of an LLD-HDRA input-port utilizing four VCs and having four slots in its buffer. The LLD-HDRA input-port is based on LLD DAMQ utilizing HDRA arbiter in its *VC- Selector* (see Figure 9.7).

EDVC and RDQ input-ports do not utilize separate *VC- Selector*. An important characteristic of high-scaled CMOS technologies (such as 15 nm technology) is that the static (leakage) power supersedes the dynamic power at 1 GHz frequency. For example, the average dynamic power of input-ports is almost 10% of their total power, which indicates that the power is more or less proportional to the hardware overhead of the design rather than its functionality. In other words, more cells consume more static power and the synthesis results given in Tables 9.2 and 9.3 also confirm this. The ViChaR-based routers are expensive in terms of hardware cost as compared to other routers. The higher cost of ViChaR is due to the large control table (even bigger than that of LLD) [15]. The LLD- and ViChaR-based routers that utilize Matrix arbiters consume more chip area but have lower critical path delay. This is due to the fact that the Matrix arbiter uses more registers than other arbiters. For example, the number of registers employed by Matrix, RoR, HDRA and IRR arbiters is 28, 8, 8 and 3 respectively. The EDVC- and RDQ-based input-ports employ optimal hardware as compared to the other input-ports with the same number of VCs and buffer slots. This is due to their lower hardware overhead, as discussed earlier in this chapter. The RDQ router architecture has extra hardware in terms of an AND gate and some registers per VC (see Figure 9.11). The registers are used in the Blocking Header table. In this way, EDVC input-ports consume less hardware than RDQ input-ports for four and eight buffer slots. However, RDQ read- and write-pointers scale with the index of slot number, as opposed to those of EDVC, that scale with the number of slots. In this way, RDQ input-ports become optimal as compared to EDVC input-ports for 16 and 32 buffer slots.

The main difference among the various arbiter structures is the RR arbiters used in SA modules. A few extra OR gates are used in RDQ and EDVC VC-block modules. For example, the *IRR_8v* arbiter utilizes an IRR arbiter in its SA module and five OR gates in VC-block modules, as compared to the *RoR_8v* arbiter that utilizes an RoR arbiter in its SA module. This trend among the arbiter characteristics can be observed in Table 9.3. The Matrix-based arbiters consume more chip area but have lower critical path delay as compared to RoR- and HDRA-based arbiters. However, *IRR_4v* and *IRR_8v* arbiters consume optimal power and chip area as compared to other arbiters. An important point relates to the identical critical path delays for 4-VC and 8-VC arbiters, as shown in the results of Table 9.3. This is due to the same RR arbiters (5-input) being used in both 4-VC and 8-VC arbiters. Figure 9.14 shows that the size of RR *output -arbiters* (used in SA) is equal to the number of input-ports of the router.

The crossbar module does not have any critical path delay as it does not utilize any register in its structure. One can observe that the critical path delays of the arbiters shown in Table 9.3 are less than those of the corresponding input-ports of Table 9.2 despite their higher area and power

TABLE 9.2

Input-port hardware characteristics

VCs	Slots	Input-Port Model	ASIC Design 15 nm NanGate Library		
			Area (μm²)	Power[a] (uW)	Critical path (ps)
4	4	port_LLD_RoR_4v_4s	376	172	111
		port_LLD_Matrix_4v_4s	382	173	97
		port_LLD_HDRA_4v_4s	377	183	109
		port_ViChaR_RoR_4s	391	202	135
		port_ViChaR_Matrix_4s	397	206	116
		port_ViChaR_HDRA_4s	392	220	130
		port_EDVC_4v_4s	290	64	70
		port_RDQ_4v_4s	301	65	53
	8	port_LLD_RoR_4v_8s	632	252	123
		port_LLD_Matrix_4v_8s	638	254	112
		port_LLD_HDRA_4v_8s	633	264	121
		port_EDVC_4v_8s	596	96	88
		port_RDQ_4v_8s	594	98	81
	16	port_LLD_RoR_4v_16s	1197	435	147
		port_LLD_Matrix_4v_16s	1202	435	133
		port_LLD_HDRA_4v_16s	1198	453	140
		port_EDVC_4v_16s	1281	175	132
		port_RDQ_4v_16s	1181	171	97
	32	port_LLD_RoR_4v_32s	2409	851	195
		port_LLD_Matrix_4v_32s	2415	847	195
		port_LLD_HDRA_4v_32s	2410	863	195
		port_EDVC_4v_32s	2714	359	168
		port_RDQ_4v_32s	2385	332	114
8	8	port_LLD_RoR_8v_8s	881	392	172
		port_LLD_Matrix_8v_8s	937	437	142
		port_LLD_HDRA_8v_8s	887	421	174
		port_ViChaR_RoR_8s	1194	741	195
		port_ViChaR_Matrix_8s	1237	777	162
		port_ViChaR_HDRA_8s	1198	771	190
		port_EDVC_8v_8s	623	102	89
		port_RDQ_8v_8s	644	105	84
	16	port_LLD_RoR_8v_16s	1424	555	179
		port_LLD_Matrix_8v_16s	1469	595	149
		port_LLD_HDRA_8v_16s	1430	586	173
		port_ViChaR_RoR_16s[b]	4578	3016	302
		port_ViChaR_Matrix_16s[b]	4814	3217	232
		port_ViChaR_HDRA_16s[b]	4589	3054	265
		port_EDVC_8v_16s	1333	185	132
		port_RDQ_8v_16s	1271	183	97

(Continued)

TABLE 9.2 (CONTINUED)

Input-port hardware characteristics

| VCs | Slots | Input-Port Model | ASIC Design 15 nm NanGate Library | | |
			Area (μm^2)	Power[a] (uW)	Critical path (ps)
	32	port_LLD_RoR_8v_32s	2667	983	202
		port_LLD_Matrix_8v_32s	2700	1021	202
		port_LLD_HDRA_8v_32s	2673	1024	202
		port_ViChaR_RoR_32s[b]	19004	13258	454
		port_ViChaR_Matrix_32s[b]	19888	14062	327
		port_ViChaR_HDRA_32s[b]	19024	13291	343
		port_EDVC_8v_32s	2820	377	169
		port_RDQ_8v_32s	2515	340	114

[a] Frequency for power estimation = 1 GHz; the static power is around 10% of total power.
[b] The number of VCs of ViChaR input-port is equal to the number of slots of input-port.

TABLE 9.3

Arbiter Hardware Characteristics

| VCs | Arbiter Model | ASIC Design 15 nm NanGate Library | | |
		Area (μm^2)	Power (uW)[a]	Critical path (ps)
4	Matrix_4v	773	436	36
	RoR_4v	701	377	58
	HDRA_4v	710	432	56
	IRR_4v	712	366	42
8	Matrix_8v	2092	877	36
	RoR_8v	2020	817	58
	HDRA_8v	2029	874	56
	IRR_8v	2031	806	42
4/8	crossbar	104	35	0

[a] Frequency for power estimation = 1 GHz; the static power is around 10% of the total power.

characteristics. For example, the critical path of arbiter *Matrix_4v* is almost half of the *port_LLD_Matrix_4v_4s* critical path. This is due to a couple of features of our IRR arbiter. First of all, the SA module is separable, which reduces the SA logic complexity. Secondly, the *input-arbiter*s of SA are accommodated in the input-port. Therefore, considering the critical path delays of input-ports, arbiters and crossbar switch modules, the critical path delays of input-ports determine the maximum operating frequency, f_{max}, of the routers as listed in Table 9.4.

The power and area characteristics of each router given in Table 9.4 are calculated by adding the characteristics of five input-ports, an arbiter, and a crossbar switch listed in Tables 9.2 and 9.3. Table 9.4 also lists the advantage rate of our RDQ-IRR router as compared to the other routers. One can observe that on average the RDQ-IRR routers have 11% less chip area, 50% less power consumption, and 103% faster frequency as compared to LLD and ViChaR routers for 4-VC router implementations. For the 8-VC buffer, our RDQ-IRR routers have at least 36% less chip area, 69% less power consumption, and 129% faster frequency as compared to LLD and ViChaR routers.

9.4.2 Performance Evaluation of RDQ-IRR NoC

Latency and throughput are the main performance parameters of NoCs, which are measured for RDQ-IRR NoC evaluation. The NoCs are implemented in System Verilog and we employ ModelSim to obtain these performance parameters. We explore and compare various NoCs mentioned earlier such as RDQ-IRR, EDVC-IRR, LLD-RoR, LLD-Matrix, LLD-HDRA, ViChaR-RoR, ViChaR-Matrix and ViChaR-HDRA for 8×8 NoC mesh topologies and for Uniform-random, Tornado and Complement traffic patterns [6, 15]. The throughput is measured by the rate of packets received to the maximum number of packets being injected at a specific time. The packet communication is based on wormhole switching where the channel width is equal to the flit size (16 bits). A packet consists of 16 flits, and each input-port has a central eight-slot buffer. There are four VCs per input-port except for ViChaR, which has eight VCs (the number of VCs in ViChaR is equal to the number of input-port buffer slots). The throughput and latency are measured for flit injection rates per time unit. For example, flit injection rate 8 means that each node (source core) injects eight flits per time unit. The maximum injection rate is considered based on the capability of delivering the flits by NoC routers. We have already mentioned that the flit arrival/departure for RDQ and EDVC routers is one cycle as compared to two cycles for LLD- and ViChaR-based routers. Therefore, if a time unit is assumed to equal 16 clock cycles, the LLD- and ViChaR-based sources can inject a maximum of eight flits, and the injection of more than eight flits is impossible. However, the RDQ- and EDVC-based source cores can inject a maximum 16 flits per time unit, and the injection of more than 16 flits is not possible. For the sake of fair comparison, we consider a maximum of eight-flit injection rate for RDQ and EDVC in our simulation.

The average latency is measured by the average time delays associated with the departure and arrival of a specific number (e.g. 2048) of packets in the NoC. The average latency estimation is formulated as follows:

$$\text{Average latency} = (\text{TRT} - \text{TST}) / \text{NSF} \qquad (9.2)$$

where:
  TRT is the total received time of flits,
  TST is the total ideal sent times of flits,
  NSF is the total number of sent flits.

TABLE 9.4

Router Characteristics and Advantage Rate

# of VCs	# of Slots	NoC Router Model	ASIC Design 15 nm NanGate Library			RDQ-IRR Advantage Rate		
			Area (μm²)	Power[a] (uW)	Delay (ps)	Area (saving)	Power (saving)	Frequency (faster)
4	4	LLD-RoR	2685	1272	111	14%	43%	109%
		LLD-Matrix	2787	1336	97	17%	46%	83%
		LLD-HDRA	2699	1382	109	14%	47%	106%
		ViChaR-RoR	2760	1422	135	16%	49%	155%
		ViChaR-Matrix	2862	1501	116	19%	52%	119%
		ViChaR-HDRA	2774	1567	130	16%	54%	145%
		EDVC-IRR	2266	721	70	-2%	-1%	32%
		RDQ-IRR	2321	726	53	NA	NA	NA
	32	LLD-RoR	12850	4667	195	1%	56%	71%
		LLD-Matrix	12952	4706	195	2%	56%	71%
		LLD-HDRA	12864	4782	195	1%	57%	71%
		EDVC-IRR	14386	2196	168	11%	6%	47%
		RDQ-IRR	12741	2061	114	NA	NA	NA
8	8	LLD-RoR	6529	2812	172	18%	51%	105%
		LLD-Matrix	6881	3097	142	22%	56%	69%
		LLD-HDRA	6568	3014	174	18%	55%	107%
		ViChaR-RoR	8094	4557	195	34%	70%	132%
		ViChaR-Matrix	8381	4797	162	36%	72%	93%
		ViChaR-HDRA	8123	4764	190	34%	71%	126%
		EDVC-IRR	5250	1351	89	-2%	-1%	6%
		RDQ-IRR	5355	1366	84	NA	NA	NA

(Continued)

TABLE 9.4 (CONTINUED)

Router Characteristics and Advantage Rate

# of VCs	# of Slots	NoC Router Model	ASIC Design 15 nm NanGate Library				RDQ-IRR Advantage Rate		
			Area (μm²)	Power[a] (uW)	Delay (ps)	Area (saving)	Power (saving)	Frequency (faster)	
32		LLD-RoR	15459	5767	202	5%	56%	77%
		LLD-Matrix	15696	6017	202	6%	58%	77%
		LLD-HDRA	15498	6029	202	5%	58%	77%
		ViChaR-RoR	97144	67142	454	85%	96%	298%
		ViChaR-Matrix	101636	71222	327	86%	96%	187%
		ViChaR-HDRA	97253	67364	343	85%	96%	201%
		EDVC-IRR	16235	2726	169	9%	7%	48%
		RDQ-IRR	14710	2541	114	NA	NA	NA

[a] Frequency for power estimation = 1 GHz; the static power is around 10% of total power.

The maximum latency in our design cannot exceed a specific limit. This is due the fact that an injection rate of more than 8 in LLD and ViChaR is impossible. In other word, considering Equation 9.2, the TRT is not changed when the flit injection rate becomes more than 8 (in LLD and ViChaR designs), but the TST can become zero theoretically (i.e. all the flits are injected in time zero). The NSF is fixed and equal to the multiplication of three parameters: the number of injected packets (e.g. 2048), the number of sources (for 8×8 NoC, it is 64), and the flits/packet (i.e. 16). Therefore, the maximum average latency is fixed and equal to TRT/NSF.

LLD-RoR, LLD-Matrix and LLD-HDRA NoCs behave similar to each other in terms of functionality at the same frequency. It is due to the fact that RoR, Matrix and HDRA arbiters behave in a functionally similar way. For instance, LLD-Matrix with faster clock rate supersedes the LLD-RoR and LLD-HDRA NoCs in terms of performance. The same conclusion can be drawn for the ViChaR-Matrix NoC. Therefore, four fast NoCs (LLD-Matrix, ViChaR-Matrix, EDVC-IRR and RDQ-IRR) were selected for evaluation, and the comparison is presented in Table 9.4. The LLD-Matrix, ViChaR-Matrix, EDVC-IRR, and RDQ-IRR run at 565, 457, 718 and 1000 MHz clock respectively with 4-VC configuration. These clock rates correspond to the critical path delays listed in Table 9.4.

The performance metrics of each NoC depend on the functional behavior of its data flow mechanism and the timing characteristics associated with the router. The critical path delays associated with the router of each NoC must be considered in the evaluation of NoC performance. Therefore, we tested these NoCs under different clock rates according to the critical path delays associated with their routers. In this experiment, the performance parameters of aforementioned NoCs are evaluated and the results are shown in Figures 9.19 and 9.20. The clock rate is in linear relation to the performance metrics. Consider n packets passing through the NoC system during t time at a clock rate of f. Then $p \times n$ packets will pass through the NoC system at $p \times f$ clock rate during t time. One can observe that the RDQ-IRR latency and throughput results presented in Figures 9.19 and 9.20 show better performance than those of others. At higher injection rates, more flits are injected, and the NoCs become populated, producing higher contention. The back-pressure mechanism associated with the blocked packets in RDQ helps to improve the performance of RDQ-IRR-based NoC as follows [15]. First, the probability of a monopoly of an input-port buffer by a growing VC is reduced. Secondly, the free packets receive more buffer-free space to pass through the NoC. Moreover, the RDQ-IRR NoC frequency employed for the results presented in this section is higher than those of the other approaches, and the flit arrival/departure time for RDQ-IRR and EDVC-IRR routers is one cycle as compared to two cycles for LLD-Matrix and ViChaR-Matrix routers. The EDVC-IRR NoC has lower performance as compared to the RDQ-IRR due to the fact that the RDQ mechanism is an improved version of EDVC [16]. The average RDQ-IRR latencies are 78%, 83% and 81% less than those of

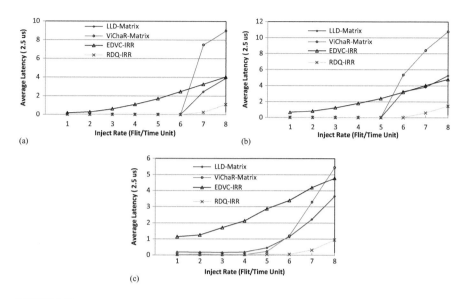

FIGURE 9.19

Latency for Tornado and Complement and Uniform-Random traffic in 8×8 mesh topology. (a) Tornado traffic (b) Complement traffic, (c) Uniform-random traffic.

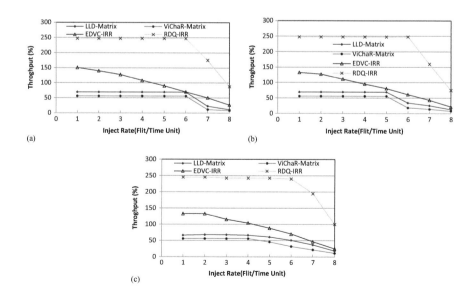

FIGURE 9.20

Throughput for Tornado, Complement and Uniform-Random traffic for 8×8 mesh NoC. (a) Tornado traffic (b) Complement traffic (c) Uniform-random traffic.

LLD-Matrix for Tornado, Complement and Uniform-random traffic patterns respectively. Average throughputs of RDQ-IRR are 74%, 75% and 75% higher than those of LLD-Matrix for Tornado, Complement and Uniform-random traffic patterns respectively.

We have also investigated the performance impact of our RDQ-IRR approach for three application-specific NoCs: an MPEG-4 decoder [32] (MPEG-4), an Audio/Video Benchmark application [32] (AV), and an enhanced Video Object Plane Decoder [33] (DVOPD) with the capability to decode two streams in parallel. The MPEG-4, AV, DVOPD applications are mapped to 3×4, 4×4, and 4×7 2D-mesh topology NoCs respectively. The core graphs of MPEG-4, AV and DVOPD applications are given in Figures 21 and 22. The communication of packets is based on wormhole routing and follows a deterministic XY routing algorithm. The arrows show the direction of packet from source cores to sinks. They also specify the number of VCs needed for each channel. For example, three arrows toward the north input channel of router #5 (Figure 9.21 left) means three packets may pass through the channel at the same time. Three packets require three VCs to service them without any blockage. The other setup conditions of NoCs are the same as the previous experiment, i.e. 8-slot/buffer, 4-VC/input-port, 16-bit/flit, and 16-flit/packet. The latency is measured through a time slice in which all the destinations receive their assigned packets, e.g. when MPEG-4 and AV destinations receive 55472 and 380128 packets respectively.

The contentions in these three NoC applications are low due to two conditions. First, we have set up 4-VC per input-port that is more than the maximum requested VCs in these applications. For instance, the maximum requested VCs in MPEG-4, AV and DVOPD are 3, 2 and 2 respectively, as one can see from Figures 9.21 and 9.22. Secondly, the packet flows in many routes are not crowded, e.g. the west input channel of router #11 of MPEG-4 carries the highest packet flow. Under the above conditions, NoC frequency

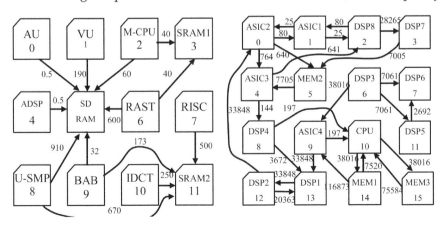

FIGURE 9.21
Mapping of MPEG-4 (left) and AV (right) core graphs to a 3×4 and 4×4 mesh topology NoC.

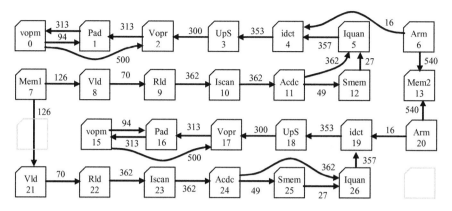

FIGURE 9.22
DVOPD core graph map in a 4×7 mesh NoC.

and flit arrival/departure have great impact on the performance of NoCs. As already mentioned, the LLD-Matrix, ViChaR-Matrix, EDVC-IRR, and RDQ-IRR run at 514, 451, 820 and 1000 MHz clock respectively, and the EDVC-IRR and RDQ-IRR routers take one cycle, as compared to two cycles for the LLD-Matrix and ViChaR-Matrix routers. The simulation results presented in Figure 9.23 support our above analysis. The results show that the average latency of RDQ-IRR is 72%, 92% and 78% less than the average latency of LLD-Matrix for MPEG-4, AV and DVOPD applications.

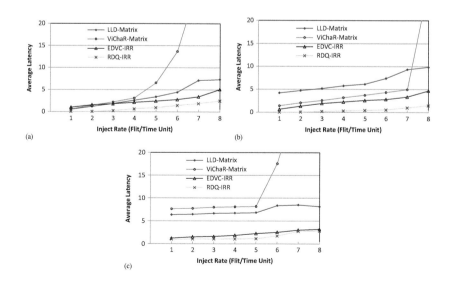

FIGURE 9.23
Latency for MPEG-4, AV and DVOPD Traffic for 3×4, 4×4 and 5×5 mesh NoCs respectively: (a) MPEG-4 traffic; (b) AV Benchmark traffic; (c) DVOPD traffic.

9.5 Conclusions

We have presented a novel NoC router micro-architecture that employs an efficient buffering (RDQ) organization along with a faster arbitration (IRR) technique. The RDQ input-port has been presented as a solution to the EDVC [15] drawbacks. The IRR arbiter employs a least- recently- served priority scheme and achieves strong fairness arbitration. The addition of these two components (i.e. RDQ input-port and IRR arbiter) has been investigated for their impact on the efficiency of NoC systems. The experimental results related to required NoC chip area and power consumption have also been presented. These verify that our RDQ-IRR NoC is low-power and economical. The RDQ-IRR routers have at least 14% less chip area, 43% less power consumption, and 83% faster frequency as compared to the LLD and ViChaR routers for 4-VC and 4-slot implementations. Moreover, the RDQ-IRR NoCs operate at on average 74%, 75% and 75% higher throughputs and 78%, 83% and 81% lower latencies than those of LLD for Tornado, Complement and Uniform-random traffic patterns respectively. The effectiveness of our RDQ-IRR NoCs has also been investigated for some applications, such as MPEG-4, AV benchmark and DVOPD. The average RDQ-IRR latency decreases by 72%, 92% and 78% as compared to that of LLD-Matrix NoCs for MPEG-4, AV and DVOPD applications.

Acknowledgments

The author acknowledges and thanks his graduate students for generating the experimental results presented in this chapter. The financial and equipment support provided by the Electrical and Computer Engineering Department & FEAS at Ryerson University, NSERC and CMC Canada.

References

1. W.J. Dally and B. Towles. (2004). Arbitration. In: *Principles and Practices of Interconnection Networks*. Morgan Kaufmann Publishers, pp. 349–362.
2. W.J. Dally and B. Towles. (2004). Buffered flow control. In: *Principles and Practices of Interconnection Networks*. Morgan Kaufmann Publishers, pp. 233–256.
3. C.A. Nicopoulos, P. Dongkook, K. Jongman, N. Vijaykrishnan, M.S. Yousif, and C.R. Das. ViChaR: A dynamic virtual channel regulator for network-on-chip routers, in *Proc. 39th IEEE/ACM Int. Symp. on Microarchitecture*, pp. 333–346, Orlando, FL, Dec. 2006.

4. M. Evripidou, C. Nicopoulos, V. Soteriou, and J. Kim, Virtualizing virtual channels for increased network-on-chip robustness and upgradeability, in *Proc. IEEE Symp. on VLSI*, pp. 21–26, Amherst, MA, Aug. 2012.

5. Y. Choi and T.M. Pinkston, Evaluation of queue designs for true fully adaptive routers, *Journal of Parallel and Distributed Computing*, vol. 64, no. 5, pp. 606–616, May 2004.

6. Y. Xu, B. Zhao, Y. Zhang, and J. Yang, Simple virtual channel allocation for high throughput and high frequency on-chip routers, in *Proc. Int. Symp. on High Performance Computer Architecture*, pp. 1–11, Bangalore, India, Jan. 2009.

7. M. Lai, Z. Wang, L. Gao, H. Lu, and K. Dai, A dynamically-allocated virtual channel architecture with congestion awareness for on-chip routers, in *Proc. 45th annual Design Automation Conf.*, pp. 630–633, Anaheim CA, Jun. 2008.

8. Y. Tamir and G.L. Frazier, Dynamically-allocated multi-queue buffers for VLSI communication switches, *IEEE Trans. on Computers*, vol. 41, no. 6, pp. 725–737, Jun. 1992.

9. M. Lai, L. Gao, W. Shi, and Z. Wang, Escaping from blocking: A dynamic virtual channel for pipelined, in *Proc. Int. Conf. Complex, Intelligent and Software Intensive Syst.*, pp. 795–800, Barcelona, Spain, Mar. 2008.

10. D.U. Becker and W.J. Dally, Allocator implementations for network-on-chip routers, in *Proc. of the Conf. on High Performance Computing Networking, Storage and Analysis*, pp. 1–12, Portland, OR, 2009.

11. G.L. Frazier and Y. Tamir, The design and implementation of a multiqueue buffer for VLSI communication switches, in *Proc. IEEE Int. Conf. on Computer Design: VLSI in Computers and Processors*, pp. 466–471, Cambridge, MA, 1989.

12. J. Park, B.W. O'Krafka, S. Vassiliadis, and J. Delgado-Frias, Design and evaluation of a DAMQ multiprocessor network with self-compacting buffers, in *Proc. Supercomputing*, pp. 713–722, Washington, DC, Nov. 1994.

13. P. Forstner. (1999). FIFO architecture, functions, and applications. www.ti.com/lit/an/scaa042a/scaa042a.pdf, Last accessed Dec. 2017.

14. J. G. Delgado-Frias and R. Diaz, A VLSI self-compacting buffer for DAMQ communication switches, in *IEEE Proc. of the 8th Great Lakes Symposium on VLSI*, Lafayette, LA, 1998.

15. M. Oveis-Gharan and G.N. Khan, Efficient dynamic virtual channel organization and architecture for NoC systems, *IEEE Trans. on VLSI Systems*, vol. 24, no. 2, pp. 465–478, Feb. 2016.

16. M. Oveis-Gharan and G.N. Khan, Dynamic VC organization for efficient NoC communication, in *Proc. IEEE 9th Int. Symp. on Embedded Multicore/Many-core Systems-on-Chip (MCSoC)*, pp. 151–158, Turin, Italy, Sep. 2015.

17. G. Jiang, Z. Li, F. Wang, and S. Wei, A low-latency and low-power hybrid scheme for on-chip networks, *IEEE Trans. on VLSI Systems*, vol. 23, no. 4, pp. 664–677, Apr. 2015.

18. Y. Ben-Itzhak, I. Cidon, A. Kolodny, M. Shabun, and N. Shmuel, Heterogeneous NoC router architecture, *IEEE Trans. on Parallel and Distributed Systems*, vol. 26, no. 9, pp. 2479–2492, 2015.

19. R. Ramanujam, V. Soteriou, B. Lin, and L. Peh, Design of a high-throughput distributed shared-buffer NoC router, in *Proc. 4th ACM/IEEE International Symposium on Networks-on-Chip (NOCS)*, pp. 69–78, Grenoble, France, May 2010.

20. D. Zoni, J. Flich, and W. Fornaciari, CUTBUF: Buffer management and router design for traffic mixing in VNET-based NoCs, *IEEE Trans. on Parallel and Distributed Sys.*, vol. 27, no. 6, pp. 1603–1616, Jun. 2016.

21. M. Oveis Gharan and G. N. Khan, Packet-based adaptive virtual channel configuration for NoC systems, *International Workshop on the Design and Performance of Network on Chip, Procedia Computer Science*, vol. 34, pp. 552–558, 2014.

22. T. Yung-Chou and H. Yarsun, Design and evaluation of dynamically-allocated multi-queue buffers with multiple packets for NoC routers, *Sixth International Symposium on Parallel Architectures, Algorithms and Programming (PAAP)*, pp. 1–6, Beijing, China, Jul. 2014.

23. A.T. Tran and B.M. Baas, Achieving high-performance on-chip networks with shared-buffer routers, *IEEE Trans. on VLSI Systems*, vol. 22, no. 6, pp. 1391–1403, Jun. 2014.

24. K.A. Helal, S. Attia, T. Ismail, and H. Mostafa, Priority-select arbiter: An efficient round-robin arbiter, in *IEEE 13th Int. Conf. on New Circuits and Systems (NEWCAS)*, pp. 1–4, Grenoble, France, Jun. 2015.

25. Z. Fu and X. Ling, The design and implementation of arbiters for network-on-chips, in *2nd Int. Conf. on Industrial and Information Systems*, pp. 292–295, Dalian, China, 2010.

26. M. Oveis-Gharan and G.N. Khan, Index-based round-robin arbiter for NoC routers, in *Proc. IEEE Computer Society Annual Symposium on VLSI (ISVLSI)*, pp. 62–67, Montpellier, France, July 2015.

27. Y. Lee, J. M. Jou, and Y. Chen, A high-speed and decentralized arbiter design for NoC, in *Proc. IEEE Int. Conf. on Computer Systems and Applications*, pp. 350–353, Rabat, Morocco, 2009.

28. F. Guderian, E. Fischer, M. Winter, and G. Fettweis, Fair rate packet arbitration in network-on-chip, in *Proc. IEEE Int. System-on-Chip Conference (SOCC)*, Taipei, Taiwan, pp. 278–283, 2011.

29. L. Shaoteng, A. Jantsch, and L. Zhonghai, A fair and maximal allocator for single-cycle on-chip homogeneous resource allocation, *IEEE Trans. on VLSI Systems*, vol. 22, no. 10, pp. 2229–2233, Oct. 2013.

30. M. Fattah, A. Manian, A. Rahimi, and S. Mohammadi, A high throughput low power FIFO used for GALS NoC buffers, in *IEEE Computer Society Annual Symposium on VLSI (ISVLSI)*, pp. 333–338, Lixouri, Greece, July 2010.

31. NANGATE. 2014. Nangate Releases 15nm Open Source Digital Cell Library. [ONLINE] Available at: www.nangate.com. Accessed December 2017.

32. M. Oveis-Gharan, and G.N. Khan, Statically adaptive multi FIFO buffer architecture for network on chip, *Microprocessors and Microsystems*, Vol. 39, No. 1, pp. 11–26, Feb. 2015.

33. A. Pullini, F. Angiolini, P. Meloni, D. Atienza, S. Murali, L. Raffo, G. D. Micheli, and L. Benini, NoC design and implementation in 65nm technology, in *Proc. 1st Int. Symp. on Networks-on-Chip*, Princeton, NJ, pp. 273–282, 2007.

10

Rapid Static Memory Read-Write for Energy-Aware Applications

**Theodoros Simopoulos, Themistoklis Haniotakis,
George Ph. Alexiou, and Nicolas Sklavos**

CONTENTS

10.1 Introduction

Low-power designs and systems are targets of great interest today, especially for energy-aware applications for current and future critical technologies [1–3]. With the miniaturization of the transistor size, as today's IC devices are implemented using deep submicron technology processes, the response of the ICs is quicker [4], with each of their semiconductor elements consuming less power. However, the density of the IC has increased, leading to the search for new design techniques to minimize further the power that is consumed [1,2]. According to the International Technology Roadmap for Semiconductors [5], 90% of the device die logic will be spent on internal static memory in the following years. It is a fact that modern processors, not excluding the ones that are used for systems concerning IOT (see [6,8]), have an increased cache size of at least 1 MB concerning the L2 cache with latency of more than 15 cycles [6–8]. So, it is an essential need to reduce this latency time in order to speed up the system and do it without paying power costs.

However, due to the size of the cache memory, the Bit Lines that form the in-memory data bus (i.e. the data bus of the core of the memory) see an increased grade of parasitic elements. This results not only from the set of the capacitances between the source/drain terminal that is directly connected to its adjacent Bit Line but also from the gate terminal of the memory cell's access transistors and the metal material area of the narrow but long Bit Lines, according to Equations 10.1 and 10.2. All these parasitic elements tend to delay the data that are transferred to and from the IC system. It could be said that sizing down the transistors firstly speeds up the core, which increases the cache latency and, secondly, makes the size of cache larger, which increases the parasitic capacitance.

$$R = \frac{\rho L}{Wt} \tag{10.1}$$

$$C = A\frac{\varepsilon_{di}}{t_{di}} \tag{10.2}$$

Applying the RC lumped model [9] to the memory's Bit Lines, we get a simplified parasitic RC model, as illustrated at Figure 10.1. The delay that is inserted due to the long Bit Lines can be estimated by the resulting equation (10.3), which verifies that the long Bit Lines insert an increased delay to the storage cell's access.

$$T_{DK} = RC + (R+R)C + \cdots + \left(\overbrace{R+R+\cdots+R}^{K \text{ times}}\right)C + R_K C_K$$

$$= (1+2+\cdots+K)RC + R_K C_K \Rightarrow \tag{10.3}$$

$$T_{DK} = \sum_{i=1}^{k} iRC + R_K C_K$$

10.2 Generic SRAM Model

The deconstruction of the single-paged SRAM model reveals its generic block diagram, consisting of three major blocks (see Figure 10.2), each one dedicated to a special memory operation. These blocks are the memory's storage cells array block, which is also the one that occupies most of the area the memory needs, the memory's data handler block and the memory's control and decoding block.

The storage cells block and the data handler block are usually implemented using dedicated cells [10], tiled and abutted together. The use of cells in the

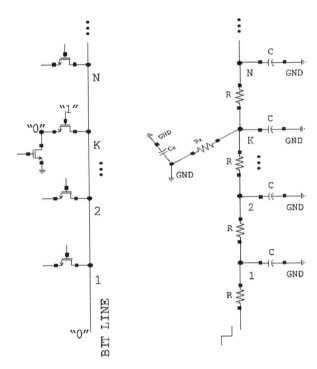

FIGURE 10.1
Bit Line simplified parasitic RC model.

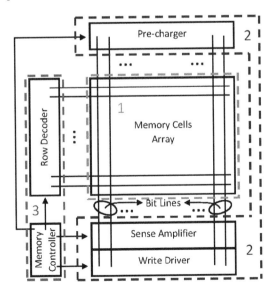

FIGURE 10.2
Basic SRAM structure: (1) Memory cells array block; (2) Memory data handler block; (3) Memory control and decoding block.

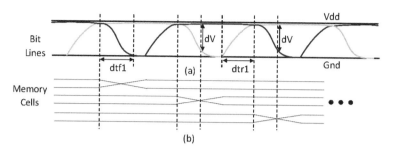

FIGURE 10.3
Conventional Read-Write: (a) Bit Lines transition; (b) Storage cells access.

creation of the in-chip memory array is a technique that is also followed in order to implement the cache memory system of commercial processors, as expressed in [11] concerning the implementation of an Itanium family processor. The control and decoding block can use any implementation method from HDL synthesis to direct layout abutment using standard cells, which are even available online [12,13].

Prior to the memory's Read or Write operation the Bit Lines need to be pre-charged, an internal operation of the memory that is performed by its conditioning circuitry – the pre-chargers, as they are also called. Taking into account the conventional SRAM implementation, the pre-charge of the Bit Lines is done at the Vdd process nominal value. The Read or Write function of the memory leads to the discharge of one of the Bit Lines through the storage cell or the Write driver accordingly. In concern of the memory's write function, only when the Bit Lines have been associated with a minimum voltage difference of dV, as shown in Figure 10.3, can the storage cell's cross-coupled inverters flip their value, while the Read function operates similarly as the Bit Lines make the small but sufficient voltage difference needed to be sensed by the Read mode's amplifier.

10.3 Conventional Storage Cells

Each of the memory storage cells which comprise the first basic block of the static memory (see Figure 10.2), consists of two inverters connected in back-to-back formation (i.e. cross-coupled) and an accessing circuitry to these inverters. Different implementations of the storage cell are possible, which concern the pull-up circuitry of the inverters or the access transistors of the cell. These cells are characterized by the number of the transistors their circuitry includes (i.e. 4T, 6T and so on) or by the number of elements that the accessing circuitry has.

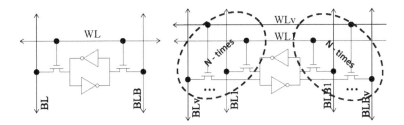

FIGURE 10.4
Conventional static memory cell: (a) Single-ported, (b) n times ported.

As regards the accessing circuitry, the conventional SRAM's storage cells can be divided into the single-ported, the dual-ported and the multi-ported ones in general. In each case, the number of access transistors that are applied to the sides of the cross-coupled inverters, which hold the cell's value, varies, as indicated in Figure 10.4. The single-ported SRAMs are those that can perform consecutive Read or Write operations. A dual-port and in general a multi-port SRAM is able to handle Read and Write cycles simultaneously. However, concerning multi-ported SRAMS, the presence of multiple address and data buses is compulsory. Therefore, multiple decoders are essential in order to serve multi-ported operations.

All the different implementations of the cells are intended to perform successfully three basic operations: the Read operation, the Write operation and the Data Retain operation. Moreover, all different implementations of the cells need the Bit Lines (BL and BLB signals) of the memory to be already precharged by the time the Read or Write function starts. This is an unavoidable task for the proper accomplishment of these particular operations, because SRAMs have an increased width and, because of their increased width, the Bit Lines lengthen as they have to span all the rows of the memory page. However, as already explained, by lengthening the Bit Lines, the area they occupy will inevitably increase, resulting, according to Equation 10.2, in an analogous increase of the line's parasitic capacitance and, therefore, an increased delay.

It is desirable to have the storage cell's size small enough in order to increase the memory's density. However, this leads to the inability of each cell to compensate for the great parasitic capacitance of the Bit Lines. It is this inability of the memory cells that demands the Bit Lines to be pre-charged to Vdd before the beginning of the Read or Write operation.

10.3.1 Conventional Cell Read Operation

Before the read operation starts, the Bit Lines, BL and BLB (see Figure 10.5), are pre-charged to logic level '1' (Vdd). The cell-to-be-read is accessed by inserting a pulse to the WL signal. In most cases, where the access transistors are of type 'n', the WL pulse should be positive. This pulse results in the

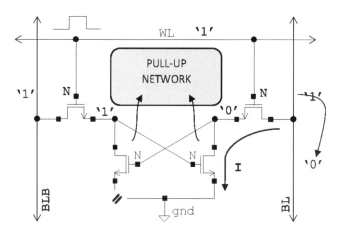

FIGURE 10.5
Cell Read operation.

enabling of the access transistor, which will cause one of the Bit Lines to discharge through the pull-down circuitry of one of the cross-coupled inverters. Due to the parasitics on the Bit Lines, the discharge is of long duration. However, a short voltage difference between them is able to be sensed and enhanced by the Read amplifiers.

The Read operation should not be destructive, which means that the memory must retain the data it held as they were before the Read function. Figure 10.5 illustrates diagrammatically the Read operation of the memory's conventional storage cell, when the cell holds data of logic value '0'.

10.3.2 Conventional Cell Write Operation

As with the Read operation, the Bit Lines, BL and BLB (see Figure 10.6), are pre-charged to logic level '1' (Vdd) before the Write operation starts. Subsequently, the Write driver generates the data-to-be-written by setting at logic '1' one of the Bit Lines and at logic '0' the other. This results in the discharge of one wireline of each Bit Line pair instead the other through the Write driver. After a suitable timing, a pulse is generated to the WL signal, enabling the access transistors of the cell. Supposing that writing the cell will flip the value it already holds, the node of the cross-coupled transistors that holds the logic '1' discharges through the Bit Line its access transistor is connected to. On the other hand, the node that holds the logic '0' is charged via the voltage divider that is formed between its access transistor and the pull-up network. Hence voltage levels of different binary logic are assigned to the nodes of the cross-coupled inverters which form the cell. Figure 10.6 illustrates the Write operation of the general storage cell, writing to it logic '1', when it already holds a value of logic '0'.

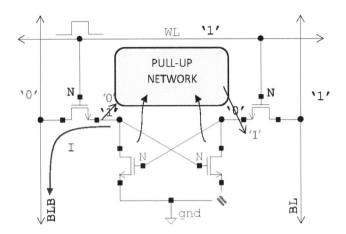

FIGURE 10.6
Cell Write operation.

10.3.3 Data Retain Operation

If the memory cell is not in the Read or Write operation state, then the WL signal is not driven by the memory's address decoder, leading to the disabling of the cell's access transistors. Therefore, in this operation state, the cross-coupled inverters are cut off from the pre-charged Bit Lines of the memory. One inverter outputs logic '1', which is driven as input to the second inverter. This input level should be above the gate threshold voltage of the second inverter. Then this inverter outputs a logic '0', which is driven as input to the first inverter. This input level should be below the gate threshold voltage of the first inverter. The operation is repeated and the cross-coupled inverter retains the data that are stored inside the cell (Figure 10.7).

10.4 Energy-Aware Applications

Mobile electronic devices, in order to ensure a long duration of at least the basic system operation, are equipped with applications that monitor and even manage their power consumption. A number of software solutions have been provided that are quite helpful in this direction, starting from different levels of abstraction. Some solutions perform on the application software abstraction level, as power proofing or management tools [14] and others are enhancements to the system software abstraction level concerning extension of the operation system [15].

Besides the reduction of the power a system consumes via software utilization, hardware enhancements are also possible, most of which target the computational and logic part of the ICs. One effort in this direction is the

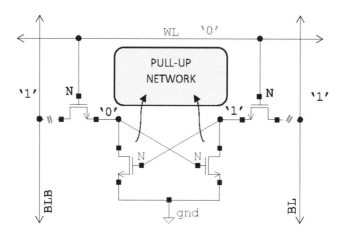

FIGURE 10.7
Cell data-retain operation.

minimization of the power dissipation via the reduction of the bit-width [16], the precision [17] or the switching activity [18] of arithmetic units, where a solution is gained by altering the functional logic that is implemented. On the other hand, using a different design methodology for special IC units is another technique that can reduce the power consumed even further. Often asynchronous implementations of autonomous IC sub-units take place [1], resulting in the possible isolation of the sub-unit when it is in the idle state. Therefore, the power that such a unit consumes can be selectively eliminated.

On the other hand, as more and more applications are loaded to the devices, it is inevitable that the memory size has to grow in order to serve the increasing amount of data that is set under process by the active applications. Therefore, despite minimizing the power dissipation on the computational and logic part of the hardware, a significant and increasing amount of energy is spent by the device's memory system. This leads to the necessity of inserting improvements in the memory system of at least the power-aware devices and thus helping the hardware to reduce the power further. The Rapid Read-Write Static Memory technique is working in this direction as its fast operations allow the transfer of the demanded data to the functioning units quicker. Therefore, in the same duration of time, the system can perform more operations. Moreover, as it is explained, the short Read-Write pulse of the Rapid SRAM leads to less power consumption by these memory micro-operations.

10.5 Rapid SRAM Read-Write Function

The block diagram of the single-paged conventional SRAM model has been extended in order to support the enhancements of the Rapid Read-Write

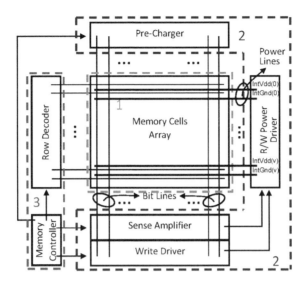

FIGURE 10.8
Rapid Read-Write block diagram.

SRAM model. A Read-Write Power Driver unit has been inserted in the memory's data handler block, which comprises of a number of power cells, one placed at every memory row, which forms the memory word. As a result, every memory word is connected to a Read-Write Power Driver unit (Figure 10.8).

With the Rapid Read-Write Static Memory technique, using a multi-voltage driver [19], it is possible to change the powering scheme of the memory, as illustrated in Figure 10.9. Via this technique the Read-Write pulse of the memory is somewhat shorter, leading to a quick transfer of data between the system and the memory. As the time needed to accomplish the transfer shortens, the energy that is consumed is less, according to the basic formula below:

$$E = P \cdot t \tag{10.4}$$

The Read-Write Power Driver unit generates the IntVdd and IntGnd internal power lines, which are directly connected to the Rapid SRAM's storage cells. Via these power lines a new powering scheme is provided to the storage cell's transistors with the Vdd voltage level their pull-up circuitry sees downgrading, and the Gnd voltage level their pull-down circuitry sees upgrading from the process nominal values. The alternation of the power voltage levels happens when a memory Read or Write operation starts. The behavior of the Read-Write Power Driver unit is shown in Figure 10.9.

Subsequently, the new Rapid SRAM conditioning circuitry pre-charges the Bit Lines of the memory to a voltage level between the IntVdd and IntGnd,

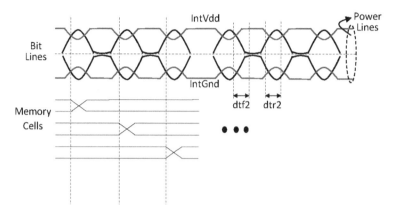

FIGURE 10.9
Rapid Read-Write SRAM signals behavior.

instead of the Vdd voltage level the conventional memories do. The time duration the Rapid SRAM needs in order to pre-charge (dtf2) or discharge (dtr2) the Bit Lines is significantly shorter than the relevant time durations of the conventional SRAM (dtf1, dtr1). The relation between the pre-charge and discharge times of the Rapid SRAM and the conventional SRAM is shown in the following formula:

$$\left\lceil \begin{array}{l} dtf2 \ll dtf1, \text{and} \\[2mm] \quad dtr2 \ll dtr1 \end{array} \right. \tag{10.5}$$

Because of the short and sharp Read-Write pulse and the new pre-charge scheme [19], the time the transistors of the Rapid SRAM's storage cells are in the transition state is shorter. Therefore, less dynamic power is dissipated in order to transfer more data between the memory and the computational-logic part of the IC.

10.6 Rapid SRAM 6T Cell Model

The Rapid SRAM storage cell is based on the 6T model, which was chosen due to its advantages of having a uniform Read and Write stability [20] and a quick respond time [21]. The cell is directly connected with the Read-Write Power Driver cell unit, which is present on the same row. The transistors of the storage cell's cross-coupled inverters are powered by these direct connections (see Figure 10.10).

The BL and BLB lines, which, most often, are present at the sides of the storage cell and form the memory's internal data bus, are driven by the Rapid

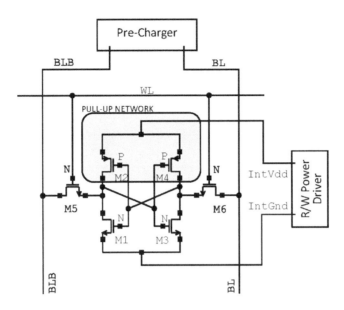

FIGURE 10.10
Rapid SRAM 6T storage cell model.

SRAM's new pre-charge scheme, when the memory bank is activated, and the storage cells are not accessed for writing or reading. The connections of the Rapid SRAM storage cell with the Read-Write Power Driver and the Pre-charger units are illustrated in Figure 10.10.

At the beginning of the Read or Write function, a logic '1' pulse is inserted in the storage cell's WL signal line. This signal line is issued from the memory's Row Decoder unit (i.e. the address decoder), which is placed at the side of the cell's array (see Figure 10.8). The WL pulse causes the access transistors to create a path from the Bit Lines to the intra-cell nodes, where the transistor gates of the opposite inverter are connected (see Figure 10.10). When the cell is either written or read, one of the Bit Lines will start to charge, while the other will start to discharge in order to accomplish the function.

10.6.1 Rapid 6T Read Operation

Figure 10.11 illustrates the reading operation of the Rapid SRAM 6T storage cell when it holds the value '1'. Before the beginning of the Read function, the Rapid SRAM Bit Line Pre-charger has already pre-charged the BL and BLB wirelines to a voltage level suitable for the Rapid SRAM. Simultaneously, the Read-Write Power Driver has altered the Vdd and Gnd voltage levels to IntVdd and IntGnd. When the IntVdd and IntGnd have reached their low and high peak values respectively, the reading function begins, inserting a high pulse at the WL signal. The cell's inverter output at the BLB side is at logic '0', causing the discharge of the BLB line through its corresponding

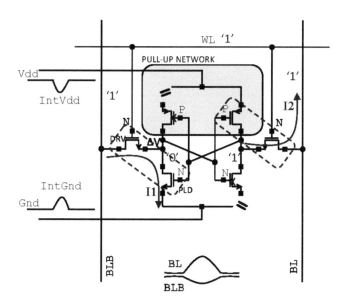

FIGURE 10.11
Rapid SRAM Read operation.

access transistor and the nmos pull-down transistor it is connected to. On the other hand, the cell's inverter output at the BL side is at logic '1', causing a charge of the BL line through its corresponding access transistor and the pmos pull-up transistor it is connected to. This BL–BLB voltage difference is quick sensed by the Read amplifier, which is placed at the bottom of the memory array (see Figure 10.8) and the Read function completes.

10.6.2 Rapid 6T Write Operation

As explained, at the Read operation of the Rapid SRAM, the new Bit Line Pre-charger that the Rapid SRAM makes use of has already pre-charged the BL and BLB wirelines to a suitable voltage level for the memory, before the Write operation starts. With the beginning of the Write operation, the Rapid SRAM's Read-Write Power Driver (see Figure 10.8) has altered the Vdd and Gnd voltage levels to IntVdd and IntGnd. When the IntVdd and IntGnd have reached their low and high peak values respectively, the writing operation begins, inserting a high pulse at the WL signal.

Figure 10.12 illustrates the writing operation of the Rapid SRAM 6T storage cell when it already holds data of logic '0' and the cell is written with data of logic '1'. The BL and BLB signals start to diverge from their pre-charged point, with the BL charging and the BLB discharging. The voltage difference between the BL and BLB signals is applied to the nodes of the cross-coupled inverters of the storage cell, immediately after its access transistors, directing the inverters to flip their value (as explained in Section 10.3.2). When the

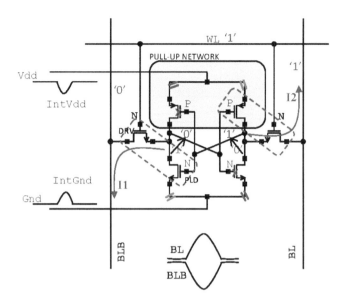

FIGURE 10.12
Rapid SRAM Write operation.

Write operation completes, the storage cell's access transistors are switched off, allowing the convergence of the BL and BLB signals to their pre-charge operating point.

10.6.3 Rapid SRAM 6T Data Retain Operation

The Rapid SRAM 6T transistor is in the data retain operation state when the WL signal is switched off. During this operation, the Read-Write Power Driver unit has raised the IntVdd and lowered the IntGnd to the technology's process nominal Vdd and Gnd voltage levels. The cell's access transistors are disabled and therefore the cell's cross-coupled inverters are cut off the memory's Bit Lines and the new pre-charge scheme. Therefore, in this operation state, the Rapid SRAM 6T storage cell behaves like the conventional 6T storage cell, as explained at section 1.3.3 above (Figure 10.13).

10.7 Conclusion

Power-aware devices can partially reduce the power they consume via software which targets the system's power management and configuration on the application and on the system level. Hardware enhancements can assist the minimization of the power consumption by reducing functionally the

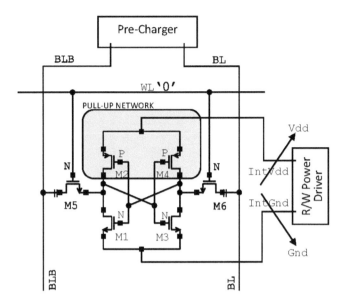

FIGURE 10.13
Rapid SRAM 6T data retain operation.

switching activity or by lowering the bit-width of the system's operations, which again results in the reduction of the signal switching activity.

However, as the size of the in-chip memory grows in order to cope with the increasing number of active applications that are installed in today's mobile devices, a significant portion of the power is dissipated by the memory's micro-operations. Rapid SRAM is a static memory implementation technique that can help to minimize the power that is dissipated by the in-chip memory system. With the Rapid SRAM hardware enhancement technique, power-aware mobile devices can extend their in-battery system's operating time, inserting a hardware solution in this direction.

References

1. N. Sklavos, A. Papakonstantinou, S. Theoharis, and O. Koufopavlou. Low-power implementation of an encryption/decryption system with asynchronous techniques, *VLSI Design, An International Journal of Custom-Chip Design, Simulation, and Testing*, Vol. 15, no. 1, pp. 455–468, 2002.
2. N. Sklavos and K. Touliou. Power consumption in wireless networks: Techniques & optimizations. *Proc. of the IEEE Region 8, EUROCON 2007, International Conference on "Computer as a Tool" (IEEE EUROCON'07)*, Poland, Sep. 9–12, 2007.

3. N. Sklavos. Cryptographic hardware & embedded systems for communications. *Proc. of the 1st IEEE-AESS Conference on Space and Satellite Telecommunications,* Rome, Italy, Oct. 2–5, 2012.

4. J. Ebergen, J. Gainsley, and P. Cunningham. Transistor sizing: How to control the speed and energy consumption of a circuit. *Proc. of the International Symposium on Advanced Research in Asynchronous Circuits and Systems,* IEEE Computer Society Press, Crete, Greece, 2004.

5. International Technology Roadmap for Semiconductors. Reference online at: www.itrs2.net

6. AMD Opteron A1100 Series SoC. Specification available online at: www.amd.com/en-us/press-releases/Pages/amd-and-key-industry-2015jan14.aspx

7. AMD Opteron A1100 Series SoC. Product brief available online at: www.amd.com/Documents/A-Hierofalcon-Product-Brief.pdf

8. Samsung Exynos 4 Dual 32nm RISC Microprocessor. User manual available online at: www.samsung.com/global/business/semiconductor/file/product/Exynos_4_Dual_32nm_User_Manaul_Public_REV100-0.pdf

9. J. Rabaey, A. Chandrakasan, and B. Nicolic. *Digital Integrated Circuits: A Design Perspective,* 2nd ed. Upper Saddle River, NJ: Prentice Hall, 2003.

10. T. Simopoulos, T. Haniotakis, and G. P. Alexiou. CeidMem: A compact static memory library. *Journal of Engineering Science and Technology Review,* Eastern Macedonia and Thrace Institute of Technology, Vol. 9, no. 4, pp. 138–141, 2016.

11. S. Naffziger, et al. The implementation of a 2-core multi-threaded Itanium family processor. *IEEE Journal of Solid-State Circuits,* Vol. 41, no. 1, pp. 197–209, Jan. 2006.

12. T. Simopoulos, T. Haniotakis, and G. P. Alexiou. Implementation of a low leakage standard cell library based on materials from UMC 65nm technology. *Proc. of the 18th Panhellenic Conference on Informatics (PCI), Hardware Design Tools and Technologies Special Session,* Athens, Greece, Proceedings by ACM, 2014.

13. Virginia Tech VLSI for Wireless Communications (VTVT) Cell Libraries. Available online at: www.vtvt.ece.vt.edu/vlsidesign/cell.php

14. J. Flinn and M. Satyanarayanan. PowerScope: A tool for profiling the energy usage of mobile applications. *Proc. of 2nd IEEE Workshop Mobile Computing Systems and Applications,* p. 2, New Orleans, LA, USA, 1999.

15. A. Weissel, B. Beutel, and F. Bellosa. Cooperative I/O – A novel I/O semantics for energy-aware applications. *Proc. of the Symposium on Operating Systems Design and Implementation (OSDI),* Boston, MA, USA, Aug. 2002.

16. A.A. Gaffar, J.A. Clarke, and G.A. Constantinides. PowerBit – Power aware arithmetic bit-width optimization. *Proc. of the International Conference. on Field-Programmable Technology,* pp. 289–292, Dec. Bangkok, Thailand, 2006.

17. Y.F. Tong, R.A. Rutenbar, and D.F. Nagle. Minimizing floating-point power dissipation via bit-width reduction. *Power Driven Microarchitecture Workshop,* Spain, in conjunction with ISCA '98, Barcelona, pp. 114–118, June 1998.

18. V.G. Moshnyaga. Reducing switching activity of subtraction via variable truncation of MSB. *Journal of VLSI signal processing systems for signal, image and video technology,* Vol. 33, nos. 1–2, pp. 75–82, 2003.

19. T. Simopoulos, T. Haniotakis, and G.P. Alexiou. Memory write speed up via multi voltage driver on CeidMem Library. *Microelectronic Engineering Journal,* Elsevier, Vol. 163, pp. 21–25, Sep. 2016.

20. E. Seevinck, F.J. List, and J. Lohstroh, Static-noise margin analysis of MOS SRAM cells. *IEEE Journal of Solid-State Circuits*, Vol. SC-22, no. 5, pp. 748–754, Oct. 1987.
21. W. Singh and G.A. Kumar. Design of 6T, 5T and 4T SRAM cell on various performance metrics. *Proc. of the IEEE International Conference on Computing for Sustainable Global Development*, pp. 899–904, New Dehli, India, 2015.

11

Application Specific Integrated Circuits for Direct X-Ray and Gamma-Ray Conversion in Security Applications

Kris Iniewski, Chris Siu, and Adam Grosser

CONTENTS

11.1 Introduction

11.1.1 Ionizing Radiation and Its Detection

Radiation strikes people all the time, but most of it, like radio waves and visible light, is not ionizing and typically does not damage human cells unless excessive heat is dissipated in small areas of the human body. However, ionizing radiation has enough energy to knock electrons out of atoms, creating electrically charged particles that can damage human body cells. At high levels ionizing radiation may lead to cancer, genetic mutations, sickness, or in extreme cases death.

Radiation damage can be classified in two classes. Deterministic effects are those for which the severity of the effect varies with the dose, therefore specific thresholds can be assigned for these effects to occur. Examples of such effects include vomiting, diarrhea, or hemorrhage. Stochastic effects are those for which only the probability that an effect will occur can be determined, rather than the severity of the effect, and therefore no specific thresholds can be assigned. Stochastic effects include cancers and genetic damage.

People are exposed to background levels of ionizing radiation every day from such sources as granite (which often contain traces of uranium, radium, and radioactive potassium), radon gas, and cosmic rays. Food and drinking water generally contains trace amounts of radioactive materials. The radiation dose from a long-distance flight is a few millirems (mrem), from a chest X-ray about 4–6 mrem; and from living at high altitudes it can be as high as 50 mrem per year. Needless to say, ionizing radiation is difficult to avoid and can clearly be dangerous to our health and well-being if we are exposed to too much of it. Control of radiation exposure is therefore a heavily regulated area of technology. For example, the Environmental Protection Agency (EPA) in the USA uses a standard of 25 mrem per year of whole-body dose. Other international agencies adopt various standards and safe dose levels, in general in the 5–100 mrem range.

One extreme example of the radiation threat is a nuclear explosion. Nuclear weapons generate massive amounts of radiation as neutrons, gamma rays,

X-rays, visible light and infrared. The resulting radiation doses can cause sickness or death in hours to months. Some less threatening materials, so-called dirty bombs, do not involve a nuclear explosion but might contain some tens of grams of radioactive material, enough to contaminate a large area. Such a dirty bomb would generate a hazardous dose of radiation over a much smaller area than would a nuclear weapon but still would contaminate the area and could be very damaging to a population exposed to such an attack. The first portion of this chapter is devoted to methods of identifying nuclear materials using gamma -ray detection.

While ionizing radiation can clearly be very harmful to humans and other living organisms it also can play a very useful role. The ability to look inside the human body using X-rays started with German physicist Wilhelm Röntgen in late 19th century and continues today with the use of modern Computed Tomography (CT) equipment. Similarly, X-rays serve as the best way to detect explosive material in baggage scanned at airports. Presently used baggage X-ray scanning equipment varies from basic machines that utilize simple X-ray imaging, similar to the one used in chest X-rays, to complex and sophisticated scanners that utilize X-ray diffraction (XRD) effects or analyze objects in 3D domain using Computed Tomography. The second portion of this chapter is devoted to methods of identifying dangerous baggage content using X-ray detection.

On the electromagnetic spectrum, X-rays have wavelengths from 0.01 nm to 10 nm, as shown in Figure 11.1. Due to wave-particle duality, it is also useful to consider X-rays as high-energy photon particles, with the electron volt (eV) as a measure of their energy that ranges from a few keV to a few MkeV. The term 'gamma rays' is used to denote that the source of them is nuclear decay.

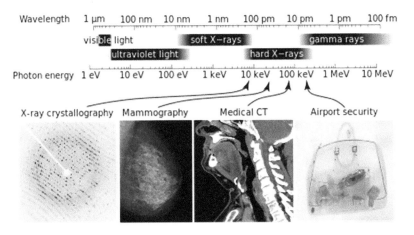

FIGURE 11.1
X-ray, gamma ray and the electromagnetic spectrum [1].

X-rays are classified as an ionizing radiation; the photon energy is suffi-cient to ionize atoms and disrupt molecular bonds, and therefore is harmful to living tissue at high doses. On the other hand, X-ray photons can also pass through objects that are opaque to visible light, and X-rays at low dosages are used for medical imaging. X-rays can be generated by a vacuum tube with a high voltage to accelerate the electrons from the cathode to the anode. When the highly energetic electrons bombard the anode (made of tungsten, for example), X-rays are emitted.

Gamma rays are emitted by isotopes of certain elements. For example, cobalt-57 (Co57) under radioactive decay will give off photons with energy at 122 keV. Another radioactive source, americium-241 (Am241) has pho-tons with an energy peak at 60 keV while cesium-137 produces 662 keV photons.

Older X-ray detectors use indirect detection, in which the conversion of photons to electrical charge is done in two steps. First, X-ray photons are converted to visible light via a scintillator material. Second, the light gener-ated by the scintillator is detected by a photodiode, converting the light to an electrical charge. Note that in an imaging application, the photodiode sensor will be pixelated. The spatial resolution and efficiency are degraded by this two-step process (Figure 11.2).

To improve the efficiency and resolution of the detection process, it is desir-able to go directly from photons to charge, and detect the resulting charge with low-noise electronics. The elemental semiconductor germanium (Ge) is an excellent material for direct detection, providing performance that is close to the theoretical limit. However, germanium must be cooled to −173 °C using liquid nitrogen for these applications. While this has been done in practice, the cryogenic cooling raises costs and reduces the reliability of these systems.

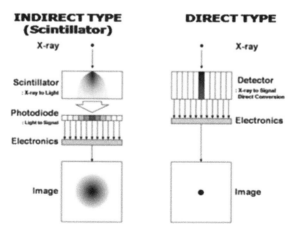

FIGURE 11.2
Indirect vs. direct X-ray detection system.

Discovery of the compound semiconductors cadmium telluride (CdTe) and cadmium zinc telluride (CZT) as direct detectors has helped greatly to lower the cost, as both materials can operate efficiently at room temperature. The direct detection process, used by Ge, CdTe and CZT sensors, offers better efficiency and resolution compared to the traditional method using scintillators described earlier.

When a photon is absorbed within the semiconductor detector, electron-hole pairs are usually generated over an extremely small volume and then drift in opposite directions in an externally applied electric field or bias. These moving charges, particularly the electrons, are sensed by a small electrode, which provides not only the energy of the absorbed photon, but also the position of the interaction site.

Signal processing of the charge generated within the sensor is typically done with an Application Specific Integrated Circuit (ASIC) that is directly attached to the semiconductor detector. Modern integrated circuit technology makes it economically feasible for each pixel to have its own low-noise electronics, creating what are known as hybrid pixel detector readout ASICs.

The block diagram of the electronics for each pixel or channel is shown in Figure 11.3.

The charge generated by the sensor enters a charge-sensitive amplifier (CSA), which converts the charge to a voltage using the feedback capacitor C_f:

$$V_{CSA} = \frac{Q_o}{C_f}$$

As an example, a typical value for C_f is 20 fF. Since a 59.5 eV photon produces of 2 fC of charge, this corresponds to a step change of 100 mV at the CSA output. The transient V_{CSA} step then enters a band-pass filter to improve the signal-to-noise ratio (SNR). Since this band-pass filter changes the shape of V_{CSA}, it is also called a pulse shaper. The peak output of the shaper is compared to a programmable threshold to determine if a valid pixel event has occurred; this comparison is done to remove false events due to noise.

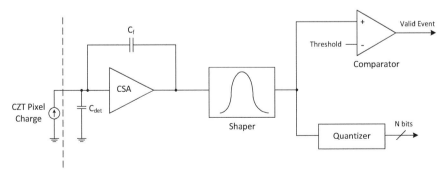

FIGURE 11.3
Charge detection electronics block diagram for CZT sensors.

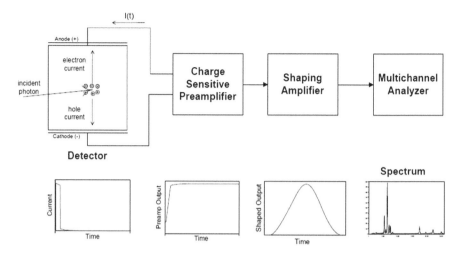

FIGURE 11.4
Typical output from each stage of the detector electronics [2].

Finally, a valid pixel event is digitalized using an analog-to-digital converter (ADC) for further downstream processing, as illustrated in Figure 11.4.

11.1.2 Nuclear Material Detection

Radioactive materials such as cobalt-60 and cesium-137 pose a major threat as even a fraction of a gram emits a large number of high-energy gamma rays; such materials can be very harmful due to their ability to damage human tissue. The dirty bomb, in which conventional explosives disperse such a radioactive material, is a significant security threat. A nuclear weapon, on the other hand, uses uranium and plutonium, which are much less radioactive. However, the processes of fission and fusion of uranium, plutonium, and other materials release vast amounts of energy. The resulting explosion produces catastrophic results, as with nuclear bombs dropped on Hiroshima and Nagasaki in 1945.

Radioactive radiation takes several forms:

- Alpha particles: Alpha particles (α) consist of two protons and two neutrons bound together into a particle identical to a helium-4 nucleus. They are generally produced in the process of alpha decay (α-decay) where an atomic nucleus emits an alpha particle (helium nucleus) and thereby transforms or 'decays' into an atom with a mass number that is reduced by four and an atomic number that is reduced by two. Because they are massive by subatomic standards, alpha particles must carry off a considerable amount of energy to escape the nucleus; at the same time, because of their mass they can only travel few centimeters in air. They are stopped by a sheet of paper or the dead outer layers of skin.

- Beta particles: A beta particle (β), is a high-energy, high-speed electron or positron emitted in the radioactive decay of an atomic nucleus, such as a potassium-40 nucleus, in the process of beta decay. Two forms of beta decay, β− and β+, respectively produce electrons and positrons. Beta particles are much less massive than alpha particles, so they can travel up to a meter in air, but are less energetic than alpha particles. Some are stopped by outer layers of skin, while others can penetrate a few millimeters.

- Neutrons: Some radionuclides decay by emitting a neutron. Neutron emission is a mode of radioactive decay in which one or more neutrons are ejected from a nucleus. It occurs in the most neutron-rich/proton-deficient nuclides, and also from excited states of other nuclides. As only a neutron is lost by this process the number of protons remains unchanged, and an atom does not become an atom of a different element, but a different isotope of the same element. Neutrons are lighter than alpha particles but much heavier than beta particles. They can travel tens of meters in air. Neutrons are typically stopped by hydrogen-containing material, such as water or plastic. Energetic neutrons can penetrate the human body.

- Gamma rays: Gamma rays (γ) are highly energetic photons released during radioactive decay arising from the radioactive decay of atomic nuclei. French chemist Paul Villard discovered gamma radiation in 1900 while studying radiation emitted by radium, a radioactive material discovered by Polish scientist Marie Curie-Skłodowska. Gamma rays have a wide range of energies, can travel kilometers in air and can penetrate the human body. causing biological damage. They can be stopped by dense material like lead.

Different materials attenuate neutrons and gamma rays in different ways. Dense materials like lead, tungsten, uranium, and plutonium have a high atomic number Z (the number of protons in the nucleus). High-Z materials attenuate gamma rays efficiently. In contrast, neutrons are stopped most efficiently by collisions with the nuclei of light atoms, with hydrogen being the most effective because it has a weight similar to a neutron. As a result hydrogen-containing material like water, wood, plastic, or food are particularly efficient at stopping neutrons.

Detection of the nuclear materials needs to overcome potential attenuation and shielding effects. Various techniques for X-rays, gamma rays and neutrons are being utilized in the security applications. Many of the detection devices use large sheets of plastic scintillator material, such as polyvinyltoluene (PVT), to detect radiation coming from unknown sources. PVT is doped with anthracene or other wavelength-shifting dopants to produce a plastic scintillator. When subjected to ionizing radiation (both particle radiation and gamma radiation), the amount of visible radiation emitted is proportional to the absorbed dose.

However, PVT cannot identify the source of radiation, and many items in everyday commerce contain radioactive material. As a result, some benign radioactive materials can produce many false alarms, which may require considerable effort to resolve and delay the flow of commerce. Newer versions of detection equipment contain direct semiconductor detectors that have built-in isotope identification capability. In the following portion of this chapter we will focus our attention on those modern detectors and their associated electronics.

As an example of using modern direct detection for nuclear needs let us use the recent announcement of the Prairie Island Nuclear power Generating Plant (PINGP) that has implemented a new and highly effective dose reduction initiative using cadmium zinc telluride (CZT) monitoring (www.neimagazine.com/features/featureprairie-island-putting-czt-on-the-map-5769805/). Two types of CZT instrument have been used: spectroscopic imaging and process monitoring. These have reduced the dose and risk associated with: forced oxidation monitoring, temporary shielding installation, hot particle identification, and shipping survey verifications. The ultimate goal is to use the monitors to support continued dose reduction efforts by providing a real-time picture of plant conditions for enhanced decision making.

The nuclear industry uses solid-state detectors with a 3D position-sensitive CZT detector to perform gamma spectroscopy and imaging. These systems have an energy resolution of less than 1.1% at 662 keV, rivaling high-purity germanium (HPGe) detectors without the need for cryogenic operating temperatures. CZT detector sub-systems are provided by companies such as H3D Inc. (www.h3dgamma.com). H3D offers the world's highest-performance imaging spectrometers. Their CZT technology is based on over a decade of groundbreaking research at the department in the University of Michigan. H3D's gamma camera, Polaris-H, was prototyped in 2012 and introduced commercially in late 2013. It has been deployed at nuclear plants around the US, including at the Fermi 2 and Cook nuclear facilities. In application in nuclear plants the spectroscopic imaging detector takes a visual image and overlays it with a radionuclide specific heat map. This produces a visual means of communicating radiation fields and can provide verification of traditional dose rate surveys. The imager was used to verify the location of the highest dose rate for a radioactive material shipping container and to evaluate the adequacy of a temporary shielding package. The evaluation of the temporary shielding package determined that there was no radiation streaming, but that additional shielding needed to be placed in the colored area of the image.

Process monitoring work has demonstrated that the monitors are versatile and that the technological fundamentals for real-time radiological characterizations are sound. Further development of CZT technologies is being pursued in the form of native incorporation of collimators and deployment procedures. Using monitors will quantify the radionuclide activities in real time. This would provide significantly more reliable information for ALARA planning and provide an avenue to reduce radiological shipping risk by changing resin beds before they reach Class-B levels. ALARA is an acronym used in radiation safety for "As Low As Reasonably Achievable." The ALARA radiation safety principle is based on the minimization of radiation doses and limiting the release of radioactive materials into the environment by employing all reasonable methods.

11.1.3 Baggage Scanning

Baggage scanning equipment needs to detect bulk, sheet, liquid, and slurry explosives. Traditional radiation detection systems use scintillator technology, as shown in Figure 11.5. X-rays produced by the X-ray tube are being used to scan the object. Ceramic scintillators detect those photons that have passed through the object and produce visible light signals. The received light signal is in turn converted by the photodiode to produce analog electrical signals that can be used to produce an image.

Most detection systems take linear projections through the luggage travelling on a conveyor belt in a so-called line scan mode. The detectors used are efficiently collimated linear arrays. For dual-energy capabilities two solutions have been developed. In the first technique, a linear scan is performed twice, once without, and then repeated with an additional X-ray filter in front of the beam. In this way an elementary technique of low- and high-energy photon separation can be obtained.

In the second technique the linear scan is performed only once but sandwich detectors have been optimized for dual-energy scanning. They consist of two layers of scintillator, separated by a metal filter. The first layer absorbs primarily low-energy photons and the second layer absorbs the remainder. An example of an X-ray spectrum produced this way is shown in Figure 11.6.

In both cases, due to the poor energy separation of those acquisition systems, and to a significant noise level resulting from the acquisition speed, the obtained accuracy at best only allows materials to be classified into broad bands such inorganic and organic [3].

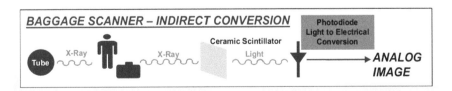

FIGURE 11.5
Scintillator-based baggage-scanning technology as typically used in standard systems.

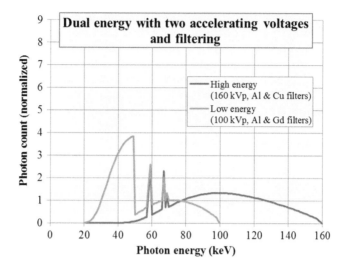

FIGURE 11.6
Simulated dual-energy X-ray spectrum that uses two accelerating voltages (100kVp and 160kVp) and various filters (Al, Cu, Gd).

Modern baggage-scanning equipment uses a more efficient, direct method of radiation detection, as shown in Figure 11.7. The two-step process involving scintillator and photodiode is replaced by a semiconductor detector that converts X-rays directly into electric charge.

Various semiconductor detectors can be used made of silicon (Si), germanium (Ge), gallium arsenide (GaAs), cadmium telluride (CdTe) or cadmium zinc telluride (CdZnTe). Silicon detectors have very poor stopping power. Germanium detectors are very expensive, while GaAs detector technology is not developed enough for commercial applications. CdTe and CdZnTe (CZT) detectors fit perfectly into baggage scanning applications due their high stopping power, reasonable cost, high stability [4] and reliability [5].

In addition, through the use of high-energy resolution semiconductor detectors, multiple energy analysis and multiple source points, the equipment might be able to detect thin sections and subtle differences in atomic

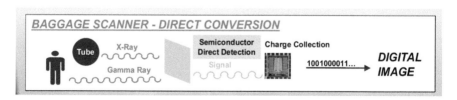

FIGURE 11.7
Direct conversion-based baggage-scanning technology as typically used in modern detection systems.

number. The equipment can also provide full volumetric reconstruction and analysis of the object being imaged. The result of using advanced CZT technology could be an exceptional ability to detect and discriminate a wide variety of threat materials, providing enhanced overall security. One of the key motivations in technological developments is to eliminate the liquid carry-on ban currently in place. Once that ban is eliminated (legislation planned in Europe) the airport operators will be forced to use better technology for liquid security detection.

An example of such technology [6] is shown in Figure 11.8. A suitcase enters the device at right of figure where spatial landmarks for registration purposes are measured by pre-scanner. In the main housing center-left of figure a primary cone-beam executes a meander scan, either of a region-of-interest or suitcase in its entirety. The underlying principle of detection is X-ray diffraction imaging or XRD [7]. XRD refers to the volumetric analysis of extended, in-homogenous objects by spatially resolved X-ray diffraction and is discussed in more detail later on. The detector technology used in such a system requires very good energy resolution and for this reason only solid-state detectors such as high-purity germanium (HPGe) or CZT qualify.

Another example of the semiconductor direct-readout technology scanner is shown in Figure 11.9. SureScan x1000 is the first TSA-certified multi-energy static gantry explosive detection system (EDS) for checked baggage screening, representing the next generation in EDS detection technology and design. With its innovative use of computed tomography (CT) and implementation of multi-energy detection for atomic number analysis, the x1000 delivers low false alarm rates, and a high level of accuracy of current and emerging threat detection. The x1000 scanner utilizes direct X-ray detection using photon-counting principles.

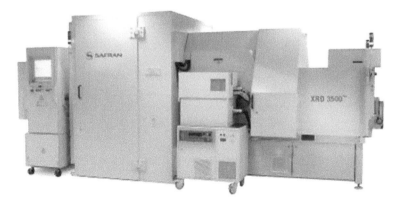

FIGURE 11.8
Direct tomographic, energy-dispersive XRD 3500 system from Morpho-Detection (www.morpho.com).

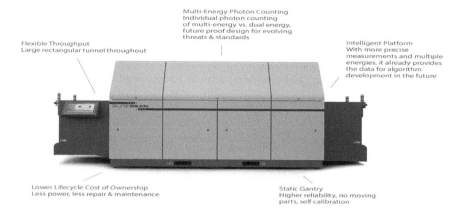

Multi-Energy Photon Counting
Individual photon counting
of multi-energy vs. dual energy,
future proof design for evolving
threats & standards

Flexible Throughput
Large rectangular tunnel throughout

Intelligent Platform
With more precise
measurements and multiple
energies, it already provides
the data for algorithm
development in the future

Lower Lifecycle Cost of Ownership
Less power, less repair & maintenance

Static Gantry
Higher reliability, no moving
parts, self calibration

FIGURE 11.9
Direct detection photon-counting system from SureScan (www.surescaneds.com).

11.2 Techniques for X-Ray and Gamma Ray Detection

Let us lay some foundation of radiation detection before specific detection techniques are described. Energetic photons such as hard X-rays and gamma rays interact with matter mainly by four basic processes:

- Elastic scattering (also known at Thomson scattering), in which photons change path without changing energy. This process is useful in forward-scattering techniques, as discussed later. Scattering direction is a function of the incoming photon's energy and the crystallographic structure and orientation of the scattering object. Commercial scanners use collimation to limit the acceptance angle of the scattered photons. Elastic scattering has been widely used for crystallography and the techniques developed for powder diffraction are now being deployed into the marketplace for explosive detection.

- Photoelectric absorption where the photon disappears after transferring all its energy to an electron. CZT is a particularly suitable absorber in the energy range of interest (40 keV–1 MeV). Many hand-held instruments have been used for security detection applications.

- Compton scattering, in which some fraction of the photon energy is transferred to a free electron in the material. The path of the photon is changed. This process can be used to detect the direction from which the photon arrived at the detector. Compton scattering can be used for back-scatter imaging. Pair production, in which the photon energy is spontaneously converted into an electron-positron pair. Pair production is dominant at the very high-energy range (well above 1 MeV) and therefore will not be discussed in this chapter.

From the above comparison it is clear that elastic scattering, photoelectric absorption and Compton scattering are physical processes that can be leveraged in radiation detection with semiconductor sensors. The last two processes lead to a complete (photoelectric) or partial (Compton) transfer of the photon energy to electron energy. The charge generation processes are not dependent on temperature. Photons of high enough energy can penetrate solid objects, but are scattered or absorbed by dense objects (or a sufficient thickness of less-dense material). This is the basis for radiography. For cargo scanning, X-rays or gamma rays are beamed through a container, and a detector on the other side records the number of photons received in each pixel.

There are several detection techniques available in the marketplace. Their principal of operation, cost, performance characteristics and ease-of-use vary broadly. We are presenting here a short summary of those techniques that have been used or are in active development.

The key issue in baggage scanning applications is the ability to identify explosive materials. The difficulty in explosive detection lies in the fact that the materials contained in luggage are unknown in number and nature, and chemically they are very close to common materials. The effective atomic number (Z_{eff}) of most explosive materials ranges from 7 to 7.7 and their density from 1.4 to 1.9. Some of the techniques discussed here are capable of overcoming that limitation.

11.2.1 X-ray Radiography

Conventional X-ray radiography (CXR), also known as advanced technology (AT), baggage-scanning is a quick way to provide an initial screening at low performance levels. These machines produce a 2D projection image of the integrated density through an object. Both photoelectric (PE) and Compton absorption vary with density and that enables some material identification. However, the material identification and contrast provided are rather poor and require additional screening methods. Despite these disadvantages, CXR is used extensively due to its low cost as it does provide visual capability for manually identifying conventional threats (knife, gun, etc.).

Adding dual-energy capability to CXR provides higher accuracy and some rudimentary capability of energy separation. In practice, the technique is implemented by scanning the object twice, once as a normal scan, the second time with a high-energy filter added. Poor energy separation might be capable of distinguishing between organic and inorganic matter but will likely not be able to detect any serious security threats. In recent deployments high-speed switching of accelerating voltage is the preferred method, as it is cheaper than dual-layer scintillator technology.

Adding more energy bins leads to so-called color X-ray technology, where multiple energy bins (typically three to five) might actually provide some useful energy separation and material identification capabilities [8]. CZT detectors are ideally suited for this application due to their ability to operate at room temperature and the fact that they have sufficient energy resolution at the count rates that are required. Adding more energy bins enables clearer visualization of the baggage content. Traditional security threats (knife, gun, etc.) can be easily detected, especially when 3D scanning is adopted. However, material identification of explosives requires more sophisticated diffraction techniques, as discussed below.

11.2.2 Computed Tomography (CT)

CXR, double-energy and color X-ray techniques provide a two-dimensional (2D) view of the object under investigation. A three-dimensional (3D) view can be obtained using Computed Tomography approach. Computed tomography (CT) is a well-known medical imaging modality used in virtually every hospital in the world. CT equipment can be tailored for use in security applications by providing a 3D image of the object under search. However, pure visualization will likely not be sufficient to detect serious security threats, including various types of explosive materials.

To be truly useful in threat detection CT-like technology needs to be coupled with energy discrimination using color X-ray concepts. Combined technology can be used to detect explosives using information of object density combined with atomic number analysis for enhanced detection. By discriminating the atomic composition of baggage contents, the enhanced CT technology can deliver high accuracy in threat detection, with very low false alarm rates (FAR). 3D CT technology with color X-ray is expensive though and airport operators may not be willing to pay for it. However, the key question is not the capital cost of CT scanners but rather the cost of ownership. When using automated threat detection the cost of handling false alarms can be substantially more than the machine's capital cost. DHS TSA estimated that the cost of reducing FAR by 1% could save the US $25 million per year in secondary manual luggage inspection costs – mostly related to transportation security officer (TSO) personnel.

11.2.3 Back-Scattering and Forward-Scattering

Due to the limited capabilities of X-ray detection in identifying explosive materials a different technology has been developed relying on analyzing scattering effects of X-rays. One class of techniques relies on back-scattering information [9]. It is more accurate although slower than CXR but still might be insufficient for explosive determination. The military has been using this technology for mine detection, because it allows for discrimination of low-density vs. high-density materials. But it provides side-only access.

The techniques based on forward-scattering or diffraction can provide ultimate performance in threat detection. XRD (X-ray diffraction) is a powerful analytical tool that has been used for the non-destructive analysis of a wide variety of materials for nearly 100 years. XRD is now widely applied in a variety of industries including metallurgy, photovoltaics, forensics, pharmaceuticals, semiconductors, and catalysis and can be used to analyze virtually any solid with a crystalline structure. It has been recently applied to security detection. The fundamental strength of the technique is its ability to characterize the periodic atomic structure present in crystalline or polycrystalline materials. It also has some capability to distinguish liquids.

In XRD analysis a sample is illuminated by a collimated X-ray beam of known wavelength. If the material is crystalline, it possesses a three-dimensional ordering or "structure" with repeat units of atomic arrangement (unit cells). X-rays are elastically scattered (i.e., diffracted) by the repeating crystal planes lattice of materials, while X-rays are randomly scattered by amorphous materials. X-ray diffraction occurs at specific angles with respect to the lattice spacings defined by Bragg's Law. Any change or difference in lattice spacing results in a corresponding shift in the diffraction lines. It is using this principle that such properties as identification (based on phase) and residual stresses are obtained.

The application of XRD in security detection provides a very powerful opportunity to detect the chemical/structural property of the material under investigation [10]. For effective baggage screening this technique is frequently coupled with CT to visualize the object in question. Traditionally this is done in a serial fashion making the scan time long and equipment expensive. With CZT it is, however, possible to perform CT-like imaging and XRD-like detection simultaneously. If you are interested in discussing this opportunity please contact authors of this document.

An example of the commercial application of the XRD technology is shown in Figure 11.10. Note the difference in obtained scattered X-ray spectra for Semtex (explosive) and some non-hazardous materials. The HPGe detector used in XRD 3500 can be substituted by CZT which offers a similar energy resolution (a few percent at 30–120 keV energy range). However, none of the scintillator materials can be used for this application as their energy resolution is too poor.

11.2.4 Hyperspectral X-ray Radiography

While X-ray diffraction technique is very powerful in material identification it places hard demands on the detector technology used. While they can be met with Ge, CdTe and CZT material to provide the required spectral response, a simpler technique is highly desirable to reduce equipment costs. Although high-purity Ge detectors provide the best energy resolution, the

▌▌▌▌▌ XRD 3500: ARRANGEMENT OF COMPONENTS

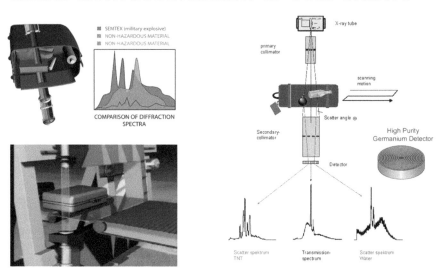

FIGURE 11.10
XRD 3500 architecture and principle of operation (www.morpho.com).

stopping power of Ge is rather low and these detectors require cooling, typically at liquid nitrogen temperatures (77 K).

It has been shown very recently that a traditional high-energy X-ray system can capture scattered X-rays to deliver 3D images with structural or chemical information in each voxel. This type of imaging can be used to separate and identify chemical species in bulk objects with no special sample preparation. Defining hyperspectral technology precisely is difficult as it is a relatively new concept that as a minimum it should contain 100 energy bins. In addition, hyperspectral technology takes advantage of measurement orthogonality.

The capability of hyperspectral imaging has been demonstrated by examining an electronic device where we can clearly distinguish the atomic composition of the circuit board components in both fluorescence and transmission geometries. Researchers were not only able to obtain attenuation contrast but also to image chemical variations in the object, potentially opening up a very wide range of applications, from security to medical diagnostics. The possibility of this technique being introduced in commercial applications does exist but has not been demonstrated yet.

11.2.5 Dirty Bomb Detection

Following the events of September 11, 2001, there has been an increased effort to develop systems to detect the terrorist threat of improvised or "dirty bombs" – where traditional explosives are detonated and used to disperse radioactive materials, causing widespread contamination. Solid-state radiation detectors are required to produce a small and portable device for detection of dirty bombs. Such equipment would be carried by security and emergency personnel and would automatically report location and background radiation levels.

Detecting nuclear radiation is relatively easy. All you need is a simple, inexpensive Geiger counter. But to be able to precisely identify both the type of radiation and where it is coming from is much more challenging. That's crucial for everything from searching for possible nuclear weapons in shipping containers to performing routine maintenance or emergency response in nuclear power plants.

To respond to new needs for hand-held detection instrumentation, several products have been launched recently. FLIR's Raider™, nano-Raider™, and identiFINDER R300™ instrument family has been developed to address the specific sensitivity, resolution, size, and weight requirements of a variety of applications (http://gs.flir.com/detection/radiation/handhelds/ nano-raider/nanoraider-benefits). Each of these instruments employs the same simple user interface, which in turn reduces the training burden. Similarly, Thermo Scientific Interceptor™ is a spectroscopic personal radiation detector (SPRD) combining the qualities of a personal radiation detector (PRD) with radio isotope identifier capabilities (RIID). These instruments provide good detection sensitivity with on-the-spot, accurate identification at an affordable price and at a very compact size. Other companies are working on similar hand-held products, most based on CZT technology.

With such hand-held instrumentations available it is possible to detect nuclear materials within a few minutes and see the obtained spectra using CZT detectors. Gamma photons related to uranium-235 source are clearly visible, as are individual X-ray peaks. While first-generation CZT-based spectrometers have been successful in securing a place within the broader radiation detection market, their current value proposition does not compel broad displacement of lower-end scintillator-based instruments (sodium iodide, lanthanum bromide) nor displacement of high-end germanium-based instruments. Various CZT detector configurations can be leveraged depending on the system needs, as discussed in the following section.

The dirty bomb detection devices are typically hand-held. They have software that can identify a radioisotope by its gamma-ray spectrum. The most capable of these devices use a crystal of high-purity germanium (HPGe), a semiconductor material, and are considered the "gold standard" of all identification devices. Such devices are heavy and delicate, and must be cooled with liquid nitrogen or by mechanical means, limiting their usability in the field. CZT technology plays a critical role in displacing HPGe as a detector material that enables cost-effective radiation detection in the field.

H3D Inc. has developed a number of CZT-based products – small, room-temperature, hand-held devices called the Polaris-H, Polaris-S and Apollo (www.h3dgamma.com). Polaris-H, was prototyped in 2012 and introduced commercially in late 2013. It has been deployed at nuclear plants around the US, including at the Fermi 2 and Cook nuclear facilities. Polaris-S and Apollo followed. As an example, Apollo is designed to exceed the ANSI N42.34 standard, provides better than 1.0% FWHM energy resolution at 662 keV, offers real-time 360° isotope-specific directionality, is ready to use in two minutes, requires no cryogenic cooling (as HPGe does), and supports energy range of interest up to 3 MeV.

In additional to faster identification (higher efficiency), these devices also provide location functionality. This feature is important, as while in a moving vehicle the device will alarm as it passes by a radioactive area (such as a warehouse or building). The device can then be hand-carried to find the exact location of the source (such as within a barrel in one of the rooms of the second floor of the warehouse).

11.3 Readout ASICs

11.3.1 ASIC Technology

Semiconductor pixelated detectors used for photon counting need to have a high level of segmented multi-channel readout. Several decades ago the only way to achieve this was via massive fan-out schemes to route signals to discrete low-density electronics. At the present time CMOS technology is used to build very dense low-power electronics with many channels which can be bonded directly or indirectly (through a common carrier printed circuit board (PCB)) to the detector.

There are different requirements for the CMOS technology used for the analog front-end signal processing, as opposed to that for the digital signal processing. For the analog part of the electronics there is a requirement for a robust technology that has low electronic noise and high dynamic range, which typically requires high-power supply voltages. Digital signal processing in turn requires very high speed and high density, which is more compatible with the more modern low-voltage supply, deep submicron processes.

There seems to be a technology optimum at around 0.35 um to 0.18 um minimum feature size for the analog requirements. The large feature size limits the complexity of circuitry that can be integrated in a pixel, but even at 0.35 um it is possible to place a million transistors on a reasonable size silicon die. In comparison, digital signal processing can benefit from the rapid development of deep submicron processes. Some selected research developments now take place using 90 nm or 65 nm process nodes. These technologies are well suited to high-speed ADC architectures and to very fast data manipulation for data specification and compression. The deep submicron technologies have their own limitations in terms of analog performance, noise and cost.

11.3.2 ASIC Attachment

Semiconductor pixel detectors require a connection from the pads on the detector material to the bond pads on the ASICs. In some cases (and ideally) the pixel pitch on the readout ASIC is the same as the detector pixel pitch. It is also possible to fan out the connections on the detector with multilevel metal routing on the detector or with the use of an interposer board. This fan-out routing has to be done very carefully because there is increased possibility of signal crosstalk. The pitch of X-ray and gamma-ray imaging systems currently ranges from about 100 um to 1 mm. For small pixel pitches bump bonding is used to connect the detector pixels to the ASICs. There are many different technologies to do this depending on the requirements of the detectors and environmental constraints.

One of the simplest electrical connections between a chip and the circuit board can be made with small balls of electrically conductive material, called bumps. A bumped die can then be flipped upside down and aligned so that the bumps connect with matching pads on the board. Flip-chip bonding has several advantages over traditional wire-bonding, including small package size and greater device speed.

Bumping can be performed by extending conventional wafer fabrication methods. After the chips are made, under-bump metallization (UBM) pads are created to connect to the chip circuitry, and bumps are then deposited on the pads. Solder is the most commonly used bumping material, although alternative materials – such as gold, copper, or cobalt – can also be used depending on the application. For high-density interconnects or fine-pitch applications, copper pillars can be used. While solder bumps spread during the joining process, copper pillars retain their shape, which allows them to be placed much more closely together.

The industry standard area bump-bonding method is to deposit solder onto under-bump metallization on the pads of the detector and ASIC, then the two are aligned and heated to reflow the solder. Various solders are used, including lead-tin, bismuth-tin, indium alloys, silver alloys, depending on the temperature to reflow and the operating temperature required. Typically these materials require 240 °C to 140 °C to reflow. Indium is used either in a lower temperature reflow process or straight compression bonding with good results, but indium cannot be used if high operating temperatures will ever be experienced.

Gold (Au) stud bumps are simply modified thermosonic wire-bond connections in which the wire has been purposely severed from the ball-bond, leaving only the Au-ball (stud bump) attached to the chip's bond- pad. A major advantage of this technique is that the thermosonic ball-bonding process scrubs through the aluminum oxides present on typical IC pads, eliminating the need to pre-apply any UBM layer.

Reliable low-impedance, metal-to-metal interconnects are a distinct characteristic of this technology and a principle reason why wire-bonding has remained in favor for so long. Solder and conductive epoxies, conversely, require a compatible metal surface for reliable connections. Silver-filled conductive epoxies perform best in contact with noble metal finishes but will not provide a reliable connection in direct contact with untreated aluminum. Gold-stud bump bonding is ideal when only a few devices need to be bumped, but is not a particularly efficient process when attempting to bump an entire wafer. Depending on the number of contacts per device, a single wafer may require several hours of bonder time. Nevertheless, the process is extremely clean and relatively economical, since no chemicals are involved and minimal material is consumed in forming the individual ball bumps.

Ensuring a uniform height for every bump within a large array of Au stud bumps requires control of the diameter of the ball bumps and amount of wire that protrudes from the top of the ball bump. Simply pulling the wire until it breaks does not ensure a repeatable process. A better method employs a shearing tool and/or coining technique immediately after the ball bump is thermosonically welded to the pad to establish a more consistent ball bump height.

A more economical bumping process when handling whole wafers is electroless-plated nickel-gold (Ni-Au) bumps. This process typically deposits small amount of Ni over the aluminum pads through a zincate process after etching away the surface oxides. A thin Au coating is then plated over the Ni to prevent oxidation of the Ni-bumps. This immersion plating process is fairly benign to most ASIC devices and rarely requires any photoresist masking step. The process is commonly employed for bumping Radio Frequency Identification (RFID) and smartcard chips for flip-chip assembly onto flexible inlets – an application that is extremely cost-sensitive. Bump height, however, is limited due to the tendency of the plated pads to mushroom outward with increased plating times.

The present trend toward all-copper interconnect at the ASIC level, instead of aluminum, may make UBM a thing of the past, since the Cu-pads are expected be finished with a thin Ni-Au layer. An alternative method, capable of producing tall columnar-shaped bumps with high aspect ratios on whole wafers, involves an electroplating process. With this technique a copper seed layer is first sputter-deposited across the wafer's surface followed by a thick photoresist layer that is then exposed and developed creating openings above the pads into which copper posts are then electroplated. The bump height is determined by the thickness of the confining photoresist. The copper seed layer is later etched away after stripping the photoresist from the wafer. High-aspect copper bumps of this type may also be plated on quartz carriers and later transferred to the wafer.

In summary, tall conductive columns may be added to ASIC chips at either the die level or the wafer level using a variety of means. These tall bumps aid in compensating for Coefficient of Thermal Expansion (CTE) mismatch between the detector and ASIC chip and lower the capacitance between surfaces. This latter characteristic has been identified as a potential advantage for improving the energy resolution of X-ray sensors in dental and industrial inspection applications.

11.3.3 ASICs for Spectroscopy

A CZT detector typically operate in a single photon detection mode where an electric charge generated by one photon needs to be collected by the readout electronics. Since the amount of generated charge is small (about 5 fC for a 122 keV photon), very sensitive analog circuitry is required to amplify that charge. In spectroscopic applications the amount of charge, which directly corresponds to the photon energy, needs to be precisely determined. In photon counting application decisions are only binary (or multi-binary) but the count rate might be very high, creating related challenges. The purpose of this section is to explain some of the design considerations that are important when building semiconductor readout electronics systems.

11.3.3.1 Analog Front End

Analog signal processing can be divided into the following steps:

- *Amplification*: The input charge signal is amplified and converted to a voltage signal using a charge-sensitive amplifier (CSA). A main characteristic of the amplification stage is equivalent noise charge (ENC), which is required to be as low as possible in order not to degrade intrinsic detector energy resolution. Another important consideration for the CSA operation is a dark current compensation mechanism. A solution that accommodates continuous compensation for dark currents up to several nAs while maintaining low ENC is desired.

- *Signal shaping*: The time response of the system is tailored to optimize the measurement of signal magnitude or time and the rate of signal detection. The output of the signal chain is a pulse whose area is proportional to the original signal charge, i.e. the energy deposited in the detector. The pulse shaper transforms a narrow detector current pulse to a broader pulse (to reduce electronic noise), and with a gradually rounded maximum at the peaking time to facilitate measurement of the amplitude. A solution that provides effective signal shaping while maximizing the channel count rate needs to be applied.

- *Pulse detection*: The input pulse, broaden by the shaping process, needs to be detected against a setup threshold value. The threshold level is a critical parameter that determines whether the event is recognized as a true event or false reading caused by noise. As a result the threshold value is typically adjustable both globally and at the pixel level. The peak detection value determines energy level information. A solution that prevents temperature drift of the peak detector needs to be used.

- *Channel multiplexing*: In the case of ASIC spectroscopy all parallel channels of the channel readout ASIC need to have their signals multiplexed at the output before being sent out to an external analog-to-digital converter (ADC). The key requirement to channel multiplexing and signal shaping is a maximum channel count rate determined by the given application.

11.3.3.2 Charge-Sensitive Amplifier (CSA)

The current signal induced in the sensing electrode can be integrated in the pixel capacitance and read out with a high input impedance stage, which amplifies the resulting voltage at the pixel node, or it can be read out directly with a low input impedance stage, which amplifies the charge Q and keeps the pixel node at a virtual ground, such as a charge amplifier. The latter is the preferred choice since, among its other advantages, it stabilizes the sensing electrode by keeping its voltage constant during the measurement and/or the readout.

In both cases, low-noise amplification is required to reduce the noise contribution from the processing electronics (such as the shaper, peak detector, and ADC) to a negligible amount; good design practice dictates maximizing this amplification while avoiding overload of subsequent stages. This low-noise amplification would also provide either a charge-to-voltage conversion (e.g., source follower, charge amplifier), or a direct charge-to-charge (or current-to-current) amplification (e.g., charge amplifier with compensation, current amplifier). Depending upon this choice, the shaper would be designed to accept a voltage or a current, respectively, as its input signal.

In a properly designed low-noise amplifier, the noise is dominated by processes in the input transistor. Assuming that CMOS technology is employed in the design, the input transistor is referred to as the "input MOSFET," although the design techniques can easily be extended to other types of transistors, such as the JFET, the bipolar transistor, or the hetero-junction transistor. The design phase, which consists of sizing the input MOSFET for maximum resolution, is called "input MOSFET optimization" and has been studied extensively in the literature.

11.3.3.3 Equivalent Noise Charge (ENC)

Equivalent noise charge (ENC) expresses an amount of noise that appears at the chip input in the absence of useful input signal and is a key chip parameter that affects the energy resolution of the system. Following the standard approach, the total ENC can be divided into three independent components: the white thermal noise associated with the input transistor of the CSA (ENC_{th}), the flicker noise associated with the input transistor of the CSA (ENC$_{1/f}$), and the noise associated with the detector dark leakage current (ENC$_{dark}$). Noise arising in other components connected to the ASIC input node such as the bias resistor is generally made negligible in a properly designed system. For a first-order shaper the ENC components can be approximately expressed as:

$$\mathrm{ENC_{th}}^2 = (8/3)kT/\left(T_{peak} * g_m\right) * C_{tot}^2$$

$$\mathrm{ENC}_{1/f}^2 = K_f/2 * WL * C_{tot}^2/C_{inp}^2$$

$$\mathrm{ENC_{dark}}^2 = 2q * I_{dark} * T_{peak}$$

$$\mathrm{ENC} = \left(\mathrm{ENC_{th}}^2 + \mathrm{ENC}_{1/f}^2 + \mathrm{ENC_{dark}}^2\right)^{1/2}$$

where g_m is the transconductance of the CSA input transistor, C_{tot} is the total capacitance at the input of the CSA, T_{peak} is the shaper peaking time, K_f is the CSA input transistor flicker noise constant, W and L are input transistor width and length, and I_{dark} is the detector leakage current. Note that C_{tot} is the sum of the detector capacitance C_{det}, the gate-source and gate-drain capacitances of the input transistor C_{inp}, and any other feedback or parasitic capacitance at the CSA input originating from the chip package, ESD diodes and PCB traces. Based on experience with ASIC design and radiation detection module manufacturing we have assumed C_{inp} to be fractions of pF while the remaining C_{tot} components to be about 1–2 pF. Clearly, particular values are strongly dependent on the chosen technology for ASIC design and packaging, as well as on the chosen connectivity scheme between the CZT detector and the chip. It can be easily shown that the optimum peaking time T_{opt} is given by the condition where ENC$_{th}$ is equal to ENC$_{dark}$, leading to the following expression:

$$T_{opt}^2 = 4kTC_{tot}^2/\left(3g_m q I_{dark}\right)$$

ENC is typically measured in the lab by measuring the output noise and referring it back to the input using the overall gain of the system. It is also possible to measure channel performance using the scope. By acquiring the channel shaper output signal on the oscilloscope at 1 MHz sampling frequency (for example 5 msec observed time on 5000 points), a fast Fourier transform (FFT) can be applied to the acquired data and the resulting spectrum calculated for each frequency. The noise expressed in mV is calculated as the standard deviation with respect to the shaper output average value, and expressed as the equivalent noise in terms of electrons (by considering the known nominal gain of the data channel, typically in the order of hundreds of mV per fC of the input charge). The result of these calculations is ENC value expressed in number of electrons, typically in a range of hundreds of electrons, depending what electronics are used, how high the count rate is, and the loading capacitance at the detector input.

11.3.3.4 Signal Shaping

The low-noise amplifier is typically followed by a filter, frequently referred to as the shaper, responding to an event with a pulse of defined shape and finite duration ("width") that depends on the time constants and number of poles in the transfer function. The shaper's purpose is twofold: first, it limits the bandwidth to maximize the signal-to-noise ratio (SNR); second, it restricts the pulse width in view of processing the next event. Extensive calculations have been made to optimize the shape, which depends on the spectral densities of the noise and system constraints (e.g., available power, and count rate).

Optimal shapers are difficult to realize, but they can be approximated, with results within a few percent from the optimal, either with analog or digital processors, the latter requiring analog-to-digital conversion of the charge amplifier signal (anti-aliasing filter may be needed). In the analog domain, the shaper can be realized using time-variant solutions that limit the pulse width by a switch-controlled return to baseline, or via time-invariant solutions that restrict the pulse width using a suitable configuration of poles. The latter solution is discussed here as it minimizes digital activity in the front-end channels.

In a front-end channel, the time-invariant shaper responds to an event with an analog pulse, the peak amplitude of which is proportional to the event charge, Q. The pulse width, or its time to return to baseline after the peak, depends on the bandwidth (i.e., the time constants) and the configuration of poles. The most popular unipolar time-invariant shapers are realized either using several coincident real poles or with a specific combination of real and complex-conjugate poles. The number of poles, n, defines the order of the shaper. Designers sometimes prefer to adopt bipolar shapers, attained by applying a differentiation to the unipolar shapers (the order of the shaper now is $n - 1$). Bipolar shapers can be advantageous for high-rate applications, but at the expense of a worse SNR.

In a typical readout system, the shaping time varies from a fraction of a µs up to several µs. The shaping time is defined as the time-equivalent of the standard deviation of the Gaussian output pulse. In the laboratory, it is the full width of the pulse at half of its maximum value (FWHM) that is typically being measured. The FWHM value is greater than the shaping time by a factor of 2.35.

The DC component of the shaper from which the signal pulse departs is referred to as the output baseline. Since most extractors process the pulse's absolute amplitude, which reflects the superposition of the baseline and the signal, it is important to properly reference and stabilize the output baseline. Non-stabilized baselines may fluctuate for several reasons, like changes in temperature, pixel leakage current, power supply, low-frequency noise, and the instantaneous rate of the events. Non-referenced baselines also can severely limit the dynamic and/or the linearity of the front-end electronics, as in high-gain shapers where the output baseline could settle close to one of the two rails, depending on the offsets in the first stages. In multiple front-end channels sharing the same discrimination levels, the dispersion in the output baselines can limit the efficiency of some channels.

11.3.3.5 Peak Detection

The peak detector (PD) is one of the critical blocks in the radiation signal detection system as accurate photon energy is determined by the detected peak amplitude. Standard PDs may be sampled or asynchronous solutions. Sampled PDs are more precise but suffer from high circuit complexity and high power dissipation. Asynchronous PDs have simpler structure but suffer from lower output precision.

11.3.4 ASICs for Photon Counting

One of the major advantages of photon-counting detectors is electronics noise rejection. A well-designed photon-counting detector allows for an ASIC electronics threshold high enough to reject noise pulses while still counting useful signals. Therefore, quantum-limited operation of the photon-counting detector can be achieved as image noise is determined only by statistical variations of X-ray photons. On the other hand, energy-integrating detectors suffer from electronics noise that is mixed with useful photon signals and separating it from statistical noise is not possible. Electronics noise rejection is important because its magnitude for currently used digital X-ray detectors is not negligible.

After converting the CZT-generated charge to voltage by the CSA and subsequent filtering by the shaping amplifier, the signal is ready for digitization. Typically, the signal is compared against user-selected threshold voltage (discriminator box in Figure 11.11) to produce a 1-bit trigger signal indicating detection of the pulse. In parallel, the value of the shaped signal is sent to an ADC converter (or time-over-threshold (ToT) processor) with n-bit accuracy. The conversion resolution n is typically between 8 and 16 bits depending on the system accuracy, noise levels and degree of signal precision achieved.

One important consideration in the practical system is CSA reset. As the feedback capacitor C_f is charged by the input signal there must be some means of discharging this capacitor in order for the CSA to be ready for the next signal. This circuitry is schematically shown as the reset block in Figure 11.11.

There are two possible implementations for the reset block: digital and analog. The digital one involves using a switch that will discharge the feedback capacitor quickly. Unfortunately, this process typically creates too much disturbance for the sensitive CSA. The analog solution involves using a resistor (or MOSFET operating in the triode region) and provides continuous discharging during the entire process. The discharge cannot be too slow (in which case the capacitor will not be fully discharged before the next event) or too fast (as that will affect signal formation).

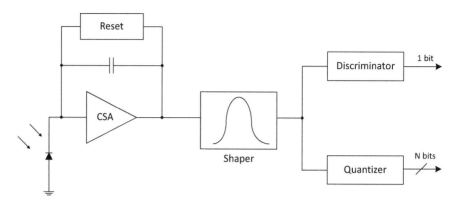

FIGURE 11.11
Photon-counting detector readout signal chain.

On a final note, while the principle of CSA signal amplification, pulse shaping and ADC conversions outlined above are fairly simple, practical implementations can be very challenging due to the very small input signals involved (below 1 mV). One has to pay particular attention to system noise, power supply decoupling, ESD protection, EMI radiation, and op-amp stability issues.

A typical photon-counting ASIC implementation contains hundreds of channels frequently implemented with multiple energy bins. One of the early 128-channel ASICs is shown in Figure 11.12. A clear advantage of the

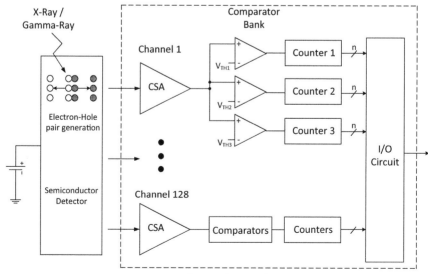

128 Channel Signal Processing ASIC

FIGURE 11.12
Block diagram of 128-channel photon-counting ASIC.

photon-counting detectors over the integrating detectors is ability to perform at higher SNR at low photon counts.

11.3.5 Photon Counting vs. Spectroscopy

There are two ways of signal processing for baggage scanning with energy-sensitive semiconductor detectors: photon counting and spectroscopy. While a precise difference between the two is hard to establish because all analog signals eventually become digital at some point in the readout system, we would like to suggest the following practical definition. Photon counting relies on energy binning with the term "binning" implying the use of comparators inside the ASIC chip. Spectroscopy preserves the analog nature of the signal representing photon energy with an A/D converter after the ASIC signal processing has been accomplished.

As a result of this architectural change, photon-counting systems can achieve very high count rate while sacrificing energy resolution (ER), while spectroscopic systems can have very good noise properties but face limitations with a maximum count rate. To achieve specific design objectives, either the count rate of the spectroscopic system is maximized or the energy resolution of the photon-counting systems is maximized. Schematic comparison between two types of signal processing is shown in Figure 11.13.

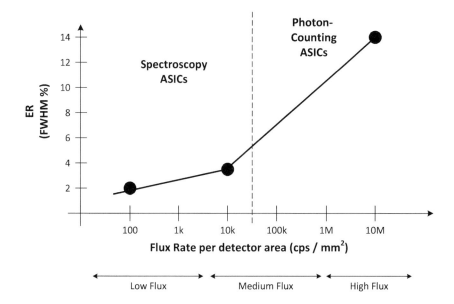

FIGURE 11.13
Schematic comparison between low-flux spectroscopic ASICs and high-flux photon-counting ASICs. Energy resolution (ER) is represented at 60 keV level.

11.4 Examples of Readout ASICs

There are literally hundreds of readout ASICs that have been published in the literature. This section summarizes the design and performance characteristics of three photon-counting and three spectroscopic devices.

11.4.1 Photon-Counting ASICs

Fast readout electronic circuits have been developed to reach count rates of several million counts per second [11–15]. These systems provide coarse energy resolution given by a limited number of discriminators and counters. This section provides some information about the most important photon-counting devices.

11.4.1.1 TIMEPIX

TIMEPIX ASIC is a CZT pixel detector developed in the framework of the MEDIPIX2 collaboration [16]. The pixel matrix consists of 256×256 pixels with a pitch of 55 µm which gives a sensitive area of about 14 mm $\times$ 14 mm. TIMEPIX is designed in a 0.25 µm CMOS process and has about 500 transistors per pixel. The chip has one threshold and can be operated in photon counting (PC), time-over-threshold (ToT) or time of arrival (ToA) modes. The principles of the different operating modes are described in detail in the literature [16].

In the PC mode the counter is incremented once for each pulse that is over the threshold, while for the ToT mode the counter is incremented as long as the pulse is over the threshold. In the ToA mode the pixel starts to count when the signal crosses the threshold, and keeps counting until the shutter is closed.

11.4.1.2 MEDIPIX-3

While TIMEPIX is a general-purpose chip, the MEDIPIX-3 is aimed specifically at X-ray imaging [16]. It can be configured with up to eight thresholds per pixel and features analog charge summing over dynamically allocated 2×2 pixel clusters. The intrinsic pixel pitch of the ASIC is 55 µm, as in TIMEPIX. Silicon die can be bump bonded at this pitch (fine pitch mode), and the chip can be run with either four thresholds per pixel in single-pixel mode or with two thresholds per pixel in charge-summing mode. Optionally the chip can be bump bonded with a 110 µm pitch, combining counters and thresholds from four pixels. Operation is possible in single-pixel mode with eight thresholds per pixel or in charge-summing mode having four thresholds and a summing charge of 220 µm $\times$ 220 µm area.

Being a very versatile and configurable chip, there is also the possibility to utilize two counters per pixel and run in continuous read/write mode where one counter counts while the other one is being read out. This eliminates the readout dead time but comes at a cost of losing one threshold since both counters need to be used for the same threshold. Finally, the charge-summing mode is a very important feature to combat contrast degradation by charge sharing in semiconductor detectors with small pixels.

11.4.1.3 ChromAIX

A multi-energy resolving ASIC called ChromAIX has been designed by the Philips corporation to support Spectral CT applications. In order to enable K-edge imaging, at least three spectrally distinct measurements are necessary; for a photon-counting detector the simplest choice is to have at least the same number of different energy windows. With more energy windows the spectrum of incident X-ray photons is sampled more accurately, thus improving the separation capabilities.

The ChromAIX ASIC accommodates a sufficient number of discriminators to enable K-edge imaging applications. Post-processing allows the separating of the Photo effect, Compton effect, and one or possibly two contrast agents with their corresponding quantification. The ChromAIX ASIC is a pixelated integrated circuit that has been devised for direct flip-chip connection to a direct converting crystal like CZT. The design target in terms of observed count rate performance is 10 Mcps/pixel, which corresponds to approximately 27.2 MHz/pixel periodic pulses. Although the pixel area in CT is typically about 1 mm², both the ASIC and direct converter feature a significantly smaller pixel, or sub-pixel. In this way significantly higher rates can be achieved at an equivalent CT pixel size, while further improving the spectral response of the detector via exploiting the so-called small-pixel effect. The sub-pixel should not be made too small, since charge-sharing effects then start to degrade the spectral performance. Very small pixels would need counter-measures as implemented in Medipix-3, the effectiveness of which at higher rates remains doubtful due to charge-sharing effects.

The ChromAIX ASIC consists of a charge-sensitive amplifier (CSA) and a pulse shaper stage, as any other photon-counting device. The CSA integrates the fast transient current pulses generated by the direct converter, providing a voltage step-like function with a long exponential decay time. The shaper stage represents a band-pass filter that transforms the aforementioned step-like function into voltage pulses of a defined height. The height of such pulses is directly proportional to the charge of the incoming X-ray photon. A number of discriminator stages are then used to compare a predefined value (i.e. energy threshold) with the height of the produced pulse. When the amplitude of the pulse exceeds the threshold of any given discriminator, the associated counter will increment its value by one count.

In order to achieve 10 Mcps observed Poisson rates, which would typically correspond to incoming rates exceeding 27 Mcps, a very high bandwidth is required. The two-stage approach using a CSA and a shaper allows achieving such high rates while relaxing the specification of its components. The design specification in terms of ENC is 400 e-, which corresponds to approximately 4.7 keV FWHM. Simulations of the analog front-end have been carried out to evaluate the noise performance of the channel. According to these simulations the complete analog front-end electronic noise (CSA, shaper and discriminator input stage) amounts to approximately 2.51 mV_{RMS}, which in terms of energy resolution corresponds to approximately 4.0 keV FWHM for a given input equivalent capacitance.

11.4.1.4 PILATUS ASIC

PILATUS is a hybrid pixel detector system operating in single-photon counting mode; it was developed at the Paul Scherrer Institute for the needs of macromolecular crystallography at the Swiss Light Source (SLS). A calibrated PILATUS module has been extensively characterized with monochromatic synchrotron radiation. The detector was also tested in surface diffraction experiments at the SLS, whereby its performance regarding fluorescence suppression and saturation tolerance was evaluated, and has been shown to greatly improve the sensitivity, reliability and speed of surface diffraction data acquisition.

The operation of the PILATUS ASIC is as follows. The incident photons are directly transformed into electric charge in the semiconductor sensor, which is transferred via the bump bond to the input of the readout pixel. A schematic of the PILATUS readout chip pixel cell is presented in Figure 11.14. The analog front-end of a readout pixel consists of a charge-sensitive preamplifier (CSA) and an AC-coupled shaper. The gain and shaping time of the CSA are adjusted with a global voltage (Vrf). An analog pulse from the

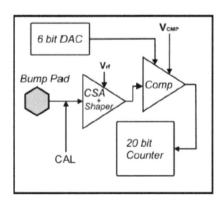

FIGURE 11.14
Architecture of the PILATUS readout cell [20].

shaper is discriminated against a threshold in the comparator (Comp) after amplification. The comparator threshold of each pixel is set with a global threshold voltage (Vcmp) and is further individually trimmed using an additional in-pixel 6-bit digital-to-analog converter (DAC). If the pulse amplitude exceeds the threshold, a digital signal is produced which increments the 20-bit counter. This detection principle is free of dark current and readout noise effects but requires precise calibration of the pixel threshold for optimum performance

11.4.2 Spectroscopic ASICs

11.4.2.1 IDEF-X

IDeF-X HD is the last generation of low-noise radiation-hard front-end ASICs designed by CEA/Leti for spectroscopy with CZT detectors [17,18]. The chip, as shown in Figure 11.15, includes 32 analog channels to convert the impinging charge into an amplified pulse shaped signal, and a common part for slow control and readout communication with a controller.

The first stage of the analog channel is a CSA based on a folded cascode topology with an inverter input amplifier. It integrates the incoming charge on a feedback capacitor and converts it into voltage; the feedback capacitor is discharged by a continuous reset system realized with a PMOS transistor. The increase of drain current in this transistor during the reset phase

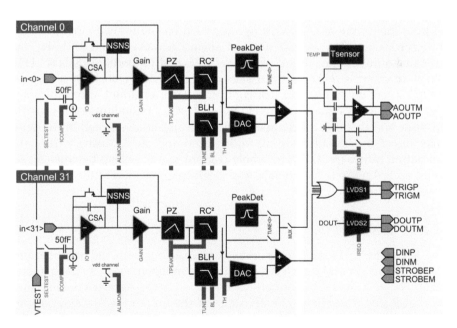

FIGURE 11.15
32-channel IDEF-X ASIC architecture.

is responsible for a non-stationary noise; to reduce the impact of this noise on the equivalent noise charge, a so-called non-stationary noise suppressor was implemented for the first time in this chip version using a low-pass filter between the CSA output and the source of the reset transistor to delay this noise.

The second stage is a variable gain stage to select the input dynamic range from 10 fC (250 keV) to 40 fC (1 MeV). The third stage is a pole zero cancellation (PZ) implemented to avoid long-duration undershoots at the output and to perform a first integration. The next stage of the analog channel is a second-order low-pass filter (RC²) with variable shaping time. To minimize the influence of the leakage current on the signal baseline, a so-called baseline holder (BLH) was implemented by inserting a low-pass filter in the feedback loop between the output of the RC² filter and the input of the PZ stage. The DC level at the output is stabilized for leakage current up to 7 nA per channel. The output of each analog channel feeds a discriminator and a stretcher. The discriminator compares the amplitude with an in-pixel reference low-level threshold to detect events. The stretcher consists of a peak detector and a storage capacitor to sample and hold the amplitude of the signal, which is proportional to the integrated charge and hence to the incident energy. In addition, each channel can be switched off by slow-control programming to reduce the total power consumption of the ASIC when using only few channels of the whole chip.

The slow-control interface was designed to minimize the number of signals and to get the possibility to connect together up to eight ASICs and address them individually. This optimization has allowed the reduction of the electrical interface from 49 pins in Caliste 256 to 16 pins in Caliste-HD for the same number of channels using low-voltage differential signals (LVDS). When an event is detected by at least one channel, a global trigger signal (TRIG) is sent out of the chip. The controller starts a readout communication with three digital signals (DIN, STROBE and DOUT) to get the address of the hit ASIC and then the hit channels. Then the amplitudes stored in the peak detectors of the hit channels are multiplexed and output using a differential output buffer (AOUT). The whole readout sequence lasts between 5 and 20 µs, according to the set delays and clock frequencies and the number of channels to read out.

11.4.2.2 VAS UM/TAT4

The VAS UM/TAT4 ASIC chip is used to read out both the amplitude of charge induction and the electron drift time independently for each anode pixel. The ASIC has 128 channels, each with a charge-sensitive preamp and two CR-RC unipolar shapers with different shaping times. The slow shaper has a 1-µs peaking time and is coupled to a peak -hold stage to record pulse amplitude. The fast shaper has a 100-ns shaping time and is coupled to simple level discriminators for timing extraction.

Of the 128 channels, 121 are connected to the pixels, one is connected to the grid, and one is connected to the cathode. Compared to the anodes, the polarity of the signals is reversed for the cathode and grid. The peak-hold properties, signal shaping, ASIC noise, and triggering procedures are included in the ASIC readout system model. The fast shaper can trigger off pulses as small as 30 keV for the anode and 50 keV for the cathode. Only the pixels with slow-shaped signals greater than a noise discrimination threshold of 25 keV are used in operation.

VAS UM/TAT4 is particularly well suited for 3D imaging and detection using thick CZT detectors (>10 mm) with high-energy photons (>1 MeV). 3D position-sensing techniques enable multiple-pixel events of pixelated CZT detectors to be used for 4π Compton imaging. Multiple-pixel events occur by either multiple gamma-ray interactions or charge sharing from a single electron cloud between adjacent pixels. To perform successful Compton imaging one has to correct for charge sharing. There is a large research effort at the University of Michigan under the direction of Professor Zhong He to resolve these complicated signal-processing issues and to re-construct the trajectory of incoming photons for dirty bomb detection (https://ners.engin.umich.edu/people/zhong-he/).

11.4.2.3 HEXITEC

HEXITEC was a collaborative project between the universities of Manchester, Durham, Surrey, Birkbeck and the Science and Technology Facilities Council (STFC). The objective of the program was to develop a new range of detectors such as CZT for high-energy X-ray imaging applications.

The HEXITEC ASIC consists of an 80×80 pixel array on a pitch of 0.25 mm [19]. Each pixel contains a 52 μm bond pad which can be gold-stud bonded to a CZT detector. Figure 11.16 shows a block diagram of the electronics contained in each HEXITEC ASIC pixel. Charge is read from each of the CZT detector pixels using a charge amplifier, which has a selectable range and a feedback circuit which compensates for detector leakage currents up to 50 pA.

The output from the each charge amplifier is filtered by a 2 us peaking circuit comprising a CR-RC shaper followed by a second-order low-pass filter, as schematically shown in Figure 11.16. A peak hold circuit maintains the voltage at peak of the shaped signal until it can be read out. Three track-and-hold buffers are used to sample the shaper and peak hold voltages sequentially prior to the pixel being read.

The HEXITEC is read out using a rolling shutter technique. A row select register is used to select the row which is to be read out. The data from each pixel becomes available on all column outputs at the same time, and at this point the peak hold circuits in that row can be reset to accept new data. The data being held on the column output is read out through a column multiplexer. The column readout rate is up to 25 MHz and the total frame rate

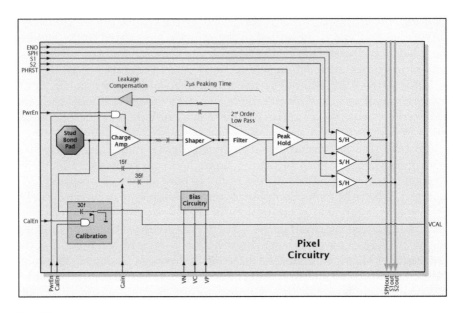

FIGURE 11.16
Block diagram of the HEXITEC architecture.

depends on the number of pixels being read out. The main limitation of the HEXITEC is a maximum count rate due to the 10 kHz frame readout scheme.

11.5 Operational Issues

11.5.1 Threshold Equalization

Before calibration of a CZT pixel detector the chip has to be equalized in order to minimize the threshold dispersion between pixels. This requirement results from the fact that the threshold that the pixel sees is applied globally but the offset level of the pixel can be slightly different due to process variations affecting the baseline of the preamplifier.

The equalization is performed with a threshold adjustment DAC in each pixel. The resolution of the adjustment DAC is usually in the range of 4 bits depending on particular ASIC implementation. The standard way to calculate the adjustment setting for each pixel is by scanning the threshold and finding the edge of the noise, then aligning the noise edges. This adjusts correctly for the offset level of the pixel but gain variations can still deteriorate the energy resolution at a given energy. To correct for the gain mismatch either test pulses or monochromatic X-ray radiation have to be used for the equalization. Equalizing at the energy of interest instead of the zero level might also be preferred.

11.5.2 Energy Calibration

Depending on the ASIC architecture there are two types of energy calibration that need to be done: calibration of the threshold and calibration of the time-over-threshold response (if applicable). For photon-counting chips such as MEDIPIX3 the only calibration required is the threshold response while in time-over-threshold ASICs such as TIMEPIX the ToT response has to be calibrated as well. Virtually all spectroscopic ASICs need to undergo energy calibration procedures.

To calibrate the threshold we need monochromatic photons or at least radiation with a pronounced peak. These can be obtained from radioactive sources, by X-ray fluorescence or from synchrotron radiation like Am241 and/or Co57 point sources. To find the corresponding energy for a certain threshold the threshold is scanned over the range of the peak, obtaining an integrated spectrum. The data is then either directly fitted with an error or sigmoid function or first differentiated and then fitted with a Gaussian function. From this fit the peak position and energy resolution can be extracted. Repeating the procedure for multiple peaks, the result can then be fitted with a linear function, and the relationship between voltage threshold setting and deposited energy in the detector is found.

11.5.3 Charge Sharing Corrections

When the pixel size is starting to approach the size of the charge cloud the input signal is subjected to charge sharing. Charge sharing creates a characteristic low energy tail and leads to a reduced contrast and distorted spectral information. To counteract this problem there are two possibilities, either to use larger pixels (reduced spatial resolution) or to implement charge summing on a photon-by-photon basis.

For lower rates and with detectors that store the energy information in each pixel either using ToT (TIMEPIX) or a peak-and-hold circuit (HEXITEC), the charge summing can be done offline. However, this requires that you do not have a second hit in the same pixel before you read the first one out. Using this approach you also lose charge that is below the threshold.

Another approach is to sum the charge in the detector as implemented in MEDIPIX3, where the analog charge is summed in a 2×2 cluster before being compared to the threshold. The advantage of this approach is that it can handle much higher interaction rates and that even charge below the threshold is summed as long as one pixel is triggered. However, since this correction has to be implemented in the ASIC architecture, it complicates the chip design and is less flexible.

11.5.4 Pile-up Effects

Given that the processing of each photon takes time, there will be problems with pile-up effects at high count rates. Pile-up happens when a second photon arrives in the same pixel before the first one is processed. Depending on the system architecture, the second photon could either be lost or added to the signal of the first photon. The result will be a deviation from linear behavior for the count rate. This deviation can be corrected for up to a certain limit in photon-counting devices, but more problematic are the spectral distortions due to pile-up that cannot be corrected for. For this reason, the operation of spectroscopic ASIC is limited to the maximum count rate that is not causing any pile-up effects. Different detectors will have different responses and it is important that the detector is characterized and suitable for the flux in a specific application. Since the flux is measured per area, smaller pixels offer the advantage of a smaller number of photons per second per pixel.

The X-ray photon events in the CZT detector occur randomly following a Poisson distribution. The probability density function between successive events is given by [4]:

$$p(t) = r \cdot \exp(-rt)$$

where r is the average incoming photon rate. When pile-up occurs, there will be a deviation from the linear relationship between flux rate and count rate; as illustrated by the PILATUS curves shown in Figure 11.17, the count rate in the ASIC begins to saturate as the flux rate increases beyond a certain limit.

An equivalent way of characterizing this limitation is via a parameter called dead time, which is the minimum amount of time that must separate photon arrivals for them to be counted separately. If successive photons arrive within this dead time window, paralysis occurs in the ASIC count, as shown in Figure 11.18.

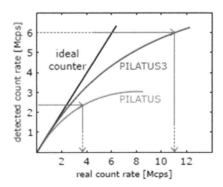

FIGURE 11.17
PILATUS3 ASIC count rate vs. incoming flux rate.

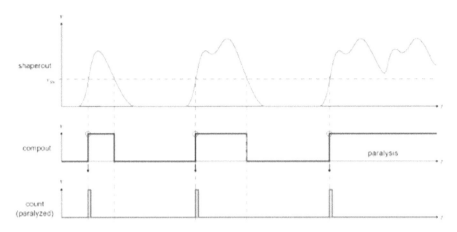

FIGURE 11.18
Signal waveforms illustrating paralyzable counting.

As illustrated in the top graph to the left, a single photon generates a pulse which is counted correctly. Subsequently, two photons arrive within the dead time window and are only counted as one event. Finally, the arrival of multiple photons causes pulse overlaps such that the pulse shaper output does not fall below the threshold, and paralysis occurs in the ASIC count.

Modern ASICs combat this problem by introducing non-paralyzable counting modes. An example is the PILATUS3 instant retrigger architecture. This is accomplished by re-evaluating the pulse shaper output after a programmable amount of time. If the output is still above threshold, it is assumed that pile-up has occurred and the counting circuit is retriggered. This technique prevents the count rate from freezing due to multiple photon arrivals.

11.6 Conclusions

Currently, most digital radiation detectors for security applications are based on integrating the X-ray quanta (photons) emitted from the X-ray tube for each frame. This technique is vulnerable to noise due to variations in the magnitude of the electric charge generated per X-ray photon. Higher energy photons deposit more charge in the detector than lower energy photons so that in a quantum-integrating detector, the higher energy photons receive greater weight. This effect is undesirable in many detection applications because the higher part of the energy spectrum provides lower differential attenuation between materials, and hence, these energies yield images of low contrast.

Direct conversion X-ray quantum-counting detectors solve the noise problem associated with photon weighting by providing better weighting of information from X-ray quanta with different energies. In an X-ray quantum-counting system, all photons detected with energies above a certain predetermined threshold are assigned the same weight. Adding the energy windowing capability to the system (i.e., counting photons within a specified energy range) theoretically eliminates the noise associated with photon weighting and decreases the required X-ray dosage by up to 40% compared to integrating systems.

Key ASIC challenges are present in spectroscopic and photon-counting systems, due to high flux and pile-up that affect both the count rate linearity and the spectral response. One way to counteract that problem is to use smaller pixels, but smaller pixels will lead to more charge sharing. In this respect, the various chips offer an interesting combination of relatively small pixels and still very good energy resolution.

X-ray systems dominate the installed base of airport baggage-scanning systems with a majority of systems being conveyers with projection line scanners. These systems can achieve a high throughput but exhibit a high false-positive alarm rate and require significant operator involvement, resulting in high operational costs. Systems employing computed tomography (CT) techniques have been also installed in large numbers. However, CT measures only X-ray absorption coefficient per voxel, which does not provide a means of specific material identification, resulting in many false positives. In addition, it is relatively straightforward to configure explosive materials so that they are undetectable by CT systems. Diffraction-based X-ray systems currently introduced in the marketplace present a potential solution to this problem. They detect and measure atomic layer spacing in crystalline and microcrystalline materials with high sensitivity and this provides means of specific material identification.

Regardless of the technology used, operators of baggage-scanning equipment are faced with multiple challenges today: difficulty in screening for liquids and home-made explosives (HME), screening personnel fatigue, false positives, and the cost of running security operations in general. Current aviation baggage-screening systems are either fast but do not generate detailed images of the object being scanned (high-speed X-ray) or they generate more detailed images but are relatively slow (CT-based systems) or are too costly for universal use (germanium-based diffraction detection systems). To solve the operational problems and maintain high operational efficiency new baggage-scanning technologies are being developed (such as X-ray diffraction).

References

1. Source: https://en.wikipedia.org/wiki/X-ray.
2. Source: www.i3system.com/eng/n_tech/tech2.html.
3. R. Redus, *Charge Trapping in XR-100-CdTe and -CZT Detectors*, Amptek, 2007.

4. R. Macdonald, Design and implementation of a dual-energy X-ray imaging system for organic material detection in an airport security application, *Proc. of SPIE*, Vol. 4301, 2001.
5. Y. A. Boucher, F. Zhang, W. Kaye, and Z. He, Study of long-term CdZnTe stability using the polaris system, *IEEE NSS-MICS*, 2012.
6. H. Chen, F. Harris, S. Awadalla, P. Lu, G. Bindley, H. Lenos, and B. Cardoso, Reliability of pixellated CZT detector modules used for medical imaging and homeland security, *SPIE Invited Paper*, 2012.
7. S. Skatter, H. Strecker, H. Fleckenstein, and G. Zienert, CT-XRD for improved baggage screening capabilities, *IEEE NSS MICS Conference*, 2009.
8. V. Rebuffel, J. Rinkel, J. Tabary, and L. Verger, New perspectives of X-ray techniques for explosive detection based on CdTe/CdZnTe spectrometric detectors, *International Symposium on Digital Industrial Radiology and Computed Tomography*, Berlin, 2012.
9. Y. Tomita, Y. Shirayanagi, S. Matsui, M. Misawa, H. Takahashi, T. Aoki, and Y. Hatanaka, X-ray color scanner with multiple energy differentiate capability, *IEEE Nuclear Science Symposium and Medical Imaging Conference*, Rome, Italy, 2004.
10. Anthony A. Faust, Richard E. Rothschild, Philippe Leblanc, and John Elton McFee, Development of a coded aperture X-ray backscatter imager for explosive device detection, *IEEE Trans. Nucl. Sci.*, Vol. 56, no. 1, Feb. 2009.
11. L. Verger, E. Gros d'Aillon, O. Monnet, G. Montemont, and B. Pelliciari, New trends in gamma-ray imaging with CdZnTe/CdTe at CEA-Leti, *Nucl. Instrum. Meth A*, Vol. 571, pp. 33–43, Feb. 2007.
12. J. Iwanczyk, E. Nygard, O. Meirav, J. Arenson, W. Barber, N. Hartsiugh, N. Malakhov, and J. C. Wessel, Photon counting energy dispersive detector arrays for X-ray imaging, *Proc. IEEE Nucl. Sci. Symp. Rec.*, 2007, pp. 2741–2748.
13. C. Szeles, S. Soldner, S. Vydrin, J. Graves, and D. Bale, CdZnTe semiconductor detectors for spectrometric X-ray imaging, *IEEE Trans. Nucl. Sci.*, Vol. 55, no. 1, pp. 572–582, Feb. 2008.
14. S. Mikkelsen, D. Meier, G. Maehlum, P. Oya, B. Sundal, and J. Talebi, An ASIC for multi-energy X-ray counting, *Proc. IEEE Nucl. Sci. Symp. Rec.*, 2008, pp. 294–299.
15. O. Tümer, V. Cajipe, M. Clajus, S. Hayakawa, and A. Volkovskii, Multi-channel front-end readout IC for position sensitive solid-state detectors, *Proc. IEEE Nucl. Sci. Symp. Rec.*, 2006, pp. 384–388.
16. J. Rinkel, G. Beldjoudi, V. Rebuffel, C. Boudou, P. Ouvrier-Buffet, G. Gonon, L. Verger, and A. Brambilla, Experimental evaluation of material identification methods with CdTe X-ray spectrometric detector, *IEEE Trans. Nucl. Sci.*, Vol. 58, no. 2, pp. 2371–2377, Oct. 2011.
17. X. Wang, D. Meier, B. Sundal, B. Oya, P. Maehlum, G. Wagenaar, D. Bradley, E. Patt, B. Tsui, and E. Frey, A digital line-camera for energy resolved X-ray photon counting, *Proc. IEEE Nucl. Sci. Symp. Rec.*, 2009, pp. 3453–3457.
18. A. Brambilla, C. Boudou, P. Ouvrier-Buffet, F. Mougel, G. Gonon, J. Rinkel, and L. Verger, Spectrometric performances of CdTe and CdZnTe semiconductor detector arrays at high X-ray flux., *Proc. IEEE Nucl. Sci. Symp. Rec.*, 2009, pp. 1753–1757.
19. L. Tlustos, Spectroscopic X-ray imaging with photon counting pixel detectors, *Nucl. Instrum. Meth. A*, Vol. 623, no. 2, pp. 823–828, Nov. 2010.
20. C. Schulze-Briese, The new PILATUS3 ASIC with instant retrigger technology, *PIXEL 2012*, Inawashiro, Japan, 2012.

12

High-Efficiency Power Amplifiers

Guillermo Velasco-Quesada, Herminio Martínez-García, and Alfonso Conesa-Roca

CONTENTS

12.1 Introduction

Analog amplifiers are usually described as small-signal circuits whose purpose is to increase the amplitude of the input signal or act as an impedance buffer between other amplifiers stages, accepting input signals over a broad range of amplitudes and frequencies. Since power amplifier stages may deliver high output power to low-impedance loads, they are significantly different from the low-power small signal amplifiers. The large-signal nature of power amplifiers requires special design considerations that may not be significant for small-signal amplifiers.

The sections below are devoted to describing amplifiers or amplification output stages that are capable of delivering high-output power levels.

12.2 Fundamentals of Power Amplifiers and Output Stages

The output stage of an amplifier must be able to deliver a significant amount of power into a low-impedance load, such as a stereo amplifier to a pair of speakers or a radio frequency amplifier to a broadcasting antenna, with acceptably low levels of signal distortion. They are also used as output stages in integrated circuits. Because high output powers may be involved, the efficiency of the amplifier to convert a low-power signal to high power becomes increasingly important. Inefficiency causes unwanted increases in transistor operating temperature and may lead to accelerated device failure.

As a consequence, it is usually required to have one or more of the following desirable properties:

1. Large output current or voltage swing
2. Low output impedance
3. Low standby power

In addition to these basic properties, an output stage is also required to have sufficiently good frequency response and thus it will not present a limitation on the rest of the amplifier circuit.

In this section, several output stage configurations will be examined with special regards to their advantages and limitations, starting with the simplest configurations and moving on to more complex designs.

12.2.1 Concepts of Efficiency and Distortion for a Power Amplifier

12.2.1.1 Efficiency in Power Amplifiers

To compare the design of power amplifiers, different figures of merit or parameters can be utilized. One of the most used is the efficiency. Notice that efficiency gives us a way to compare two different designs because it indicates how well an amplifier converts the DC input power to AC output power. The higher the efficiency, the better the amplifier is at converting the DC to AC power. This issue is important in battery-operated equipment because high efficiency means that the batteries last longer.

In order to determine the efficiency of a power amplifier, firstly it is necessary to determine the DC power supplied by the DC power source to the amplifier, P_{DC}. The power supplied by a DC voltage source, V_{CC}, is given by:

$$P_{DC} = V_{CC}I_{DC} \tag{12.1}$$

where I_{DC} is the current provided by the DC voltage source. On the other hand, it is necessary to consider the active power, P_{out}, provided to the load connected at the amplifier output terminals. Thus, the efficiency is defined by:

$$\eta = \frac{P_{out}}{P_{DC}}100\% \tag{12.2}$$

This equation says that the efficiency equals the AC output power divided by the DC input power. It is important to highlight that since all transistors and resistors (without considering the load resistor) of any power amplifier waste power, the efficiency is always lower than 100%.

12.2.1.2 Distortion in Power Amplifiers

When the output signal waveform of an amplifier differs in general shape from the input signal waveform, the output is said to be distorted. In particular, if a single-frequency signal input to an amplifier results in an output signal composed of the input frequency (fundamental component) and other frequencies, the amplifier has distorted the signal. The creation of these additional frequencies, known as Fourier harmonics components, is typically the result of non-linear distortion.

Large input signal to power amplifiers causes the amplifier to yield distorted output signals. In fact, in large-signal operation, the signals overtake the limits of the transistor forward-active region, causing distortion at the output. The relative amplitude of the fundamental component decreases with respect to the Fourier component as distortion increases.

Distortion is defined in one of several ways depending on the particular application of the circuit. However, the most important, commonly used for audio circuits, is the one named as total-harmonic distortion (THD).

If we define the input signal, $v_{in}(t)$, to a linear power amplifier as a sinusoidal input:

$$v_{in}(t) = A_{in} \cos \omega t \qquad (12.3)$$

where A_{in} is the input amplitude, it has an *ideal* transfer characteristic described by:

$$v_{out}(t) = A_0 + A_v v_{in}(t) = A_0 + A_v A_{in} \cos \omega t \qquad (12.4)$$

where:
$v_{out}(t)$ is the output voltage
A_0 is the DC offset voltage at output
A_v is the voltage gain.

Notice that there is not harmonic distortion since the single output frequency (apart from the DC component) matches the input component ω.

However, if the *real* output voltage is defined as

$$v_{out}(t) = A_0 + A_1 \cos \omega t + A_2 \cos 2\omega t + A_3 \cos 3\omega t + A_4 \cos 4\omega t + ..., \qquad (12.5)$$

there is harmonic distortion, and A_0, A_1, A_2, A_3, A_4... are the Fourier coefficients of the output voltage. The THD is commonly given as a percentage and is expressed as the ratio of the RMS (root-mean-square) values of all the harmonic terms to the effective value of the fundamental:

$$\text{THD} = \frac{\sqrt{A_2^2 + A_3^2 + A_4^2 + ...}}{A_1} \cdot 100\% \qquad (12.6)$$

As noted above, THD is used extensively in audio amplifier specifications. As an example, audio amplifiers that incorporate negative feedback to compensate for non-linearity typically have THD of less than 0.003% at low frequencies and low output power levels.

12.2.2 Power Amplifiers Classification

In order to classify an amplifier as a high-power amplifier, it must be capable of handling large-signal amplitudes where the current and voltage swings may be a significantly large fraction of the bias value. Then, this classification can be carried out according to the portion of the period of the output waveform during which the circuit transistors conduct. The conduction angle of each transistor in the circuit, assuming a sinusoidal input, determines the

TABLE 12.1

Classification of Power Amplifiers and Output Stages as
a Function of the Transistor Conduction Angle

Amplifier Class	Individual Transistor Conduction Angle
A	360°
B	180°
AB	180°–360°
C	< 180°

designation of the amplifier. The amplifier classification by conduction angle
for sinusoidal inputs is shown in Table 12.1. The output (collector or drain)
current through a transistor for class-A, -B, -AB, and -C power amplifiers is
shown in Figure 12.1.

In Figure 12.1a, named as class-A operation, the current flows through the
transistor over the whole period. In class-B operation, shown in Figure 12.1b,
each transistor only conducts over half the period of the sinusoidal waveform. Class-AB operation, shown in Figure 12.1c, illustrates current flow
through each transistor for greater than a half-cycle but less than the full
cycle of the input sinusoid. Figure 12.1d shows class-C operation, where each
transistor conducts over less than a half-cycle of the input sinusoid.

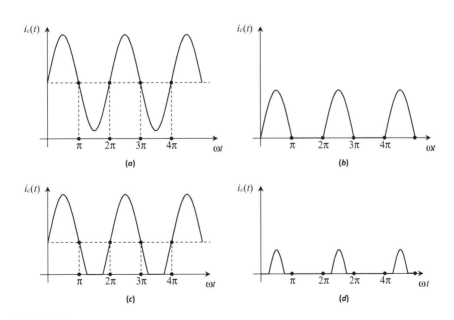

FIGURE 12.1
Transistor current for different amplifier classes: (a) Class-A amplifier, full current flows;
(b) Class-B amplifier, half-period current flow; (c) Class-AB amplifier, greater-than-half-period
current flows; (d) Class-C amplifier, less-than-half-period current flows.

Class-A operation is the only configuration that will yield low distortion signals. Class-B and class-AB power amplifiers assure signal continuity by making use of arrangements of two transistors that allow each transistor to share portions of the input signal conduction angle. Class-C amplifiers provide single-frequency sinusoidal output by driving resonant circuits over a small portion of the cycle. The continuity of the sinusoidal output is assured by the tuned circuit. This kind of amplifier is used for narrow-band signal applications.

12.2.3 Class-A Output Stages

In a class-A amplifier, the output stage is biased in the active region such that there is an uninterrupted flow of current during the entire cycle, and at no time does the transistor go into its cut-off or saturation regions. As a consequence, the amplifier operates in the active region during the entire 360° cycle of the input signal.

Consider the power amplifier shown in Figure 12.2 with a power supply voltage of $+V_{CC}$. The maximum peak-to-peak swing of the output voltage cannot exceed V_{CC} volts. Even more, in most practical cases it will be limited to 1 or 2 V less than V_{CC}, and sometime more, in part due to distortion considerations.

Considering Equation 12.2, it is easy to demonstrate that the maximum power conversion efficiency, η_{max}, for a class-A amplifier is given by:

$$\eta_{max} = \frac{P_{out,max}}{P_{DC}} = \frac{1}{4} \quad \text{or} \quad \eta_{max} = 25\% \tag{12.7}$$

Therefore, of the total DC input power, a maximum of 25% can be converted to AC power and delivered to the load, and the rest will be dissipated as heat by the amplifier.

It is important to note that the 25% power conversion efficiency is approached only when the amplifier is driven hard enough such that the

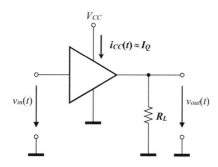

FIGURE 12.2
Class-A power amplifier.

maximum output voltage and current swings are obtained. If the input drive conditions are different from this, the efficiency is correspondingly less. In any case, it is not really possible to achieve an efficiency of 25% in practice due, in part, to other circuit losses. To limit distortion to an acceptable level, the output swing is often limited to no more than about 50% to 80% of the total active region voltage and current spans. As a result, the actual power conversion efficiency values of class-A power amplifiers generally fall in the range of 10% to 20%, and many times even lower if very low distortion is required.

Finally, it is important to highlight that this efficiency can be improved significantly using an output transformer to connect the load (transformer-coupled class-A power amplifier). In this case, for the ideal amplifier the maximum efficiency is equal to 50%. Nevertheless, the main disadvantage of this kind of class-A power amplifier is that the use of an output transformer in their design increases the price and the construction complexity, and involves transformer losses.

12.2.4 Class-B Output Stages

Practical applications of a class-B amplifier have an output stage comprised of two transistors whose outputs are combined in such a way as to reconstruct the full 360°-waveform cycle. Each transistor operates in the class-B mode and conducts during alternate half-cycles of the input signal so that by combining the two outputs an amplified replica of the input signal will be obtained. The two transistors of the class-B output stage are usually connected in a push-pull arrangement, a simple example of which is shown in Figure 12.3a.

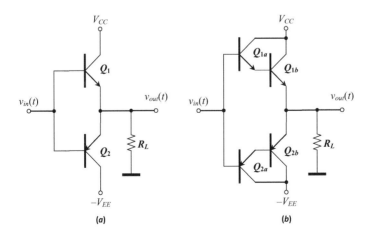

FIGURE 12.3
(a) Complementary emitter-follower class-B push-pull output stage; (b) Complementary emitter-follower class-B push-pull Darlington output stage.

In this figure, transistors Q_1 and Q_2 constitute a complementary push-pull emitter-follower output stage operating in the class-B mode. The transistors are a complementary pair: one is an NPN transistor and the other is a PNP transistor. Both bases are fed from the same point, and when the base voltage goes in a positive direction, Q_1 will be turned on and conduct while Q_2 will be off. Conversely, when the base voltage goes negative, Q_1 will be turned off and Q_2 will be biased into conduction.

Transistor Q_1 conducts during the positive half-cycles of the input voltage applied to the Q_1-Q_2 output stage and will source or "push" current into the load. On the other hand, transistor Q_2 conducts during the negative half-cycles and will sink or "pull" current from the load. As a consequence, each transistor conducts for one half of the entire cycle, and the two outputs are combined to give the full-cycle (360°) sinusoidal output current. Notice that, under quiescent conditions (that is, $V_{out} = 0$ V and $I_{out} = 0$ A), both transistors are off and the power dissipation is negligible. In the class-B mode of operation, the output transistor is biased at or very near the cutoff region, such that conduction occurs for only one half of the input waveform cycle, or 180° for a sinusoidal type of input signal. In this mode of operation, the quiescent current is essentially zero and, as a result, the class-B mode of operation offers the possibility of a much higher power conversion efficiency than does the class-A mode.

For greater current gains in the output stage, Q_1 and Q_2 can be Darlington compound transistors, as shown in Figure 12.3b.

In a similar way to a class-A amplifier, it is easy to show that the *maximum* efficiency of a class-B output stage, η_{max}, is given by:

$$\eta_{max} = \frac{P_{out,max}}{P_{DC}} = \frac{\pi}{4} \approx 0.7854 \quad \text{or} \quad \eta \approx 78.54\% \tag{12.8}.$$

Since this is the power conversion efficiency under the conditions of maximum power output, this will be the maximum possible power conversion efficiency for a class-B amplifier. In fact, the value of η that is actually obtained in practice is generally substantially less than this due to other circuit losses, and most of all due to the fact that the peak output voltage swing is always less than the supply voltage, typically by several volts. Part of this is due to the saturation voltage drop of the transistors, but it also results from distortion considerations. To limit distortion to an acceptable value, the output voltage swing is purposely limited to something considerably less than the full extent of the active region between cutoff and saturation.

Figure 12.4 shows a generalized graph of the power dissipation of the amplifier, P_d, versus P_{out}. In this plot we have two interesting points: Point '*A*', when maximum output power is reached, and point '*B*', when the maximum power dissipation in the amplifier transistors is achieved. Therefore, Table 12.2 shows the energy balance-sheet in ideal class-B power stages for these two points presented in Figure 12.4: In the case of maximum output

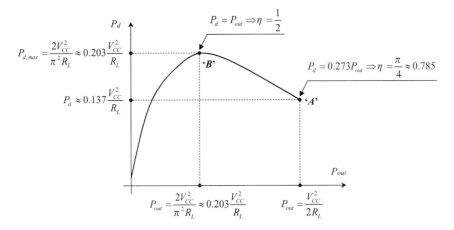

FIGURE 12.4
Power dissipation (P_d) versus output power (P_{out}) for a class-B push-pull amplifier supplied with a symmetrical power supply equal to $+V_{CC}$ and $-V_{EE}$.

power (point 'A'), and in the case of maximum power dissipation in the amplifier transistors (point 'B'). Notice that all expressions are given as a function of the maximum power reached on the output load.

The ratio of output power to power dissipation is limited to a maximum value of 0.33 for ideal transformerless class-A amplifiers, and unity for transformer-coupled class-A stages, compared to a ratio of 3.66 for class-B amplifiers. As a consequence, class-B amplifiers offer a substantial advantage over class-A stages. However, push-pull class-B amplifiers suffer an important type of distortion known as *crossover distortion* (Figure 12.5). This type of

TABLE 12.2

Energy Balance-Sheet in Ideal Class-B Power Stages for the Points 'A' and 'B' Presented in Figure 12.4: In the Case of Maximum Output Power (point 'A'), and in the Case of Maximum Power Dissipation in the Amplifier Transistors (point 'B')

Maximum Output Power When: $V_{out,peak} = V_{CC}$		Maximum Transistors' Power Dissipation When: $V_{out,peak} = 2V_{CC}/\pi$	
Total power that is drawn from the DC power supply: $P_{DC} = 1.274 P_{out,max}$	Load power: $P_{out,max} = \dfrac{V_{CC}^2}{2R_L}$	Total power that is drawn from the DC power supply: $P_{DC} = 0.812 P_{out,max}$	Load power: $P_{out} = 0.406 P_{out,max}$
	Q_1's power dissipation: $P_{d,Q1} = 0.137 P_{out,max}$		Q_1's power dissipation: $P_{d,Q1,max} = 0.203 P_{out,max}$
	Q_2's power dissipation: $P_{d,Q2} = 0.137 P_{out,max}$		Q_2's power dissipation: $P_{d,Q2,max} = 0.203 P_{out,max}$

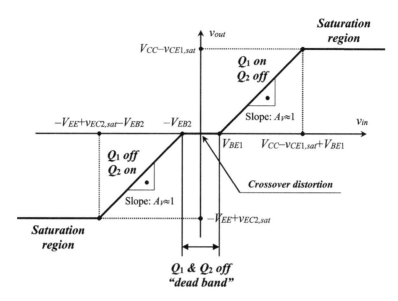

FIGURE 12.5
Transfer characteristics of the complementary push-pull emitter-follower output stage with crossover distortion.

distortion occurs because a small base-to-emitter voltage is required before collector current will flow, and is due to the very low or almost non-existent gain of the transistors in the cutoff region.

Class-B power amplifiers have the following advantages compared to class-A stages:

- Class-B power amplifiers have very small no-signal power dissipation in transistors. However, class-A stages causes maximum power dissipation of the transistor under no-signal conditions.
- If transistors with the same power rating are used for class-A and class-B amplifiers, the seconds will deliver more power to the load.
- Class-B power amplifiers have higher theoretical maximum efficiency than their class-A stage.

However, class-B power amplifiers also have some important drawbacks compared to class-A stages:

- Typical class-B stages require two or more transistors (or even two center-tapped transformers).
- Class-B amplifiers require NPN and PNP transistors with reasonably similar characteristics.
- The reduction of crossover distortion in class-B power amplifiers requires additional circuitry.

When technological limitations did not allow high-quality PNP power transistors, a balanced center-tapped input transformer, which splits the incoming waveform signal into two equal halves and which are 180° out of phase with each other, could be used in order to obtain a class-B power amplifier. Another center-tapped transformer on the output is used to recombine the two signals, providing the increased power to the load. The transistors used for this type of transformer push-pull amplifier circuit are both NPN transistors with their emitter terminals connected together. Obviously, two of the main disadvantages of this kind of class-B power amplifier are that it uses balanced center-tapped transformers in its design, making it expensive and difficult to construct, and an increase of stage losses.

12.2.5 Class-AB Output Stages

In order to avoid crossover distortion, it is necessary to bias the transistors with a small quiescent current at a point that is slightly into the active region. This can be done by applying a small DC bias voltage between the bases of the two transistors. The total bias voltage required between the two transistor bases is in the range of 1.0–1.5 V, approximately.

Although there are different alternatives in order to bias the transistors, Figure 12.6 shows a typical example of a simple class-AB complementary push-pull emitter-follower output stage based on bias diodes. Transistor Q_3 operates as a common-to-emitter gain stage with a current source active load of strength I_Q. The push-pull output stage is comprised

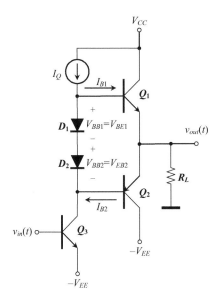

FIGURE 12.6
Class-AB push-pull output stage.

of transistors Q_1 and Q_2. The voltage drop across diodes D_1 and D_2 is the bias voltage for a class-AB operation of Q_1 and Q_2. In fact, in integrated circuit applications, these two diodes will actually be diode-connected transistors. The active areas of these transistors are scaled so as to obtain the desired standby or quiescent current through the output transistors Q_1 and Q_2. The voltage drop across D_1 and D_2 decreases with temperature, but this temperature variation is exactly what is needed to compensate for the negative temperature coefficient of the base-to-emitter voltage Q_1 and Q_2, avoiding thermal runaway in these power transistors.

Since Q_1 and Q_2 are now no longer biased in the cutoff region, but slightly into the active region, the conduction angle for each transistor will be greater than 180° (one half of a cycle), although it will still be considerably less than 360° (a full cycle). Therefore, strictly speaking, the mode of operation is no longer class-B, nor is it class-A. In fact, the mode of operation is called class-AB, and it combines the low distortion attribute of class-A operation with the high power conversion efficiency characteristics of class-B operation. Since the mode of operation is generally considerably closer to class-B than to class-A, the high power conversion efficiency available from class-B operation can still be obtained.

Along with a high power conversion efficiency, a low percentage of distortion is obtained. Indeed, with a well-balanced push-pull amplifier, the even harmonic distortion components will cancel out, leaving only the odd-numbered harmonics. Since the amplitude of the different harmonics decreases rapidly with increasing harmonic number, the cancellation of the second harmonic term, which will be the largest harmonic component, along with the other even harmonics, can lead to a very great reduction in the distortion.

For high-output power values, the circuit in Figure 12.6 can be completed substituting transistors Q_1 and Q_2 with a Darlington configuration (or by a Sziklai pair, also known as a complementary Darlington).

Finally, power amplifiers, especially those working in class-B, are usually operated under closed-loop conditions with either an internally or externally connected negative-feedback loop. The closed-loop gain, A_{CL}, is related to the open-loop gain, A_{OL}, by $A_{CL} = A_{OL}/(1 + f \cdot A_{OL})$, where f is the feedback factor. Since the closed-loop gain is generally very much less than the open-loop gain, any variation in the open-loop gain results in smaller variation in the closed-loop gain. As a consequence, the amplifier gain is basically fixed by the feedback factor selected by designers and essentially independent of the characteristics of the transistors used in the amplifier implementation.

12.2.6 Class-C Output Stages

Class-C amplifiers are used for amplification of a single frequency (tone) or a very narrow frequency band. In class-B output stages, the output current ideally flows for exactly 180° of the input sinusoidal waveform. If the duration of the output current is less than one half-cycle, class-C operation takes place. The periodic output current generates a sinusoidal output voltage by flowing through a resonant circuit tuned to the fundamental frequency or one of the harmonic components. Therefore, with class-C it is always necessary to use a resonant circuit for the load. This is why almost all class-C amplifiers are tuned amplifiers. This makes them ideal for amplifying radio and television signals because each station or channel is assigned a narrow band of frequencies on both sides of a center frequency.

The DC collector current is the only current drain in a class-C amplifier because it has no biasing resistors. Indeed, in a class-C amplifier, most of the DC input power is converted into AC load power because the transistor and coil losses are small. For this reason, a class-C amplifier has high stage efficiency.

As the conduction angle of output current approaches zero degrees, the circuit efficiency approaches 100%. Unfortunately, the output power also tends toward zero in this instance. Some compromise between good efficiency and high power output is necessary under normal conditions, with a resulting typical efficiency of 90%.

A simple class-C amplifier is shown in Figure 12.7. The ideal input voltage, collector current, collector and output voltage waveforms are shown in Figure 12.8.

Since collector current is periodic, it will contain a fundamental component of current and higher harmonics in addition to a DC or average value. The tank circuit or parallel resonant circuit will ideally present zero impedance to DC and all harmonics except the fundamental. A resistance R_P will be presented to the fundamental frequency. This resistance is related to the circuit Q and other tuned-circuit parameters. The fundamental frequency components of current will develop an AC voltage across the tuned circuit that is in proportion to this component. The AC output voltage is given by:

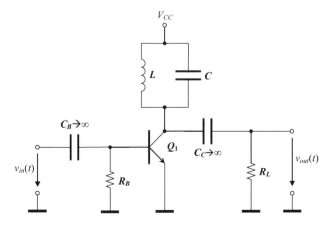

FIGURE 12.7
A basic class-C stage.

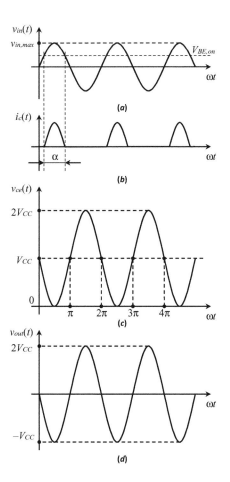

FIGURE 12.8
(a) Input voltage, (b) collector current, (c) transistor collector-to-emitter voltage, and (d) output voltage waveforms of the ideal basic class-C stage.

$$v_{out}(t) = -R_P I \cos \omega t, \tag{12.9}$$

where I is the magnitude of the fundamental current component. For an ideal tank circuit, there will be no distortion of the voltage waveform since load impedance exists only for the fundamental frequency. Nevertheless, in practice, the higher components may appear in the voltage waveform due to non-zero impedances presented to harmonics by the tank circuit.

Some class-C stages are driven from cutoff through the active region to the saturation region. Therefore, additional distortion is introduced as the transistor output impedance drops drastically when saturated. In addition, if harmonic generation is required, the tank circuit is tuned to the desired frequency and develops a voltage only at this particular frequency.

In order to calculate the input power, we must find the DC collector current that flows. Output power can be found only if the magnitude of the fundamental frequency component is found. For transistor stages, the collector-current waveform closely approximates a portion of a sinusoid. A Fourier analysis of this waveform yields the two quantities of interest. The circuit efficiency is given by:

$$\eta = \frac{\dfrac{\alpha}{2} - \dfrac{\sin\alpha}{2}}{2\sin\dfrac{\alpha}{2} - \alpha\cos\dfrac{\alpha}{2}} \qquad (12.10)$$

where α is the transistor's angle of conduction, being $0 \leq \alpha \leq \pi$ rad. Notice that, on the one hand, when α equals π, the angle of conduction is 180° and the amplifier operates in the class-B mode. Therefore, the efficiency calculated from the previous equation is 78.56%, which agrees with previous results obtained for a class-B amplifier. On the other hand, when the conduction angle decreases, the stage efficiency increases. Thus, an efficiency of 100% is approached as α tends toward zero. However, the output power also tends toward zero, resulting in a practical lower limit on α. Figure 12.9a shows plots of efficiency as a function of α. In applications with a variable input voltage, the output voltage will depend on both the amplitude of the input voltage and α.

The development of maximum output power requires the output voltage magnitude to equal V_{CC}. When α decreases, the fundamental frequency component of collector current will decrease in magnitude. Thus, the resonant tank circuit impedance must be increased to develop the maximum output signal. As α approaches zero, a very small fundamental current component is present. Consequently, an extremely large value of R_P is required to develop the maximum output voltage. It is reasonable to expect little output power

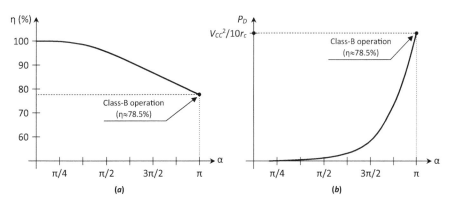

FIGURE 12.9
(a) Circuit efficiency, and (b) power dissipation of the transistor as a function of α.

under this condition. A reasonable compromise between circuit efficiency and output power often puts α in the range of $4\pi/6$ rad to $8\pi/9$, approximately, giving a conduction angle of 120° to 160°.

The power dissipation of the transistor, P_d, depends on the conduction angle α. As shown in Figure 12.9b, the power dissipation increases with the conduction angle up to 180°. The maximum power dissipation of the transistor can be derived by:

$$P_d = \frac{\text{MPP}^2}{40r_c},$$ (12.11)

where
    MPP is the maximum output, ideally given by $2V_{CC}$
    r_c is the AC resistance seen by the collector at resonance (approximately equal to the equivalent parallel resistance of the inductor in parallel with the output load).

This equation represents the worst case. A transistor operating as class-C must have a power rating greater than this or it will be destroyed. Under normal drive conditions, the conduction angle will be much less than 180° and the transistor power dissipation will be less than the value given by the previous equation.

12.3 High-Efficiency Power Amplifiers and Output Stages

As mentioned above, linear amplifiers, such as class-A, -B or -AB, have limited efficiency because the transistors of their output stages operate in their linear region. In order to improve the efficiency of power amplifiers based on linear operation, two different technological approaches can be considered. However, the output stages obtained in both cases are classified as high-efficiency power amplifiers.

This section is devoted to describing the operating principles of these amplifiers, to introduce their advantages and drawbacks, and to present some commercial application areas.

12.3.1 Introduction and Context

The first approach to reducing the energy losses in the power transistors used in analog output stages is to avoid the operation of these components in their linear region, thus eliminating the power losses due to devices' biasing and linear operation. As a consequence, the power transistors must operate by switching between saturation and cutoff regions, and then the power losses can be attributed basically to devices' on-resistances and switching operation.

The high-efficiency amplifiers based on this approach are usually classified as class-D, class-E and class-F, being also referred to as switching amplifiers or digital amplifiers.

The second approach to high-efficiency amplifiers is based on linear amplifiers (power transistors operating in the linear region) using an optimized power supply system. In this regard, the voltage supply of the linear amplifier is automatically adapted in accordance with the amplifier output voltage level. If the voltage of the power supply is adequately adjusted, the power losses of the linear amplifiers due to linear operation can be greatly reduced.

High-efficiency amplifiers based on linear amplifiers with adjustable power supply are classified as class-G and class-H, according to the strategy used for the voltage supply adjustment.

Because the power losses in power amplifiers are always dissipated as heat, the use of high-efficiency technologies in power amplifiers implies that the cooling requirements (typically large heat sinks or fans to blow/extract air over/from amplifier) can be greatly reduced or, in some cases, eliminated. As a result, these architectures increase the power density of power amplifiers, saving space and cost, these being two items of great importance in consumer products.

Consequently, they become an attractive solution for battery-powered applications and portable systems (hearing aids, headphones, smartphones and notebooks) where power efficiency is a key factor in extending the life of batteries.

On the other hand, high-power applications (such as audio amplifiers, servo motor drivers, push-pull amplifiers and radio frequency power amplifiers) can also take advantage of the high efficiency of these amplifiers' topologies.

12.3.2 Switching Amplifiers: Class-D Amplifiers

The first class-D amplifier was proposed in 1958 by R.L. Bright and G.H. Roger and it was protected under US patent reference 2,821,639 and titled "Transistor Switching Circuits", but the first commercial class-D amplifier came in 1964 from the British company Sinclair Radionics. The X-10 audio amplifier was developed by Clive Sinclair and Gordon Edge and was marketed as 10 W, but in reality it was capable of only about 2 W. In 1966 the replacement of the X-10 and X-20 amplifiers was launched by Sinclair Radionics as the Z12 audio amplifier. The Z12 was a reasonably successful product and, with it, the race of class-D amplifiers was started. History shows us, once again, that current solutions are based on old ideas.

The basic diagram of a closed-loop class-D amplifier is shown in Figure 12.10, where a pulse-width modulation (PWM) scheme is utilized. The PWM is achieved by comparing the error signal (the input signal in open-loop amplifiers) to an internally generated triangle-wave or sawtooth signal, which acts as a sampling clock and fixes the system switching frequency. The right choice of this frequency must be a balance between increasing the

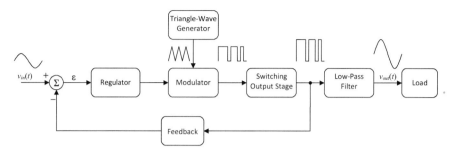

FIGURE 12.10
Closed-loop class-D amplifier block diagram.

amplifier bandwidth (BW) and decreasing the switching losses in the output stage transistors. Increasing the switching frequency increases the amplifier bandwidth, which can be at most half of the switching frequency, but switching losses will also increase, because they are proportional to the switching frequency.

The resulting duty cycle of the square- wave obtained at the modulator output is proportional to the level of the input signal. The output stage switches between the positive and negative power supplies in accordance with the PWM signal to produce a train of higher-voltage pulses. This waveform is suitable for a low-losses operation, since the output transistors have no current when they are open and have low voltage drop when they are closed, thus leading to a small amount of power dissipation. The theoretical efficiency of these amplifiers is 100% and in practice reaches values close to 98%.

Finally, the PWM waveform of the output stage is fed to a lossless inductor-capacitor (LC) low-pass filter in order to extract the amplified signal and minimize electromagnetic interference (EMI) conduced to output load. If the cutoff frequency of the low-pass filter is set properly (at least an order of magnitude lower than the switching frequency), the filter output voltage is equal to the average value of the PWM voltage and so proportional to the input signal.

These amplifiers present a low power-supply rejection ratio (PSRR) because the transistors of the output stage connect the low-pass filter with the power supplies through a low resistance path. The use of a feedback loop from the low-pass filter input will improve the PSRR and also attenuate all non-low-pass filter distortions, reducing the amplifier total-harmonic distortion (THD).

For higher PSRR and lower THD operation, a feedback loop including the low-pass filter and load can be utilized. Well-designed closed-loop class-D amplifiers can achieve PSRR values greater than 60 dB and THD values lower than 0.01%.

Pulse-width modulation (PWM) in the most common technique used for the control of class-D amplifiers, but it is not the only one possible. In order to mitigate possible EMI emissions, alternative modulation schemes can be used. Some possibilities are schemes based on pulse-density modulation (PDM), random modulation technique and click modulation technique.

Circuit designs of class-D amplifiers can be subdivided into two types. Circuits which switch voltage are called voltage-mode class-D amplifiers (VMCD) and circuits which switch current are called current-mode class-D amplifiers (CMCD). The term inverse class-D, also written as class-D^{-1} is also used, but not widely, to refer current-switching class-D amplifiers.

A voltage-mode class-D power amplifier (VMCD) circuit consists of two active devices connected in a cascade configuration. The common junction between the devices is connected to a series output filter to reconstruct a sinusoidal load signal from the pulse train, as is shown in Figure 12.11a. In this design, the drain voltages are similar to the gates' input pulse trains (square-shape waveforms) and the current through the switches is a portion of a sine wave, so that the sum of these two currents provides a sinusoidal current through the load.

The current-mode class-D amplifier (CMCD) architecture in Figure 12.11b is fed by a DC voltage source (V_{DD}) and an RF choke that has an ideal behavior similar to a DC current source. In this circuit the waveforms have been inverted if compared with the VMCD amplifier. The current through the switches is now a square-shape waveform (also similar to the gates' input pulse train) and the voltage across switches is sinusoidal. The resonant output circuit is tuned at the frequency of the switches' operation in order to achieve a sinusoidal waveform on the resistive load. This is because the higher-order harmonics are short-circuited by this parallel output network.

VMCD and CMCD amplifiers can be in low-frequency applications such as audio amplification and switch mode power supplies, but in radio frequency (RF) applications CMCD amplifiers are generally found more suitable. This is due to zero-voltage switching (ZVS) can be achieved because the voltage waveform across transistors is sinusoidal, the parasitic transistor output capacitance can become part of the resonant output filter, and the connection

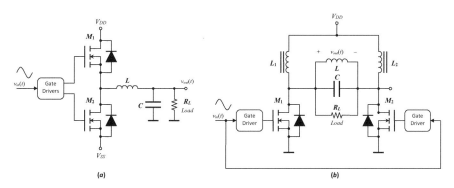

FIGURE 12.11
Class-D amplifiers: (a) Voltage-mode, and (b) current mode.

to the circuit ground (Gnd) of the transistors' source terminals simplifies the design of gate driver circuits.

VMCD amplifiers, sometimes referred to simply as class-D amplifiers and the only ones described with some detail in this text, can be classified into two topologies according to the configuration of the output stage. These topologies are the half-bridge and the full-bridge output stage, and they can be also referred as single-ended load (SEL) and bridge-tied load (BTL) topologies respectively. Each topology has its own set of advantages and drawbacks, and they will be summarized in the following sections.

12.3.2.1 Half-Bridge Output Stage

Figure 12.11a shows the open-loop class-D amplifier topology based on a half-bridge output stage.

This topology is potentially simple when compared with topology based on a full-bridge output stage. It requires a voltage power supply system which normally uses positive and negative rails of equal magnitude ($V_{SS} = -V_{DD}$) and two power switches connected between them. The load is tied between the common switches point and the system ground node through a low-pass filter.

This topology can also be powered from a single supply source, with the negative supply terminal (V_{SS}) used for ground. As a consequence, the output voltage swings between V_{DD} and ground and remains inactive at 50% duty cycle of the PWM signal. This imposes a DC offset across the output load, equal to $V_{DD}/2$, unless a DC-blocking capacitor is added.

In any case, the power supply system of this output stage must absorb the energy pumped back from the amplifier, especially when large reactive loads are driven. This effect, referred to as the "bus-pumping effect," results in bus voltage fluctuations which might become severe when the amplifier drives low-frequency voltages to the load. As a consequence, the output distortion increases since the gain of these amplifiers is directly proportional to the supply voltage.

12.3.2.2 Full-Bridge Output Stage

Figure 12.12 shows the open-loop class-D amplifier topology based on a full-bridge output stage.

The full-bridge topology is a more complex topology. It requires two half-bridge amplifiers with the load tied between their center points through an LC low-pass filter. However, the differential structure of the full-bridge output stage allows the utilization of three-state PWM strategies and the consequent cancellation of even-order harmonics and the possible DC offset present in the output voltage.

On the other hand, this output stage can achieve twice the output signal swing when compared to a half-bridge output stage with the same supply

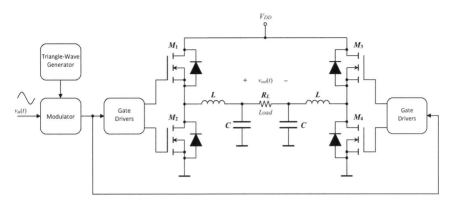

FIGURE 12.12
Output stage of bridge-tied load class-D amplifier.

voltage. This is because the load is driven in differential mode, as mentioned above. This leads to a theoretical fourfold increase in the output power when this topology is compared with a half-bridge class-D amplifier operating from the same voltage supply.

A full-bridge class-D amplifier requires twice as many transistors as a half-bridge topology. This can be considered a drawback since more switches typically mean more switching and conduction losses. This is true when high-output power amplifiers are considered, and it is due to the high output currents and supply voltages involved. Due their slight efficiency advantage, half-bridge class-D amplifiers are typically selected for high-power applications.

12.3.3 Switching Amplifiers: Class-E Amplifiers

The class-E amplifier concept was initially introduced by Nathan O. Sokal and Alan D. Sokal in the early 1970s and has recently received more attention due to the growing interest in high-efficiency amplifiers for transmitters in wireless communication systems.

The class-E amplifier is a single-transistor structure in switching operation with a 50% duty cycle at frequency f_0. The load network has a series resonant circuit (L_S and C_S) and a capacitor in parallel with the transistor (C_P), as shown in Figure 12.13a.

The series resonator circuit is used to block the DC and high-frequency harmonic components, forcing the current through the load to be a sinusoid with frequency f_0. If the V_{DD} choke is assumed as ideal (that is, it only conducts DC current), the current through the transistor-C_P parallel association must then be an offset sinusoid, and this current is commutated between these two components according to the state of the switch.

If the values of reactive components are appropriately adjusted, the voltage and the derivative of the voltage in the capacitor C_P (and transistor) can

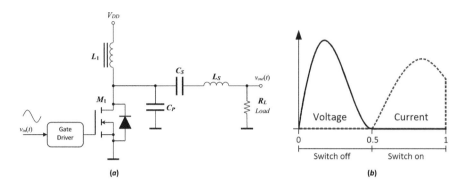

FIGURE 12.13
Class-E amplifier: (a) Structure, and (b) transistor waveforms in ZVS condition.

then be zero when the switch closes. In this condition the transistor voltage is driven to zero prior to turn-on, avoiding the overlap of voltage and current through the transistor during switching. This technique is identified as zero-voltage switching (ZVS). The transistor voltage and current waveforms for class-E amplifiers operating in ZVS conditions are presented in Figure 12.13b.

One advantage of class-E amplifiers is the easy incorporation of the transistor output capacity into the circuit topology, since C_P is the association of all parasitic capacitances and the exterior capacitance added to the circuit to allow adjustment of resonances.

12.3.4 Switching Amplifiers: Class-F Amplifiers

Class-F amplifiers utilize harmonic resonators in the output networks in order to reduce losses in the transistors, thus increasing the amplifier efficiency. Their basic principles of operation were patented by Henry J. Round in 1919, and suitably formalized by V. J. Tyler in a document published by the *Marconi Review* in 1958 entitled "A New High-Efficiency High-Power Amplifier".

The class-F approach has been developed in order to increase the efficiency of class-AB or class-B amplifiers. It is usually based on a standard class-B amplifier wherein a parallel resonant circuit (L_0 and C_0) is used to force the output voltage to be sinusoidal. Additional tuned networks are added in series with the load/resonator combination, as shown in Figure 12.14a, in order to open-circuit the transistor drain at the low-order odd harmonics. The amplifier shown in Figure 12.14a includes only two tuned networks at 3rd and 5th harmonics, but there is no theoretical limit to the number of tuned networks to be used in cascade.

By the use of these tuned networks, the transistor drain voltage waveform will begin to increasingly resemble a square wave as the generated odd harmonics will tend to flatten the top and bottom of the waveform, as can be

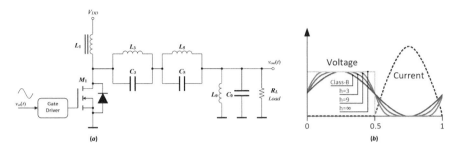

FIGURE 12.14
Class-F amplifier: (a) Structure, and (b) waveforms in the transistor drain.

seen in Figure 12.14b. This voltage shape decreases close to zero the drain voltage during the time for which the current is maximum through it and, as a consequence, the amplifier efficiency is increased. This efficiency increases quickly beyond class-B amplifiers if lower-order harmonics are tuned but the efficiency rising ratio decreases as the number of tuned harmonics increases. For instance, the maximum efficiencies for tuning up to the 3rd, 5th and 7th harmonics will be 88%, 92% and 94% respectively. When the number of tuned harmonics approaches infinity, the voltage waveform is a square wave and the efficiency limit is 100%, as in the class-D and class-E amplifiers.

The inverse class-F or class-F^{-1} amplifier is the dual tuning of a class-F one. Where class-F amplifiers short-circuit even harmonics and open-circuit odd harmonics, class-F^{-1} amplifiers open-circuit the tuned even harmonics and short-circuit the tuned odd harmonics. This has the effect of interchanging the transistor voltage and current waveforms. As a consequence, the voltage waveform resembles a half-sinusoid and the current waveform resembles a square wave.

As is done in this text, it is usual to classify the class-F and class-F^{-1} amplifiers in the switching amplifiers category since they have similar voltage and current waveforms at the transistor drain and achieve similar efficiencies. However, other authors classify these amplifiers as saturated transconductance amplifiers (or saturated amplifiers) with harmonic tuning.

12.3.5 Switching Amplifiers Comparison

Class-D amplifiers are mainly used in low-frequency applications; it is usual to find this topology in audio-frequency amplifiers operating with a switching frequency between 500 kHz and 2 MHz. For operation at higher frequencies the amplifier efficiency is limited by the switching losses due to the transistor parasitic capacitance discharge in the turn-on commutation. In this regard, class-E amplifiers integrate the transistor capacitance in the topology (ZVS tuning network), allowing a higher-frequency operation without significant efficiency losses. Additionally, class-E amplifiers may be implemented with a relatively simple circuit.

On the other hand, one of the main advantages of the class-D topology is the low voltage swing across the transistors, which is equal to the supply voltage. This is a great drawback in the case of class-E amplifiers where the voltage swing across the transistor is larger (nearly four times the supply voltage). Thus, compared with class-E topology (and also class-F) the class-D amplifier is more suitable for high-voltage applications.

Operating frequencies in class-F amplifiers are generally higher than for class-E amplifiers, but the efficiency limitations due to the lack of a simple implementation of tuned circuits for almost ideal switching conditions makes this design a poor alternative for frequencies where a class-E amplifier can be implemented.

Nevertheless, the waveforms achievable by class-F and class-F^{-1} amplifiers should allow performance benefits over class-E topology. In this regard, class-F and class-F^{-1} topologies reduce peak voltage present in the amplifier transistors (about twice the supply voltage) and, in the case of class-F^{-1} topology, RMS current trough amplifier transistors.

12.3.6 Amplifiers with Adjustable Power Supply Voltage

High-efficiency amplifiers based on switching topologies improve the overall amplifier efficiency by avoiding the overlap between voltage and current through the output stage transistors. As a consequence, power losses in the output transistors are reduced to those produced in switch-on and switch-off processes, as long as conduction losses can be neglected.

The second approach to high-efficiency amplifier topologies is based on class-AB (or class-B) amplifiers and reduces the losses in output stage transistors by adjusting the power supply voltage value in accordance with the amplifier output voltage. A reduced value of collector-emitter voltage, which is the difference between power supply voltage and amplifier output voltage, implies low conduction losses caused by wasted heat in the output stage transistors and the consequent rise of the amplifier efficiency.

The following sections are devoted to describing this second kind of high-efficiency amplifiers, which usually are known as adjustable power supply amplifiers or adjustable (moving) rails amplifiers. There are two different approaches for their implementation: Class-G topology focuses on voltage rail switching, whereas class-H topology utilizes a rail-modulation strategy. Although there are multiple references to these amplifiers in the specialized literature, technically speaking neither class-G nor class-H amplifiers are officially recognized.

12.3.6.1 Class-G Amplifiers

As was mentioned previously, the class-G topology optimizes the amplifier power supply using, as a minimum, two different bipolar supply rails, but it is possible to find architectures based on two different strategies. The first one is based on switching the supply voltage of the class-AB amplifier between the available power rails, and the second one uses the rail-boosting strategy. In any case, the objective is adapting the supply voltage to the output voltage value (in order to prevent output clipping effect) using more than one power supply rail.

In 1977, Hitachi introduced a range of audio amplifiers under the denomination of *Dynaharmony* which was based on class-G amplifiers using the rail-boost operating principle. In these amplifiers a 100 W (into an 8-Ω load) class-AB amplifier and two power rails of ±40 V and ±95 V were used. A basic structure of amplifiers based on this strategy is shown in Figure 12.15a. In this topology, two sets of power transistors operate in series connection, referred to as the inner pair (Q_1 and Q_2, connected to the load and driven from the low voltage rails: $\pm V_L$), and the outer pair (Q_3 and Q_4, connected to the high voltage rails: $\pm V_H$).

As long as the output voltage (V_{out}) remains between the low-voltage rails, the outer pair remains in the cutoff region and the collector voltages of the inner pair remain equal to the low rails voltages ($\pm V_L$). Once the output voltage is close to exceeding the low rails voltages the outer transistors boost the collector voltages, allowing the output voltage to swing up to the full high rails voltages ($\pm V_H$). Figure 12.15b shows a possible output voltage and the voltage at the collectors of inner transistors (the boosted voltage). During the boost voltage operation, the voltage in the inner transistors is maintained two or three volts above the output voltage, depending on the Zener voltage value of the diodes used as level-shifters.

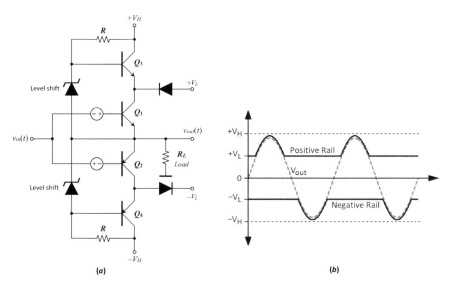

FIGURE 12.15
Class-G amplifier: (a) Boost structure, and (b) voltage rail boost effect.

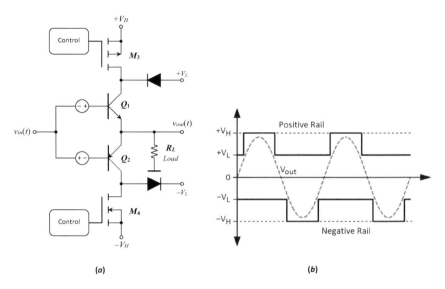

FIGURE 12.16
Class-G amplifier: (a) Switch structure, and (b) switching operation between power rails.

In 1981, Carver Corporation introduced a domestic amplifier, known as the *Carver Cube*, based on a refinement of Hitachi's amplifiers. This amplifier uses three bipolar power rails of ±25 V, ±48 V and ±80 V and claimed 500 W output when driven with music signals. The operation of this amplifier was based on switching between the three available power rails. Figure 12.16a shows a basic structure of one amplifier based on this strategy but using only two bipolar power rails. In this structure the inner pair of transistors (Q_1 and Q_2) operate as a normal class-AB amplifier and the outer pair (M_3 and M_4) operate as switches, allowing the modification of the collector voltages applied to the inner pair.

This amplifier operates from the lower supply voltages ($\pm V_L$) until output headroom (difference between output voltage and power supply voltage) becomes an issue. At this point the control system switches the power supply of the output stage to the higher supply rails ($\pm V_H$) in order to avoid the imminent output voltage clipping. When the output voltage drops below a prefixed level, the control system switches back to the lower rails ($\pm V_L$). This operative is shown in Figure 12.16b. In this regard, there are several methods to control the switching operation between the power rails but the feedback from the output voltage (V_{out}) or the input voltage (V_{in}), once pre-amplified, are normally used.

The design of any of these class-G amplifier topologies must achieve some compromise between circuit complexity and efficiency improvement. Two different power voltages minimizes the complexity of the power supply design, maintaining a reasonably high efficiency if the voltage values are properly selected. Additional rails may reduce power losses but increase the power supply complexity and reduce the overall system reliability. Amplifiers with two or three power rails have been reported in the literature, but not those using four or more rails, two-rail amplifiers being the most frequently used.

In this regard, most professional class-G audio power amplifiers use dual-voltage supply topology where the low-voltage supply is selected between 40% and 50% of the high-voltage value, but there is no optimum percentage because there are too many variables to be considered: Type of input signal, amplifier usage, power level demand and type of load. As a consequence, class-G amplifiers' efficiency depends largely on the same factors. With a high-amplitude sine wave as input signal there is no efficiency improvement compared to class-AB amplifiers operating under similar conditions. If the amplitude of the input signal remains at a level where the class-G amplifier operates from the low-voltage rails, then power efficiency does increase compared with class-AB architecture, which can only operate from the high-voltage level.

In any case, when the two operating strategies of class-G amplifiers are compared, the efficiency in the switch structures is potentially slightly higher than in the boosted structures, because the outer transistors operate in switching rather than linear mode. Consequently, the outer transistor losses are very small, but the peak dissipation of the inner transistors is increased. Nevertheless, the electric noise due to supply-rail commutation in switched structures can be translated to audible output glitches.

12.3.6.2 Class-H Amplifiers

Class-H amplifiers have a similar structure to class-G amplifiers and it can be difficult to decide into which category some structures of amplifier should be classified. Class-H is often described as topology that uses an externally modulated power supply. In other words, these amplifiers use only one power supply whose voltage value is modulated in accordance with the amplifier output voltage value to ensure reasonable headroom and thus avoid the output voltage clipping. Figure 12.17a shows the basic structure of a class-H amplifier which is based on a class-AB amplifier and an adjustable power supply system.

It is possible to find in the specialized literature two different strategies for generating the supply voltages of these amplifiers. The first one uses a buck or boost DC/DC converter as power system to generate two different voltage values from a single external supply voltage. The operation is similar to that in class-G switch structure shown in Figure 12.16b. This strategy is used by the company Maxim Integrated in their integrated headphone amplifiers.

The second strategy also utilizes a DC/DC converter as a tracking power supply system which monitors the amplifier output voltage and adjusts the supply voltage accordingly. When the amplifier output voltage increases above a threshold value, the power supply system tracks the peak level of the

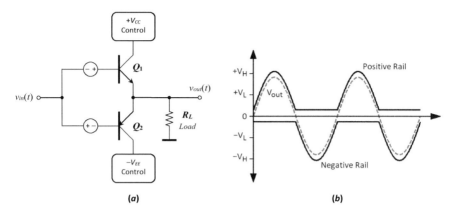

FIGURE 12.17
Class-H amplifier: (a) Structure and (b) voltage of power rails.

output voltage and impose a supply voltage only slightly higher (ΔV) than the instantaneous value of the output voltage, as is shown in Figure 12.17b. When the amplifier output falls below the threshold value, the power supply system returns to its nominal output voltage value. As a consequence, power dissipation is greatly reduced compared to conventional class-AB amplifiers. This second strategy is used by the company Texas Instruments in their integrated amplifiers for piezoelectric actuators and ceramic speakers. These amplifiers are also referred to by some authors as amplifiers with tracking power supply, and their main operating principle is used in envelope-tracking amplifiers devoted to RF applications.

When class-G and class-H amplifiers are compared, the need for multiple supplies used in class-G output stages may be a problem. The use of a multiple tap transformer is a good solution in amplifiers powered from the mains, but in the case of portable systems fed from batteries, it is an important drawback. On the other hand, the use of DC/DC converters as power supply in class-H amplifiers implies an inevitable ripple in the supply voltage. This perturbation in the power supply voltage is minimized by the voltage headroom utilized (ΔV) in the power supply system and also reduced by the PSRR of the class-AB stage. However, it inevitably affects the amplifier quality by increasing the THD value. Therefore, class-H offers a similar quality to class-AB amplifiers with efficiency that more or less matches the efficiency of class-D topologies.

12.3.7 Other High-Efficiency Power Amplifier Topologies

It should be mentioned that, in addition to the major classes of power amplifiers that were treated in previous sections, there are other documented structures of high-efficiency power amplifiers, usually based on a combination or association of stages used in the major classes. Therefore, in the remainder of this section they are briefly considered.

12.3.7.1 Class-FE and class-E/F Amplifiers

Class-FE and Class-E/F were recently introduced (the first patent dates from 2000) as power amplifier topologies which incorporate the best performance aspects of class-E and class-F or class-F^{-1} topologies, avoiding their main drawbacks but increasing the complexity of the amplifier design.

Zero voltage and zero voltage-derivative conditions (soft-switching operation) corresponding to the class-E amplifier can be used to eliminate discharge loss of the transistor capacitance, and harmonic tuning can be provided by using resonant circuits tuned to selected harmonic components, realizing class-F or class-F^{-1} mode with improved collector waveforms.

Class-FE and class-E/F amplifiers incorporate the transistor output capacitance into the tuned circuit, as in the class-E topology, and also minimize the peak voltage, as in the class-F and class-F^{-1} topologies. Additionally, class-E/F amplifiers minimize the RMS current trough in the amplifier transistors, as occurs in the class-F^{-1} topology.

12.3.7.2 Class-DE Amplifiers

The class-DE amplifier topology (first proposed in 1975) is one of the optimized voltage-mode class-D (VMCD) topologies, where both of two transistors satisfy class-E switching conditions. Class-E switching conditions mean that both the switching voltage and the derivative of switching voltage are zero when each transistor turns on. This functionality is achieved by the addition of a shunt capacitor to each transistor and the addition of two dead-time intervals on the transistors' driving pattern, when both transistors are switched off. Since the shunt capacitors must be discharged at that exact time, an additional series inductor with optimum value must be included in the load network.

Achieving the class-E switching conditions in the output stage allows class-DE amplifiers to operate with a very high power-conversion efficiency at high operating frequencies (MHz-order operation).

12.3.7.3 Class-AB/D Amplifiers

As mentioned above, the main problems of class-D amplifiers are related to the switching ripple at the output and the inherent difficulty of achieving a low output distortion. Class-AB amplifiers do not suffer from these problems, but they have a low efficiency. There are two basic topologies that allow the combination of these two amplifiers: By connecting them in series or in parallel.

In the series connection, the class-D amplifier generates the supply voltage for the class-AB amplifier. This topology is usually referred to as a class-H amplifier and it was previously described in Section 12.3.6.2.

On the other hand, the main goal of connecting linear (class-AB) and switching (class-D) amplifiers in parallel is to maximize the part of the output current provided by the efficient class-D amplifier and to minimize the output current of the class-AB amplifier. The class-AB amplifier operates as voltage source and determines the output voltage while the class-D amplifier operates as a current source, usually based on a half-bridge switch with a coil in series to the output.

In this configuration, the linear part controls the output voltage, while most of the output current is provided by the switching amplifier. In consequence, the power dissipation is minimized, when compared with class-AB amplifiers, and the output distortion is also reduced, when compared with class-D amplifiers.

12.3.7.4 Class-DG Amplifiers

Class-DG amplifiers are based on a proprietary Maxim Integrated output stage. As the company claims in the datasheets of its products, they offer higher efficiency over a greater output power range than previous amplifier topologies. These amplifiers combine class-D switching output efficiency and class-G supply-level shifting with a multilevel output modulation scheme.

Their operating principle is simple; the output stage uses PWM, a rail-to-rail output signal with variable duty cycle, to generate the output voltage as in a class-D amplifier. The magnitude of the output voltage is sensed and the supply rails are switched as needed to more efficiently supply the required power. For a low-output voltage swing requirement (below the external supply rail V_{DD}), the output range is between V_{DD} and ground. When output voltage swing above V_{DD} is required, an internal inverting charge-pump circuit generates negative rail (V_{SS}), replacing ground as the lower supply. The high-output voltage swing range is then V_{DD} to V_{SS}, approximately double the low swing range. This approach efficiently manages power consumption by switching the operating rails as needed according to the output voltage swing requirements.

12.4 Comparison and Results

This chapter concludes with some results in order to present a global comparison between power amplifiers or output stages, including the most important conclusions concerning the different classes presented in this chapter. It is important to highlight that a global comparison between power amplifiers is not an easy task for commercial equipment. Considering only the comparison for audio amplifiers, there are hundreds of products and manufacturers on the world market. However, comparing audio amplifiers with other kinds, such as radio frequency amplifiers, WiFi or Bluetooth amplifiers could be inappropriate, considering the lack of rigour due to the enormous differences between all of these sorts of equipment in terms of characteristics and performance (for instance, power level, spectral frequency, point of load like antennas, loudspeakers, etc.). Unfortunately, the information provided by different distributors and manufacturers of audio amplifiers is not always homogeneous. Therefore, for comparison purposes, it is necessary to define a merit marker or figure to make a comparison between different classes of technology in power amplifiers. What is more, the comparison must be carried out in the same conditions, and only the case of output power over a 4 Ω load has been considered.

Generally, apart from amplifier output power, information concerning dimensions and weight of the equipment is included on the specifications. Thus, an interesting and commonly used merit figure in specialized bibliography is output power with respect to the weight of the amplifier (expressed in W/kg).

However, the volume of an electronic device or equipment is important too, the compactness of the amplifier being an important target in our "electronic world". Therefore, the compactness of the amplifier (expressed as (volume·weight)$^{-1}$) can be defined as an idea of how "integrated" the equipment is. In this case, on the one hand, a low value assumes a heavy and bulky amplifier, and, on the other hand, a high value corresponds to a light and a small amplifier. Obviously, the compactness is a target in the design of the equipment, and depends on the manufacturer's skill, but is mainly fixed by the technology or class of the amplifier. In addition, this consideration may be conditioned by the use of normalized rack units in professional equipment.

More than 40 amplifiers of different manufacturers (including Skytone, Altair, Focal, Extron Electronics, Beilarly Audio, SP Audio, Yamaha, Peavy, RCF, Earthquake Sound, Xibon, Samson, BKL, Electrocompaniet, D.A.S., MTX AUDIO) have been evaluated. The results obtained are presented in the next figures.

Figure 12.18 shows the power/weight merit figure (expressed in W/kg) with respect to power (in W). Marks in this figure represent several commercialized amplifiers on the market. Only one element is presented from class-A (cross plot mark), most being from class-AB (square plot mark), class-H (triangle-up plot mark) and class-D (diamond plot mark) power amplifiers. The lines presented in the figure are the interpolation of the represented marks. This figure shows that class-D amplifiers (solid line) present the better ratio of this merit figure; secondly, we have class-H stages (dash-dot line), and, finally, class-AB amplifiers (dotted line). This conclusion obtained from commercial information is in accordance with the analysis presented in previous sections of this chapter.

Figure 12.19 shows a distribution of compactness (expressed in (m^3·kg)$^{-1}$) related to power (in W). Marks in the figure correspond to several commercial audio amplifiers, in a similar way to Figure 12.18. As can be seen, the regular tendency (interpolation lines on the figure) for all the analyzed equipment

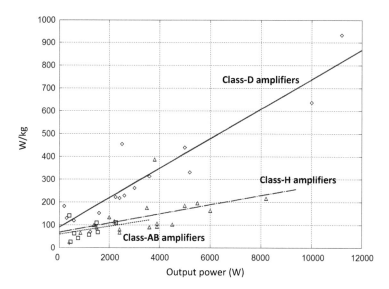

FIGURE 12.18
Power/weight indicator with respect to output power.

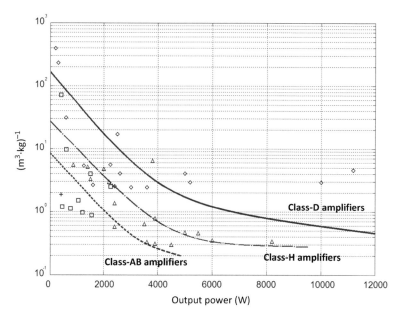

FIGURE 12.19
Compactness indicator with respect to output power.

shows higher values of compactness for class-D stages (solid line). The difference with respect to the next in order (class-H, in dash-dot line) is around five times. This difference is approximately the same as that obtained when we compare class-H and class-AB (dotted line).

Considering these two merit figures, we can conclude that the natural tendency of manufacturers will be to focus on the development of class-D power amplifiers and stages. Their advantages concerning weight and dimensions with respect to older technologies such as class-A and class-AB, guaranteeing excellent performance as regards "sound quality", will persuade consumers to use them and ensure their subsequent expansion in the market. In the opinion of the authors, class-H amplifiers will continue to be an excellent alternative in the market for low and medium power levels, to provide certain specialized customers with classic linear amplification and a good compromise in terms of weight and volume of equipment.

12.5 Would You like to Learn More?

As the reader can imagine, this section only overviews briefly some alternatives to classical high-efficiency amplifier topologies. In this sense, it is possible to find in the specialized literature more topologies of amplifiers with special relevance in RF applications or in audio systems. More information about these topologies can be found in the following references:

12.5.1 On Fundamentals of Power Amplifiers and Output Stages

- **Active and Non-Linear Electronics**. Thomas Schubert, Jr. and Ernest Kim. John Wiley & Sons, Inc. 1996. ISBN: 0-471-57942-4.

  In chapter 7 ("Power Amplifiers and Output Stages") this book discusses design principles of class-A, class-B, and class-AB power amplifiers and output stages.

- **Advanced Electronic Circuit Design**. David Comer and Donald Comer. John Wiley & Sons, Inc. 2003. ISBN: 0-471-22828-1.

  In chapter 2 ("Fundamental Power Amplifier Stages") and chapter 3 ("Advanced Power Amplification"), this book reviews and analyzes in depth power amplifier design principles and focuses on linear power amplifiers, especially transformer-coupled and transformerless class-A stages, class-B and class-AB transformer stages, transformerless class-B and class-AB, class-C and class-D output amplifiers.

- **Design and Applications of Analog Integrated Circuits**. Sidney Soclof. Prentice-Hall International, Inc. 1991. ISBN: 0-13-033168-6.

In chapter 12 ("Power Amplifiers") this book discusses in depth the design of power amplifiers, especially those devoted to micro-electronic implementation in integrated circuits (especially class-A, class-B, and class-AB).

- **Electronic Circuits. Electronic Circuits: Analysis, Simulation, and Design**. Norbert R. Malik. Prentice-Hall International, Inc. 1995. ISBN: 978-0023749100.

    In chapter 10 ("Power Circuits and Systems") this book carries out an introduction to class-A, class-B, class-AB and class-D power amplifiers.

- **Electronic Principles**. Albert Paul Malvino and David J. Bates. 8th ed. McGraw-Hill Education, Inc. 2016. ISBN: 978-0-07-337388-1.

    In chapter 10 ("Power Amplifiers") this book carries out an introduction to class-A, class-B, and class-AB power amplifiers. In addition, the authors present an interesting analysis in this chapter of class-C power amplifiers.

12.5.2 On High-Efficiency Power Amplifiers

- **Switchmode RF Power Amplifiers**. Andrei Grebennikov and Nathan O. Sokal. Elsevier Inc. 2007. ISBN: 978-0-7506-7962-6.

    This book reviews the design principles of power amplifiers and focuses on switching mode power amplifiers (class-D, class-E and class-F). Chapter 8, titled "Alternative and Mixed-Mode High-Efficiency Power Amplifiers," explains the class-DE, class-E/F, class-E_M and class-E^{-1} topologies.

- **Switchmode RF and Microwave Power Amplifiers**. Andrei Grebennikov, Nathan O. Sokal and Marc Franco. Elsevier Inc. 2012. ISBN: 978-0-12-415907-5.

    As the previous one, this book reviews the design principles of power amplifiers and focuses on switching mode power amplifiers. Chapter 9, also titled "Alternative and Mixed-Mode High-Efficiency Power Amplifiers," includes an overview of the class-EF amplifiers and the amplifiers known as outphasing power amplifiers.

- **High Efficiency Audio Power Amplifiers; Design and Practical Use**. Ronan A. R. van der Zee. Universiteit Twente. 1999. ISBN: 90-36512875.

    This PhD thesis reviews the measurement and prediction of amplifiers' dissipation, and the topologies of linear and switching amplifiers including class-AB, class-G, class-H and class-D amplifiers. It also focuses on series and parallel combinations of linear and switching amplifiers, which are usually known as class-AB/D amplifiers.

- **Audio amplifiers, class-T, class-W, class-I, class-TD and class-BS**. Paul Rako. EDN Network. 2009. Published online: https://www.edn.com/electronics-blogs/anablog/4309722/Audio-amplifiers-class-T-class-W-class-I-class-TD-and-class-BS.

  The author of this technical note comments that in the last few years several companies have been inventing amplifier classes, not as a legitimate architecture class, but as a marketing trick. This document lists and links these amplifiers to the different companies which produce them.

Finally, the datasheets provided by amplifier manufacturers are another source of information about these amplifier classes that are not officially recognized. This is the case for class-DG topology, produced by Maxim Integrated and described in detail in the datasheet of any of their integrated amplifiers based on this output stage (MAX98307 and MAX98308).

Index